Renewables

Renewables

The Politics of a Global Energy Transition

Michaël Aklin and Johannes Urpelainen

The MIT Press
Cambridge, Massachusetts
London, England

This book was set in ITC Stone Serif Std by Westchester Publishing Services. Printed and bound in the United States of America.

Library of Congress Cataloging-in-Publication Data

Names: Aklin, Michaël, author. | Urpelainen, Johannes, author.
Title: Renewables : the politics of a global energy transition / Michaël Aklin and Johannes Urpelainen.
Description: Cambridge, MA : The MIT Press, 2018. | Includes bibliographical references and index.
Identifiers: LCCN 2017028648| ISBN 9780262037471 (hardcover : alk. paper) | ISBN 9780262534949 (pbk. : alk. paper)
Subjects: LCSH: Renewable energy sources--Political aspects. | Energy industries. | Globalization.
Classification: LCC TJ808 .A454 2018 | DDC 333.79/4--dc23
LC record available at https://lccn.loc.gov/2017028648

10 9 8 7 6 5 4 3 2 1

Contents

List of Figures

List of Tables

Preface

As we give this book the finishing touches it needs, renewables are already everywhere. From the rooftops of Brooklyn to the temples of Lucknow in India, from the fields of Liguria to the steppes of Mongolia, solar and wind power are transforming our landscapes. Wind farms supply ever-larger quantities of electricity in Europe, Asia, and the Americas, and solar panels have brought renewable power to the most remote rural communities of the world.

The triumph of renewables is recent, however. When we started working on the politics of renewable energy almost a decade ago, the world's energy landscape looked very different. Environmentalists and clean technology enthusiasts aside, few people thought that modern renewables—wind and solar power—would in just a few years be the most dynamic component of the global power sector. As we travel around the world for our research, this difference is impossible to miss. We wrote this book because renewables were always bound to be one of the key stories of the 21st century. Had renewables failed to grow, they would have epitomized the world's failure to move away from polluting fossil fuels. Had renewables exploded, they would have been hailed as the human civilization's great triumph.

What transpired was something much more complex and interesting. Renewables grew because profound challenges to an energy system based on fossil fuels created an urgent need to experiment with alternatives, and governments in forerunner countries soon saw political gains in investing vast sums of money in alternative energy. These investments created powerful advocacy coalitions in favor of a sustainable energy transition, thus consolidating public policies that enabled growth over time. Economies of scale and learning by doing brought about dramatic decreases in the cost of renewable power. That's why we now see renewables everywhere, and not just in wealthy industrialized countries.

We write this book at the early stages of an exciting global energy transition, a glimmer of hope for anyone worried about climate change. The world energy system is in flux, and a future reader of our book—we hope this strange creature will thrive and prosper—will undoubtedly find much that is amiss. Yet we are confident that our analytical framework and empirical scholarship capture something important about the growth of renewable energy.

We have spent the past seven years working on renewables, revising our analysis, and accumulating new evidence in response to both changing circumstances and vigorous but constructive criticism from our colleagues. We have by now lost count of all the seminars, but we must begin by acknowledging the generous support of the University of Pittsburgh and Steve Finkel in organizing a book workshop for what we thought was an advanced draft. Frank Laird and Lyle Scruggs read through the entire manuscript and challenged us to work much harder on the theory and the evidence. As painful as it was, the book is much better for the revision.

Seminar audiences at Brown University, George Mason University, the University of Oregon, the University of Pennsylvania, the University of Tampere, the National Chengchi University, the University of Macau, the Carnegie-Tsinghua Center, the City University of Hong Kong, the Asan Institute for Policy Studies, Yonsei University, and the annual conventions of the American and Midwest Political Science Associations all made their unique contributions. We deeply appreciate these contributions.

Besides Frank and Lyle, a few individuals went far beyond the call of duty in helping us with this project. Marion Dumas read several chapters and gave detailed comments. Patrick Bayer has worked with us extensively on renewable energy and offered detailed comments on the theory. Jeff Colgan helped us see the broader relevance of renewables in the world of energy policy and has used our draft manuscript in his teaching. Joel Tarr provided insightful comments on the history of energy transitions and the history of environmental regulation in the United States. Finally, we want to thank Beth Clevenger, our editor at MIT Press, for her dedication to this project and wonderful insights.

This book is dedicated to the millions of people around the world working for a more sustainable energy future. Some protest on the streets and some formulate government energy policy. Others develop new technologies in the laboratory, and yet others invest money in clean technology companies. Our gratitude for these people is profound because they are the heroes of what we hope to be an inspiring story.

I The Argument

1 Introduction

Humans have used renewable sources of energy, such as wind and biomass, for thousands of years. Until the end of the 20th century, however, the contribution of renewable energy to modern industrial economies remained negligible. With the exception of large dams, renewables were completely dominated by fossil fuels in modern energy production. Oil, natural gas, and coal covered almost 90% of the global energy mix in the year 2000, and there was no end in sight to this state of affairs. At the time, the International Energy Agency (IEA) published its annual *World Energy Outlook*, predicting the continued irrelevance of nonhydroelectric renewables:

> Other renewables, including geothermal, solar, wind, tide, wave energy and combustible renewables (commonly known as biomass) and waste, are expected to be the fastest growing primary energy sources, with an annual growth rate averaging 2.8% over the outlook period. Despite this rapid growth, the share of renewables climbs to only 3% by 2020 from the current 2%. Power generation in the OECD countries accounts for most of this increase. Concerns over climate change will encourage the deployment of renewables, although they remain expensive compared to fossil fuels. (IEA 2000, 24)

The IEA's authoritative view was that renewable energy would play no role in climate mitigation or energy security for two decades to come. Without major changes, such as a global climate treaty, renewable energy would remain on the sidelines of the world energy system.

As we write this introduction in 2017 it is hard to believe that there was a time when renewables were considered a negligible power source. Already by 2010, the IEA prediction of a 3% share for modern renewables had been exceeded 10 years earlier than expected, with nonhydroelectric renewables accounting for as much as 5% of global energy consumption. The IEA's dismissive views are especially hard to reconcile with the rapid growth of renewables in the electricity sector. As the upper panel of Figure 1.1 shows, much of

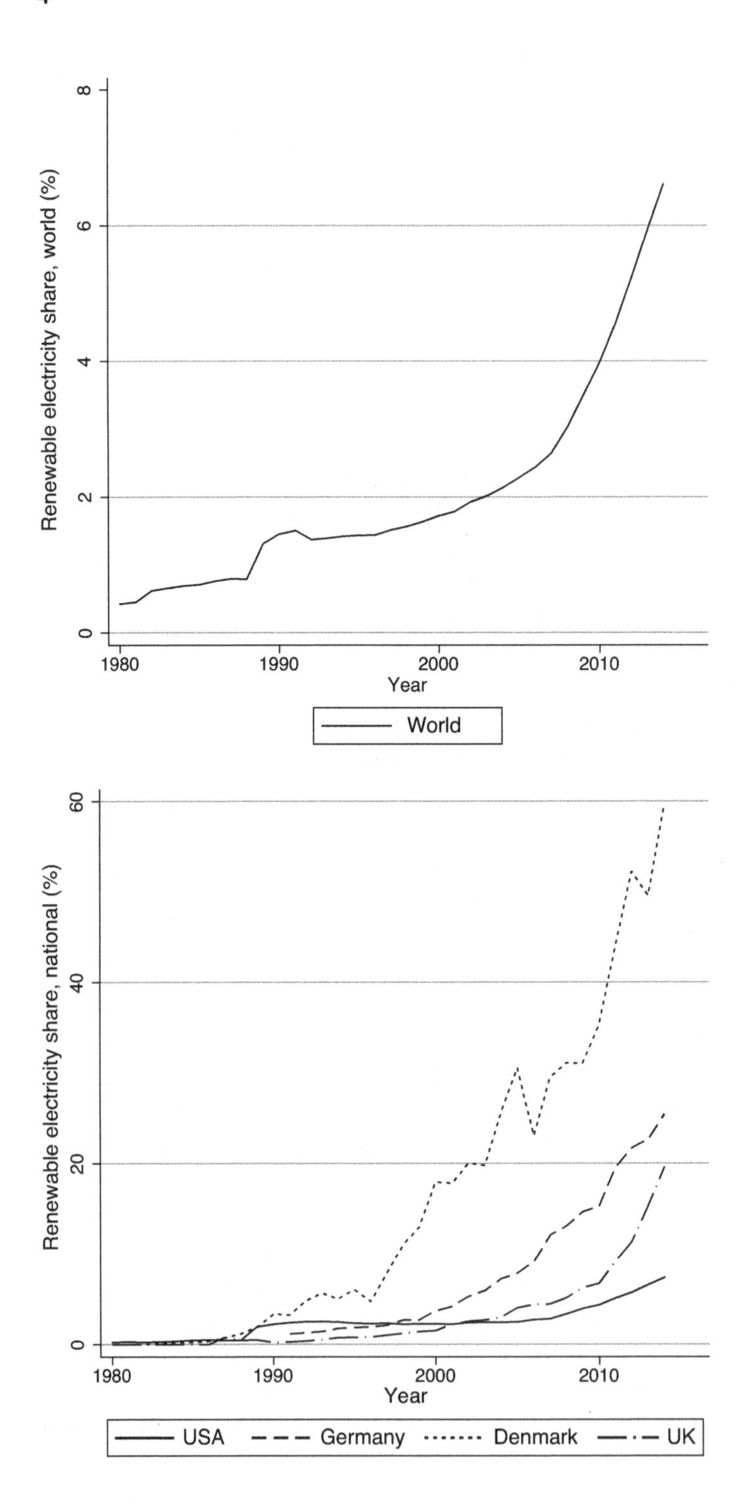
8
6
4
2
0
Renewable electricity share, world (%)
1980
1990
2000
2010
Year
World
60
40
20
0
Renewable electricity share, national (%)
1980
1990
2000
2010
Year
USA
Germany
Denmark
UK

the dynamism of modern renewables stems from the power sector, in which wind and solar power moved forward in leaps and bounds. Whereas renewables produced less than 2% of global electricity in 2000, their share had quadrupled to about 8% by 2013. In recent years, this aggressive growth has been fueled by hundreds of billions of dollars in annual investment by companies and governments (REN21 2012; Pew 2013; BNEF 2015).

The impressive growth of renewables in the power sector hides important variation across countries. Some countries have made much more progress in developing their renewable sources of energy than others, and the reasons for this variation are not obvious. Consider the lower panel of Figure 1.1,[1] which shows the growth of renewable electricity in four industrialized countries. Among these four countries, Denmark has made the most out of renewables in the power sector. Germany has also achieved impressive growth rates, whereas renewable energy industries in the United States and the United Kingdom have grown at a slower pace. Natural resources are a poor explanation for this variation because the United Kingdom has an excellent geography for wind energy, and Germany is not particularly suited for renewables.

The growth of renewables is expected to continue in the future. Although renewables are still far from overtaking fossil fuels in the global energy system, their growth is fast and contributes to a nascent and uncertain energy transition with major socioeconomic effects. BP, a major oil company, expects nonhydro renewables to overtake both hydro and nuclear power over the next 25 years (BP 2014). Bloomberg New Energy Finance predicts that renewables, including hydro, will be about half of all electricity capacity by 2030 (BNEF 2014a). Developing countries such as India are eagerly hoping that renewables will solve their energy issues. China is already the world's largest national market for renewable energy, with each national energy plan more ambitious than the previous one. Even if these predictions turn

Figure 1.1
Share of renewables in electricity generation. **Upper Panel:** Share of renewables in world grid electricity generation (%), excluding hydroelectric facilities.
Source: EIA (2017a). **Lower Panel:** Share of renewables in electricity production (%), excluding hydroelectric facilities, in the United States, Germany, Denmark, and the United Kingdom. Note the difference in the scale of the y-axis between the two figures. The increase in the United States in 1989 is largely due to changes in measurement by the EIA in order to account for nonutility electric plant data.
Source: WDI (2017).

out to be too optimistic, renewables are already changing the way our economies are powered.

The rapid but uneven growth of renewable energy is one of the most important and least understood developments of our time. Although multilateral efforts to combat climate change remain nascent and uncertain (Victor 2011), renewable energy has made more progress than even the most optimistic commentators expected only a decade ago. The 2015 Paris Agreement to mitigate climate change allows countries to choose their own emissions targets for other parties' review, and efforts to replace fossil fuels—especially coal—with renewables in electricity generation play a central role in the plans of key countries such as China and India. As the World Economic Forum put it in a public statement at the June 2016 annual meeting of the New Champions, "After Paris climate agreement, a world of renewable energy is emerging."[2]

How did renewable energy go, in less than three decades, from an obscure hobby of a motley crew of enthusiasts to the most rapidly growing kind of energy in the world economy and a business worth of hundreds of billions of dollars? Why did renewable energy continue to grow in the 1990s, at a time of exceptionally low convention energy prices? What were the primary drivers of this transformation? Why are some countries outperforming others? What does the future hold for renewable energy? We attempt to answer these questions in this book.

The answers are important for anyone interested in climate change, energy security, or economic development. Renewables are a promising approach to climate mitigation (Rabe 2004; Sandén and Azar 2005; Karapin 2012), and they also enhance the energy security of countries that depend on foreign imports of fossil fuels (Müller-Kraenner 2010).[3] Some scholars also argue that they hold promise for the creation of economic opportunities, export industries, and jobs (Lewis and Wiser 2007). Indeed, because the December 2015 Paris agreement on climate change emphasizes decentralized, bottom-up action by individual countries, national policies play a critical role in determining the success of climate mitigation in the coming decades.[4] At the same time, the promotion of renewable energy is expensive. Available evidence suggests that government policy has played a key role in the development of renewable energy. By 2012, more than 60 countries worldwide had adopted a feed-in tariff (FIT) to support the production of renewable electricity.[5] According to Bloomberg New Energy Finance, these policies "have driven 64% of global wind and 87% of global photovoltaic capacity" (UNEP 2012, 5). Consumers have paid for the growth of renewable energy through higher electricity prices and taxpayers through a heavy subsidy burden.[6] In many countries, the cost of renewable energy has sparked a heated public debate. In Germany, for example, the generous incentives for solar and wind electricity generation have drawn severe criticism by experts who argue that they amount to an income transfer from the poor to the wealthy.[7]

Renewables have also provoked a debate among scholars and practitioners. On the one hand, a group of energy scholars have for decades expressed enthusiasm about renewable energy (Lovins 1976; Jacobsson and Bergek 2004; Mendonça 2007; Scheer 2007; Sovacool 2008; Lund 2011; Dangerman and Schellnhuber 2013). These authors emphasize that renewables can be a cornerstone of a comprehensive strategy to solve the problems of climate change and energy poverty, arguing that obstacles to increasing the use of renewable energy can be surmounted. Others are more cautious, warning about potentially high costs and the difficulty of expansion (Collier and Venables 2012; Baker et al. 2013; Levi 2013). Although the IEA's prognosis in 2000 proved too pessimistic, these authors note it remains to be seen how much more renewables can grow, given how important supportive government policy has been for renewables during the past decade.

1.1 Renewable Energy: Political and Economic Challenges

Before summarizing our argument, we briefly discuss the political economy of renewable energy. Although most of this discussion is found in the next chapter, here we define terms and explain the key obstacles to the growth of renewable energy. This discussion motivates the research question and sets the stage for a summary of our argument.

Renewable energy is defined by the IEA as "energy that is derived from natural processes (e.g. sunlight and wind) that are replenished at a higher rate than they are consumed."[8] The US Energy Information Administration (EIA) offers a similar definition when it states that, "[u]nlike fossil fuels, which are exhaustible, renewable energy sources regenerate and can be sustained indefinitely."[9] An energy source is thus *renewable* if it can be used in such a way that its stock will or can be restored despite continued use. Global attention in the past decade has focused on solar and wind energy.[10] However, their contribution so far is dwarfed by other renewable sources: hydropower and biomass. Indeed, the history of renewables is as old as the discovery of fire by humans some 400,000 years ago (Price 1995). For a long time, biomass—a renewable source of energy—was the main source of energy for human beings. Its dominance only began to be challenged in the 19th century with the explosive growth of fossil fuels, a dynamic that triggered unprecedented socioeconomic changes.[11] The end of the 20th century, in a sense, marks the return of renewables thanks to wind and solar power.

Renewable energy can be used in many ways. First, renewables are used to generate electricity. Solar panels installed on rooftops or wind turbines are designed to do precisely that. Some of these devices are then connected to the regular electricity grid, whereas others are so-called off-grid installations that are directly connected to a house or factory. Second, renewable energy is also used as a liquid fuel. Biofuels, such as ethanol, are derived from biomass. Their advantage is to reduce the demand for fossil fuels and generally emit fewer greenhouse gases than fossil fuels. However, the case of biofuels is particularly interesting because it underscores that renewable energy is not synonymous with clean energy. For example, biofuels incentivize farmers to convert forests to cropland, possibly increasing the total amount of greenhouse gases released into the atmosphere (Searchinger et al. 2008).

Although renewables have many uses, this book emphasizes the use of low-carbon renewables in electricity generation—in particular, wind and solar power. This choice is motivated by four observations. First, electricity generation is one of the main sources of global warming and air pollution, as fossil fuels continue to dominate the sector. By focusing on low-carbon renewables in this sector, we identify a critical challenge that any efforts to halt global warming must meet. In contrast, other renewables, such as biomass, have not seen as significant technological breakthroughs and may produce high carbon dioxide emissions.[12] Second, electricity generation is based on a mix of fuels, which allows us to examine the development of renewable energy in the context of competition with fossil fuels. In contrast, for example, modern transportation is almost entirely based on petroleum products, and renewable energy has not emerged as a serious challenger to the status quo. Third, the most important modern developments in renewables have been solar and wind power for electricity generation. Whereas solar and wind power remain minor components of the total energy consumption at the global level, their importance for electricity generation has increased rapidly in recent years. In 2005, biomass and waste generated 227 billion kWh of electricity and wind only 104 billion kWh; but in 2014, the numbers were 488 billion kWh and 714 billion kWh, respectively (EIA 2017a). Thus, the share of biomass—a source with potential for environmental degradation—among nonhydroelectric renewables has decreased rapidly in recent years and cannot explain the renewables boom in electricity generation. Finally, although we do not want to discount the importance of renewable energy in heating and transportation, the most important developments in the political history of renewable energy have, from a global perspective, been found in the power sector.[13]

Modern renewables offer an appealing solution to many problems stemming from the use of fossil fuels and nuclear energy in electricity generation. Fossil fuels are a major source of greenhouse gases (GHGs), which in turn contribute to climate change (IPCC 2013). Nuclear energy, although not a major source of GHGs,[14] can lead to serious environmental and public health problems, as the Chernobyl and Fukushima episodes attest. Furthermore, reliance on traditional energy sources may exacerbate violence, both directly (e.g., when countries go to war for oil; Billon 2001) and indirectly (e.g., through environmental shocks; Burke et al. 2009). Finally, fossil fuels are exhaustible and cannot provide a reliable supply for our economies in the long run (Smil 2010). Despite the problems with fossil fuels, the energy sectors of modern industrial economies reflect their high dependence on fossil fuels. As Unruh (2000) argued, the large share of fossil fuels in the energy mix since the industrial revolution has caused a "carbon lock-in," which in turn has made it difficult for cleaner alternatives to emerge. In the conventional model, fossil fuels—especially coal, oil, and natural gas—and uranium are extracted from the ground for combustion or nuclear reactors. In the power sector, this process has been strongly centralized, with most power plants having a generation capacity of hundreds or thousands of megawatts. In the transportation sector, fuel is consumed by individual units such as private automobiles, but the extraction and refining of the primary transportation fuel, oil, remains highly centralized in the hands of large oil companies (Victor, Hults, and Thurber 2012).

According to Unruh (2000, 818), this energy economy has perpetuated itself despite a heavy environmental cost:

> This condition, termed *carbon lock-in*, arises through a combination of systematic forces that perpetuate fossil fuel-based infrastructures in spite of their known environmental externalities and the apparent existence of cost-neutral, or even cost-effective, remedies. Rational corrective policy actions in the face of climate change would include removal of perverse subsidies and the internalization of environmental externalities arising from fossil fuel use.

Fossil fuels maintain a tight grip on energy generation across the world. As of 2010, oil, coal, and natural gas comprise 81% of the total primary energy supply (Dangerman and Schellnhuber 2013). If we add nuclear energy, the total reaches 87%. This share has remained stable since 1973, although the total amount of energy generated from these sources doubled over this period.

The carbon lock-in presents three problems for renewable energy. The first is that decades of investments have transformed what used to be energy sources with uncertain payoffs into cheap fuels. Fossil fuels benefit from all the trials and errors of past generations. The development of ways to extract energy from fossil fuels required patience and large investments. Once these investments were made and all the difficulties of setting up these production systems were ironed out, the competitiveness of fossil fuels was greatly enhanced. Investing in traditional energy sources provided immediate returns, unlike investments in renewables. The benefits of renewables were, at least initially, uncertain, and the process of transition from fossil fuels to renewables remains fraught with risks (Jacobsson and Johnson 2000). Even investors who would dare to try their hand in renewables would face the problem of intellectual property rights. Competitors could build on technological breakthroughs funded and achieved by them (Victor 2011, 135). Fossil fuels have had centuries to solve these problems, and they continue to do so as new technologies, such as hydraulic fracturing for shale gas extraction, become available. Renewables begin the competition with fossil fuels from a position of weakness, and the fossil fuel industry is investing heavily in developing new technologies to tap previously unavailable resources.

The second issue is negative environmental externalities. The generation and use of traditional energy sources leads to pollution (I. Dincer 2000). This pollution can be local, as in the case of increased air pollution, or global. Burning oil and coal produces carbon dioxide, a potent GHG and cause of global warming. Climate change has various negative effects on human societies. The fossil fuels consumer does not have to consider the suffering imposed now and in the future on others. Most of the billions of victims are far away, both geographically and temporally speaking. This remoteness reduces the consumer's incentive to protect the planet by not consuming fossil fuels, and so the consumer price of electricity does not reflect its true societal cost. Renewables would be much more competitive if fossil fuels came with a tax based on the damage caused by carbon dioxide emissions, but such carbon taxes remain rare. Consequently, the world uses too many fossil fuels and too little renewable energy for its own good.

The third obstacle is political. Traditional energy industries have accumulated a lot of political capital over time. In contrast to the largely decentralized nature of renewable energy generation, most of the existing infrastructure is highly centralized and in the hands of major oil, gas, and coal companies. This centralization raises barriers to entry in the energy market as well as the political market for policy. In countries where energy is produced by private corporations, a small number of large and wealthy firms tend to dominate the market due to economies of scale and high levels of industry concentration (Hirsh 1999; Victor, Hults, and Thurber 2012; Cheon and Urpelainen 2013). These conditions are ideal for a powerful lobby to emerge (Olson 1968), and in many countries this has been the outcome of centuries of energy sector development (Victor 2011, 130). Indeed, the majority of fossil fuels have been extracted by fewer than a hundred "carbon majors" since the industrial revolution (Heede 2014). The dominance of traditional energy interests gives them a major advantage in the policy-making process, allowing them to slow down a transition to renewables that could be financially costly for them.

In summary, although renewables offer a promising solution to the problems caused by conventional energy sources, they face many obstacles. Yet their growth over the past years has been fast. Their progress over the past four decades goes against the conventional wisdom about the carbon lock-in of industrialized societies. Investing in them is a risky endeavor that may not pay off for the individual investor. The renewable industry learns by doing and from research over time, but the process is a slow and difficult one. Given these challenges, the puzzle that motivated us to write this book becomes even more striking. Why have renewables been much more successful than anyone thought? Why is there so much variation in their importance across countries?

1.2 Summary of the Argument

To make sense of the political history of renewables, we must first characterize the key features of the problem at hand. The first feature is the presence of a stable equilibrium. A society in a carbon lock-in simply cannot survive without fossil fuels, and initially the prospects of change appear grim. There is widespread agreement across the energy industry regarding the stability and desirability of the status quo. Any efforts to disrupt the equilibrium would require decades to be successful, and few stakeholders see the urgency of exploring the possibility of transformational change. This convergence of views among key stakeholders—government, industry, civil society, and the public—strengthens the existing equilibrium.[15] Any shift to an alternative equilibrium depends on growing demand for alternatives, such as renewable energy. Such demand, in turn, depends on widespread concern about the cost of fossil fuels to the society and uncertainties surrounding the sustainability of the lock-in.

The second feature of the analytical problem is the political difficulty of a systemic change. Even if some social entrepreneurs saw the need to break out of the carbon lock-in, they would face incumbent interests that profit from the extraction and use of fossil fuels. Thus, initial efforts to bring about systemic change would give rise to political conflict and opposition. Although prospective clean technology companies, environmentalists, or advocates of appropriate technology are individual winners, their views are not initially shared by most stakeholders. In contrast, the views of incumbent interests, such as the fossil fuel industry and heavy industries, inform policymaking and energy planning. As the challenge to the status quo grows strong, the vested interests that benefit from the old order begin to oppose change. This political struggle is dynamic and can last decades as elections, technological innovations, and global developments shape the relative cost of renewables and fossil fuels.

These broad features—a stable equilibrium that can only be broken after a long and intense political struggle—inform our theory of the politics of renewable energy. As the previous review of renewable energy showed, breaking the carbon lock-in is inherently difficult. As long as governments, investors, and other decision makers expect fossil fuels to remain the foundation of the industrial society, systematic change is simply not possible. We thus begin by arguing that, given the historical centrality of fossil fuels for industrialization, an *external shock* is necessary for renewable energy to even begin to grow. By an external shock, we refer to a major, abrupt event that reveals the weakness of current policy and is not the direct product of a government's own policy. Without some kind of an external shock, the carbon lock-in that has characterized industrialized societies in the 20th century simply does not allow alternative energy sources to prosper. If the rationality and inevitability of fossil fuels as the cornerstone of the energy economy is called into question, however, renewables have a fighting chance.

Abrupt changes in oil prices and accidents that raise concerns about nuclear power are external shocks that create political demand for policies that reduce national dependence on fossil fuels. On the one hand, an immediate and concrete threat to the society draws public attention to the systemic risks of relying on finite, polluting fossil fuels. On the other hand, the same threat also underscores the need to find alternatives and leads to a widespread acceptance of the need to explore alternative energy futures. In some countries, the government reacts to the shock by investing in renewables as a rational response to changing circumstances. The government's activities reduce the cost of renewable energy production through economies of scale, learning by doing, and scientific research.

Over time, the growth of renewable energy will generate a political backlash. From coal miners to heavy industry and fiscal conservatives, various societal interests expect to lose from generous government support for renewable energy. In countries that see an initially successful renewable energy industry, the opposition mobilizes to *politicize* renewable energy. In this case, politicization refers to attempts to cast doubt on the wisdom of investing in renewables. Opponents of renewable energy challenge the technical, economic, and social merits of these alternative energy sources. As a result, renewables become the subject of a political debate and struggle.

We argue that the political strength of renewable energy depends on factors such as public opinion, partisan ideology in the government, and the political-economic clout of various industries.[16] We provide a detailed review of the key factors and explain why they are more important than plausible alternatives. Based on these factors, we then explain and predict the development of renewable energy in different countries over time. Given that the advocacy coalition for renewable energy is in a position of weakness in the beginning, a highly favorable constellation of factors is necessary for it to be politically successful. Environmentalists, clean technology entrepreneurs, and green parties can only succeed if political institutions give them access, public opinion is favorable, and the alliance of fossil fuel producers and heavy industry is vulnerable to political challenge. Where fossil fuels continue to thrive due to favorable economic and political conditions, the growth of renewable is slow. The explosion of oil and gas production in the United States is probably the best example of the limited success of renewables when the fossil fuel lobby is sufficiently powerful and benefits from political institutions that make policy change hard.

The process of politicization generates a variety of outcomes in different countries. In some cases, such as the United States, renewable energy is suppressed by unfavorable political-economic conditions. The American fossil fuel and heavy industries had political clout, public opinion was not overly concerned about nuclear power or climate change, and the coalition of environmental groups and clean technology industries remained weak. In the 1990s, growing attention to climate change did not translate into federal renewable energy policy because a powerful industry coalition led by large energy corporations opposed policies that would encourage the growth of low-carbon energy generation.

In other countries, including Germany and Denmark, the advocacy coalition for renewables emerges triumphant and the society undergoes a *renewable energy lock-in*. The fossil fuel and heavy industries were weak relative to the broad-based coalition of environmentalists, clean technology industries, and small renewable energy producers that directly benefited from policies such as investment subsidies and FITs. The success of this renewables advocacy coalition also reflected public concerns about nuclear power—as evidenced by the importance of the Chernobyl nuclear accident for German energy policy—and later climate change. The resulting lock-in is not identical to carbon lock-in, yet it shares some fundamental features that help us understand why renewables have been so successful. The forerunner countries, in which renewables undergo a lock-in, drive down the cost of renewable energy production while providing a collection of variable policy experiences and lessons for other countries.

As renewable energy continues to grow within the boundaries of forerunners, it becomes increasingly attractive for other governments as well. In the United States, rapid decreases in the cost of wind and solar power, together with a federal production tax credit and state portfolio standards, have created an investment boom. Although societal resistance to renewable energy sources has not disappeared, over time technological progress and improved economic competitiveness have tilted the scales in favor of wind and solar power. Even more striking, however, is the rapid progress of renewable energy outside the traditional industrialized countries. From emerging economies such as China and India to least developed countries such as Kenya and Tanzania, renewable energy has grown rapidly and has begun to play a critical role in powering economic growth and contributing to energy access for the population.

Renewables, we argue, have become a global boom industry because of the various benefits they provide to a wide range of governments in an equally wide range of national settings. Decreasing energy production costs, the availability of tried and proven policy models, and the positive image of renewables together make renewables an attractive solution to a variety of socioeconomic woes. These woes include fossil fuel dependence, energy security, environmental problems, and rural electrification. Although renewables, along with natural gas and nuclear power, have helped China and South Africa begin a transition away from coal in electricity generation, in India renewables allow the government to expand electricity generation capacity for the people, industry, and services. In Kenya, off-grid products such as solar home systems have spread rapidly and now contribute to domestic electricity access for millions.

These three stages of development constitute an analytical model of the politics of renewable energy. From external shocks to intense politicization and a period of rapid global growth due to a political-economic lock-in, government policy plays a decisive role in the promotion of renewable energy. The policies and outcomes that emerge are not economically rational, but their political logic is impeccable. This argument can explain why renewable energy has grown so fast since the 1973 oil crisis and why it has done so at a high economic cost in a process of fits and starts. The rapid global growth of renewable energy reflects an auspicious combination of initial external shocks, successful political maneuvering under fire, and diffusion upon initial success in frontrunner countries. In today's globalized world, positive political and economic experiences with renewable energy spread across borders rapidly, turning renewables into a truly global business. The process is significantly more complicated and involved than previously envisioned, but the fundamental principles of political economy can shed light on the underlying causal mechanisms.

This theory can explain why renewables have grown so rapidly during the past decade and why there is so much variation across countries. The pioneer countries that initially chose to invest in renewables—Denmark, Germany, and the United States—did so in response to the 1973 and 1979 oil crises, as well as the nuclear accidents of the time. By 1981, President Reagan's policies ended the progress of renewables in the United States. In Denmark and Germany, however, the advocates of renewable energy emerged victorious from the political contest with their detractors. In Denmark, the pioneering 1976 Tvind turbine was an instant success and drew thousands upon thousands of visitors every year, making the small Nordic country a global leader in renewable energy for decades to come. In Germany, a broad coalition of farmers, engineers, environmentalists, and ordinary citizens worried about nuclear power and climate change launched a dynamic progress toward replacing nuclear power and fossil fuels with renewables. Thanks to the early investments made by pioneers, a rapid decrease in the cost of renewable electricity generation enabled interested governments to use renewable energy. Over time, many other countries, including Spain and China, chose to exploit the decreased price of renewable energy to pursue a variety of socioeconomic goals. Even in countries that initially showed little enthusiasm for renewables, such as France and the United Kingdom, the industry is now growing due to the positive reinforcement mechanisms that have strengthened renewables over time.

1.3 Ideas, Implications, and Contributions

Ours is the first book to provide an overview of the rise of modern renewables in the power sector, first in a few pioneer countries and then globally from the perspective of political economy. A growing body of literature now exists on energy transitions in general (Fouquet 2008; Smil 2010), but this line of inquiry focuses on techno-economic considerations and treats policies and politics as an explanatory factor of secondary importance. The number of case studies has grown steadily in the past years (Heymann 1998; Toke 2002; Agnolucci 2007; Lipp 2007; Sovacool 2008; Laird and Stefes 2009), however, and their results emphasize the importance of government policy for the growth of renewable energy. Espen Moe's (2015) *Renewable Energy Transformation or Fossil Fuel Backlash* describes the progress of renewables in six countries and discusses the role of vested interests, such as the fossil fuel industry. However, the author does not analyze the dynamics of renewable energy growth over long periods of time or offer an integrated model of the roles of policymakers, interest groups, and public opinion. Karapin's (2016) *Political Opportunities for Climate Policy* compares climate and energy policies across the United States and over time, but his research lacks an international dimension and does not consider the global spread of climate policies.

Adopting a global perspective that spans four decades, we provide the most comprehensive analysis of the politics of renewable energy to date.[17] We offer case studies of the three key forerunners (United States, Germany, and Denmark) and contrast their trajectories with three countries that did not make early investments into renewable energy (United Kingdom, France, and Finland). We also offer an overview of the recent globalization of renewable energy, going beyond case studies of industrialized countries and discussing developments across all continents, from China and India to Brazil, Kenya, and South Africa. We show that the origins of the globalization of renewable energy can be found in the initially experimental and later increasingly ambitious policies enacted by a small number of frontrunner countries.

In the literature on sustainable energy transitions, political economy analysis is becoming increasingly common (Jacobsson and Lauber 2006; Lipp 2007; Klick and Smith 2010; Cheon and Urpelainen 2012; Aklin and Urpelainen 2013b; Moe 2015; Karapin 2016). This book brings together several strands of thought, including the role of carbon lock-in (Unruh 2000), fossil fuel dependence (Cheon and Urpelainen 2012), interest group politics (Sabatier and Jenkins-Smith 1993; Aidt 1998; Cheon and Urpelainen 2013; Moe 2015), and public opinion (Klick and Smith 2010; Aklin and Urpelainen 2013a). We show that national politics play a complex role in the evolution of renewable energy, with variation in their effects and importance at different stages of this development. Because of a carbon lock-in, for most of the 21st century, renewable energy was a negligible political issue. The 1973 and 1979 oil shocks did not result in the politicization of renewables because few societal interests considered modest investment in alternative energy sources a threat to their status. Over time, however, politicization followed the unexpected growth of some renewables, such as wind power. A coalition of environmentalists, renewable technology industry, and small-scale producers of renewable energy grew so powerful in countries such as Denmark and Germany that, riding on the wave of public concern about nuclear power and climate change, they induced successive governments to offer generous support to renewable energy. Our analysis captures these developments and provides an analytical framework that can be applied to investigate sustainable energy transitions. The analytical framework explains why renewables were able to grow over time in pioneer countries even at times of low fossil fuel prices, such as the 1990s, as the combination of rapidly decreasing generation costs and expanding advocacy coalitions made renewables a political winner in Germany and Denmark—and later elsewhere.

Beyond the political economy of renewable energy, this book contributes to the study of external shocks in politics. Although our empirical application focuses on renewable energy, the analytical framework offers new insights to a wider range of political-economic processes. Renewable energy is but one example of a potentially disruptive technology that struggles to establish itself in a previously stable system. Electric vehicles and open-source software are other examples of potentially transformative technologies that could be studied under our analytical framework. In the case of climate policy, Karapin (2016) draws on the notion of "focusing events" to understand federal- and state-level developments in the United States. Calder (2012) investigates the role of "critical junctures" in the Eurasian geopolitics of energy. Our theory can be applied as long as the basic scope conditions, such as the existence of a dominant, status quo technology and the centrality of government policy for the growth of the new technology, are met.

Besides the study of technology, our analytical framework can be applied to other questions in comparative and international political economy. Consider financial crises. Before 2007, few people worried about the stability of the financial system, just like few people worried about oil scarcity before 1973. Nobel Prize laureate Robert Lucas (2003, 1) famously quipped that the "central problem of depression prevention has been solved." Then came the crisis, and demand for financial reforms grew in many countries. In the United States, President Obama signed the Dodd-Frank Act in 2010. Eventually, however, an intense political struggle ensued as opponents of financial regulation called for the repeal of the Dodd-Frank rules.[18] This storyline bears uncanny resemblance to the political history of renewable energy, as a major shock was necessary to break a lock-in and initiate a political struggle over the future of financial regulation. Another example is urban renewal (Evans 2002; Fitzgerald 2010; Seto, Sánchez-Rodríguez, and Fragkias 2010; Romero-Lankao and Dodman 2011). Changing the trajectory of an urban social system is difficult because of vested interests. An external shock may create a shared understanding of a need to change, but systemic transformation is a slow process that requires sustained infrastructural investments and will give rise to political conflicts in the urban sphere (Campbell, 1996). Other possible applications include transportation infrastructure (Cowan and Hultén 1996; Åhman 2006; Frantzeskaki and Loorbach 2010) and regulatory reforms in the governance of infrastructure (Horowitz 1989; Levy and Spiller 1994; Hirsh 1999; Vogel 1999; Victor and Heller 2007). In all of these cases, systemic change is difficult because of lock-in effects, and governments have little hope of changing outcomes rapidly without long periods of political conflict and clashes between proponents and opponents. Our analytical framework provides social scientists with tools to understand and solve such problems.

In theoretical terms, we solve an important puzzle for the field by showing that external shocks can slowly engender policy responses, producing a pattern that differs starkly from the conventional "punctuated equilibrium" theory of policy change (Gersick 1991). Although the initial policy responses to oil and other shocks did not leave a mark, they enabled other, more ambitious policies in the future. This longitudinal analysis shows how external shocks can trigger initially invisible, slow-moving processes that over time create new opportunities for deep societal changes.

Finally, the results also speak to the question of when and how governments can consistently enact and implement policies over long periods of time. Political economists often argue that politicians have short time horizons (Blinder 1997). They navigate from one election to the other, with little interest in policies that furnish benefits over time (A. Jacobs 2011). In the case of renewable energy, it is initially puzzling that governments continued policies even after the initial energy shocks had waned. The solution we offer is the growth of the renewables advocacy coalition over time due to supportive policy. If initial policies strengthen and create constituencies that support a certain outcome, such as renewable energy, these policies expand the coalition of their supporters. This is why consistent policy formulation is possible even after the initial window of opportunity seems to have already closed. This constituency creation through policy solves the classical problem of "time inconsistency" (Kydland and Prescott 1977). Governments often cannot credibly commit to continuing certain policies over long periods of time, but if these policies create supportive constituencies, the initial promises are credible because various interest groups and constituencies see that the policies create a politically robust support base for themselves.

1.4 Plan of the Book

This book has eight chapters. The introduction is followed by a summary of the development of renewable energy over time, emphasizing both technological progress and government policy, in chapter 2. The chapter highlights both the abrupt increase in the growth rate of renewables during the past decade and the slower growth of the industry during the preceding three decades. An experienced analyst of renewables can briefly skim this chapter and move directly to the theory chapter that follows.

Next, chapter 3 provides a theory of how policy responses to external shocks, such as abrupt oil price increases and concerns about climate change, allowed renewables to grow. The purpose of the theory is to structure the empirical analysis and identify the relevant explanatory and dependent variables for the study. The chapter explains why renewables were able to initially break the carbon lock-in identified by Unruh (2000) and, in some but not all frontrunner countries, prevail in the political conflict that followed their initial success. The final section of the chapter details the global implications of the success of renewable energy. Throughout the chapter, we propose a series of predictions that can be subjected to stringent empirical evaluations.

Chapters 4 to 6 use tools of political-economic analysis to offer an analytical account of the history of renewable energy. Chapter 4 applies our theory to the birth of renewable energy as a policy issue in the aftermath of the 1973 oil crisis. Comparative case studies of several key countries show that the oil crisis is key to understanding why renewable energy started to grow in a small number of pioneer countries. In addition to examining the cases of frontrunner countries, we investigate why some dogs did not bark. The number of comparative cases is high relative to the earlier research cited above, and so we are able to draw new inferences on the early years of the Odyssey of renewable energy.

In chapter 5, we examine the politicization of renewable energy during the decades that followed. We show that the initial success of renewable energy raised concerns and provoked opposition to renewables among a variety of social interests, ranging from fossil fuel producers to heavy industry and fiscal conservatives worried about the cost of renewable energy policy. Even in the strongest advocates of renewables, Germany and Denmark, politicization was unavoidable due to tremendous pressure. In many of the frontrunner countries, however, renewables had already generated enough enthusiasm and vested interests that the policies supporting them survived the attack of the opposing coalition. In these countries, renewables began to grow at a rapid pace.

Chapter 6 first investigates how renewable energy prevailed and underwent a technopolitical lock-in in the frontrunner countries. The chapter identifies the conditions that allowed renewables to prosper as the political controversies around them started to subside. Equally important, the chapter generates expectations about the global spread of renewable energy policy and capacity, as various governments outside the frontrunner countries start to use renewables to address their core concerns in the energy sector. In addition to industrialized countries, renewable energy has become a key strategy in emerging and least developed economies, which goes against the expectation that the price tag of renewables would limit their scope to wealthy societies. Together these three empirical chapters allow us to test the theory and offer new insights into the causes and consequences of the growth of renewable energy. We offer detailed quantitative and qualitative evidence using a large number of available data sources, both old and new.

In chapter 7, we look into the future and discuss some of today's pressing policy challenges. We begin with an analysis of national policy, applying the lessons from our analysis to contemporary policy challenges. We identify a set of policy design principles that allow renewables to weather political storms and grow without excessive economic cost. We then shift from national to international policy, analyzing current efforts to engage in international cooperation on renewable energy. We argue that international organizations can promote the growth of renewables especially in the least developed countries, in which governments struggle to formulate effective policies to solve their difficult problem of energy poverty.

The concluding chapter brings together different aspects of the argument and relates them to big questions in political science and economics. We discuss the implications of the study for energy research, comparative and international political economy, and the future of renewable energy in the world economy. The concluding chapter connects the argument to broader debates in political economy, arguing that the story of renewables can be best understood through the methodical application of the first principles of political economy. At the same time, our analytical account of the politics of renewable energy can inform many other debates on governments, policy, and economic development.

2 Renewable Energy: Past, Present, and Future

Compared with fossil fuels, modern renewables are still a rather immature source of energy, but their growth has been rapid in the past decade or so. How did renewables grow from an obscure pursuit of a few enthusiasts to a key and growing component of the modern energy economy? This chapter lays the foundation for answering this question by tracking the historical evolution and development of renewable energy.

The development of renewable energy can be illustrated in the American context. The EIA has kept track of monthly renewable electricity generation since January 1973. As shown in Figure 2.1, the trend is clear. Nonhydroelectric renewable sources of energy have grown from zero to more than 9% of national net electricity generation. In combination with hydroelectric power, this means that renewables are now more important for the US energy economy than nuclear power. As the figure also shows, the growth has occurred despite no growth at all in hydroelectric generation—the hydroelectric share has decreased because total electricity generation has increased without added hydroelectric capacity. This pattern is not driven by bioenergy either, with wood and waste accounting for only 20% of these nonhydro renewables (down from 80% in the 1990s). Wind power has grown rapidly, and now solar power also appears to be on the verge of taking off. This development is all the more remarkable given that the United States is often considered a laggard in renewable energy development (Laird and Stefes 2009).

At first sight, these numbers may appear underwhelming to some. Can we talk of an energy transition, even one that is ongoing and uncertain, if the share of nonhydroelectric renewables in electricity generation is still only 9% and the sum of hydroelectric and nonhydroelectric 15% in the United States? From a systemic perspective, the answer is yes. First, energy transitions do not mean that traditional sources disappear. A transition is

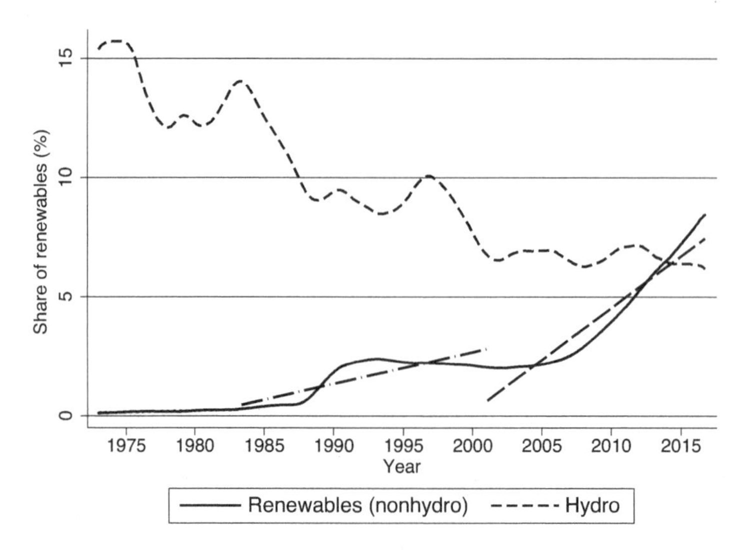

Figure 2.1
Monthly share of renewables in electricity generation in the United States, from January 1973 to October 2016. For nonhydro renewables, the share is computed as the sum of electricity generated from wind, solar, geothermal, waste, and wood divided by total electricity generation. The time series are smoothed for easier reading. The dotted lines indicate the trends in the three phases of renewable energy policy in the United States (see chapter 3).
Source: EIA (2017b), January 2017: Table 7.2a ("Electricity Net Generation: Total (All Sectors)").

not defined by the complete replacement of one energy source with another. In this case, a transition instead implies that an energy system, although designed to facilitate the use of fossil fuels and shaped by the carbon lock-in, sees a rapid increase in the use of renewable electricity. Second, renewable energy can be disruptive even at low levels of output. Stock markets have noticed it for a long time. In Germany, the value of utilities has collapsed as renewables have begun to drive down the price of electricity.[1] Even accommodating 10% of electricity from renewables has serious consequences for the valuation and profitability of other sources. New projects may suddenly be threatened. For instance, the decrease of wholesale electricity prices in Europe due to German investments in renewable energy has killed new hydroelectric projects in Switzerland.[2] A transition is characterized by the transformation of the energy market, energy institutions, and energy infrastructure. Once a transition is completed, these sunk costs make it difficult (but not impossible) to revert back to the old model.[3] Because renewables are growing fast, regulators, utilities, and energy producers are forced to reform their rules and business models to accommodate renewables.

To give but one example, the growth of renewables requires significant changes in the electricity generation system because utilities must find solutions to the issue of intermittent solar and wind power (Jacobsen and Zvingilaite 2010; E. Baker et al. 2013). The sun only shines during the day and wind speed varies both between seasons and over any given day. According to Joskow (2011), accounting for intermittency—something that standard methods based on levelized costs fail to do—significantly increases the cost of renewable electricity in general and wind power in particular. This intermittency creates technical difficulties because electricity is not necessarily available when there is demand for it. For example, solar power must be stored if it is to be used at night. Similarly, wind power generated in the middle of the night presents a problem because demand for electricity is often low at that time.

The end result of a successful energy transition is a lock-in. As explained in the introductory chapter, until now, our economies have been under a carbon lock-in (Unruh 2000). The conventional energy system is suited for fossil fuels and nuclear power.[4] For decades, this energy system has raised high barriers for renewable energy sources. It is hard for competitors to survive when fossil fuels have benefited from years of systems development and aggressive investment in infrastructure for fossil fuels and nuclear power.

In many ways, then, the ongoing energy transition to modern renewables is more significant and complex than any past transition. Past energy transitions were largely completed with little regard for negative environmental externalities. This is not to say that earlier transitions were unregulated or that the government did not play a role (Nye 1998). It doesn't even mean that people were completely oblivious to some of the downsides of new fuels, notably pollution-intensive coal. For instance, in the late 19th century, some referred to hydropower as "white coal" because it provided electricity without the pollution (Landry 2012, 9). American cities such as Pittsburgh or St. Louis, which suffered from coal-related smoke, tried to regulate air quality around the same time (they mostly failed partly because courts tended to side with polluters; Tarr and Zimring 1997; Tarr and Clay 2014). Gifford Pinchot, an early leader of the American conservationist movement and governor of Pennsylvania, advocated for the development of hydropower over coal, urging his fellow citizens to save the latter (Pinchot 1910). However, the primary driving force was productivity and efficiency, and the use of coal grew over time together with industrial production of goods

(Smil 2010). As Andrews (2008, 34) notes, for example, in 19th-century America, advocates of coal consider the move away from biomass into fossil fuels a momentous transformation: "[w]hile organic economies remained subject to the limited flow of solar energy, mineral-intensive economies could draw on the accumulated capital of eons past—the fossilized plant matter that had piled up over many millions of years in Britain and Colorado alike." Modern renewables, in contrast, faced roadblocks because their value did not come primarily from more or cheaper power; instead, it came from fewer negative externalities—and, at least initially, at the expense of abundance and affordability.

How can we make sense of the ongoing renewable energy transition? We begin with a technical discussion of the challenges of renewable energy, adding precision to the general characterization of the challenge given in the introduction. Following this discussion, we describe how renewable energy was for centuries the main resource that humankind could use for heating, cooking, and other endeavors. This state of affairs changed with the industrial revolution as coal and later oil offered an abundance of energy that enabled the development of heavy industries. The reliance on fossil fuels only became a policy issue in the aftermath of the 1973 oil crisis. The era of abundant and cheap energy was over, and policymakers around the world had to rethink how to power their economies. Some remained stubbornly dependent on fossil fuel, often for political reasons. Others, however, soon laid the foundation for the deployment of renewable energy.

2.1 Renewable Energy: Opportunities and Challenges

To make sense of the growth of renewables in electricity generation, it is important to understand their status today and key problems associated with their growth. The review in this section sets the stage for the political analysis in the subsequent chapters.

To begin with, different types of renewables are at different stages of their development. Some, such as wind and solar energy, are in some areas already competitive with conventional electricity sources (Foxon et al. 2005). The example of wind power is telling (Ackermann and Söder 2002). In 1989, a single wind turbine had a capacity of 300 kW. A decade later, that number had climbed to 1.5 MW while prototypes with a capacity of 2 MW were being tested. By 2012, the size of the largest wind turbine reached 7.5 MW.[5] To give some context to these numbers, current nuclear power plants have a capacity of about 1,000 MW on average.[6] The gains are reflected in capacity as well as prices. The cost of wind electricity from

American plants dropped from $4.30 per watt in 1984 to $1.90 per watt in 2009 (IPCC 2011). Similar improvements were observed in Europe. In Denmark, for instance, the price of a watt of wind electricity dropped to $1.40 per watt over the same period. As shown later, these learning and scale effects are the outcome of aggressive efforts by governments to promote renewable energy.

Solar power has made huge strides, too. Although the sizes of solar parks vary considerably, their average capacity is about 22 MW in Germany and 24 MW in China. Spain has been at the forefront of these developments, with parks that have a capacity of up to 60 MW (F. Dincer 2011). This progress is staggering given that the first 1 MW park was opened in 1982, only three decades ago (Arnett et al. 1984). At the same time, the price of solar electricity has massively decreased and in some places in the United States has reached grid parity, with retail prices below $1.00 per watt for large parks; the average cost was $2.43 per watt in the United States as of 2011 (Quiggin 2012). Other estimates suggest that the price of solar energy dropped even further from $76 per watt in 1977 to $0.74 per watt in 2013.[7] In other words, solar electricity is now competitive at the margin with traditional sources in a number of regions. Overall, these technologies have shown that they can be a reliable source of energy at an affordable cost.

In contrast, progress in renewables outside the power sector has been much less impressive. Worldwide, about 58% of all fossil fuels are used in the transportation sector (Nigam and Singh 2011). Given the effects of fossil on public health, the environment, and economic stability, biofuels initially seem to be a promising solution, and their production grew from 315 barrels per day in 2000 to 1,901 in 2012 (EIA 2013b). In the United States, one of the major supporters of biofuels, production increased from 106 to 940 barrels per day. Even these numbers remain small when compared with the production of oil. In the United States alone, production of oil over the same period increased from 9,058 to 11,119 barrels per day. Since then, it has increased to almost 14,000 barrels per day. Partly, this is due to the different technical challenges faced by renewables in the electricity and liquid fuel sectors (Nigam and Singh 2011). First, from a technical perspective, biofuels must be easy to store, easy to distribute, and efficient in combustion (Roman-Leshkov et al. 2007). As of now, even ethanol, the most advanced type of biofuel, comes up short. Although increased research investments may reduce these concerns (Chu and Majumdar 2012), the second problem is more fundamental. Biofuels are typically generated from biomass, such as sugar cane or corn, and so the resource is not freely available. Consequently, biofuels must operate in a different kind of market.

The technical challenges that renewables face in the power sector are quite different. To understand them, it is necessary to consider how the electricity market works in general. It can be divided into three parts: generation, transmission, and distribution. Electricity is generated in a power station and then transmitted to a substation transformer, which distributes it to consumers (typically households or firms). This basic setup is managed by national and subnational operators, generally in close cooperation with other neighboring systems. For instance, continental European grids are for the most part connected with each other and placed under the umbrella of the European Network of Transmission System Operators for Electricity. In contrast, electricity pricing is generally managed at the national level.

In the old days (prerenewables), the system was based on two pillars. Some electricity sources provided a base load, which corresponds to the minimum demand for power. In many countries, the base load is generated by sources such as nuclear power that can supply a steady stream of electricity with few unanticipated disruptions. Consumers, however, do not need a constant quantity of electricity. For instance, heating is in higher demand in the winter. Variations also occur within a single day. People turn on lights and the television when they return from work. In the summer, some may use air conditioning. To satisfy variations in demand, base load plants are augmented by peaking power plants. Some sources, such as hydropower or gas turbines, are particularly suitable for responding to these seasonal and short-term fluctuations in demand.

The rapid growth of renewables disrupts this well-oiled machine. Most modern renewable power sources are highly variable, a problem called "intermittency" (Sovacool 2009). Turbines, for instance, do not work below a certain wind speed, referred to as the "cut-in" speed. Modern wind turbines require a minimum speed of 5 meters per second (about 11 miles per hour) to start generating electricity, although the exact number varies by turbine (Wan, Ela, and Orwig 2010; Vanvyve et al. 2015). When the wind is too strong or above the "cut-out" speed, the turbine is put to rest to avoid damage to the rotor. Therefore, wind power can only be generated when the wind blows within a certain range. Wind patterns are somewhat predictable because wind blows more strongly in the winter than in the summer (Taylor McSharry, and Buizza 2009; Heide et al. 2010). Even then residual variation is difficult to model and predict. Solar power is subject to similar limitations (Zahedi 2011). Besides night time, the duration of which varies over the year, cloudy weather can also negatively affect the power generated by solar photovoltaic systems (Nguyen and Lehman 2006; Singh

2013). Unlike wind, however, solar power is mostly available during the summer and can thus counterbalance seasonal variations.

To summarize, the intermittency problem has two faces. Renewables such as solar can predictably be unavailable during certain times. On top of that, resources such as wind can unpredictably fall short. The lack of predictability means that grid operators may need to rely on other plants to provide power when wind generation falls (Kavasseri and Seetharaman 2009). If they are unable to do so, power shortages could ensue. This short-run uncertainty makes accurate weather forecasting vital, even if only minutes ahead (Anderson and Cardell 2008; Tewari, Geyer, and Mohan 2011). Economically, intermittency is also a source of uncertainty for the actual return that turbines will generate. Wind turbines are a risky asset in regions that experience high variance of wind strength. For similar reasons, intermittency may increase the variability of electricity prices. For example, some studies suggest that this has been an important issue in Germany (Ketterer 2014).

One possible solution to the issue of intermittency is to store power (Beaudin et al. 2010). One approach uses batteries. According to various estimates for Europe, an amount of power equivalent to 10% to 15% of total generation must be stored if all electricity were generated by wind and solar power (Heide et al. 2010, 2011; Rasmussen, Andresen, and Greiner 2012). Whether this is feasible under present technology is unclear (Rasmussen, Andresen, and Greiner 2012). Another solution is to create larger electric systems that can balance excess power in one region with shortages elsewhere (Heide et al. 2010). For instance, excess wind in Denmark can be used to offset unexpected low winds in Germany. According to Heide et al. (2010), this approach can significantly reduce the need for storage.

The intermittent nature of renewable electricity generation causes additional economic complications. Renewables generate power at a low marginal cost. Solar panels can be expensive to build and install, but once they are there, they can be used at a low cost. This is problematic because of the way electricity prices are determined. Under the current system, spot prices are defined by aligning all possible sources of supply from the least to the most costly. The price is defined as that suggested by the last supplier needed to meet the demand. Because renewables are extremely cheap, they push the more expensive suppliers out of the market and, in doing so, depress electricity prices. As more and more units of electricity are generated by renewable sources, their own profitability decreases (Hirth 2013).

In other words, renewables do not simply substitute for conventional sources. Rather, they transform the market by pushing spot prices down. Holding demand constant, the aggregate revenue generated in the power sector goes down. In fact, various studies suggest that renewables such as wind and solar power undercut prices exactly when they need to be high enough to make them worthwhile (Joskow 2011; Edenhofer et al. 2013; Ueckerdt et al. 2013). In turn, decreased revenues reduce investments and profitability. Ignoring this factor makes it easy to overestimate the competitiveness of renewables. At some point, renewables may hit a ceiling of their own making. According to Jesse Jenkins, an energy analyst, "wind may eventually be able to provide on the order of 25-35 percent of a power systems' electricity, while solar may top out at 10-20 percent in most regions."[8] Whether these predictions are accurate remains to be seen, but finding a sustainable model of funding is a major challenge to increased use of renewables in the power sector.

A third challenge comes from potential new competitors. Possibly the most prominent of these competitors is unconventional natural gas, especially in the United States. Within a few years, two technological developments, the advent of horizontal drilling and hydraulic fracturing, transformed a dormant industry into a powerful new energy provider (Schrag 2012). Gas that seemed beyond the grasp of humankind could suddenly be extracted in a cost-effective manner, and abundant quantities could now be used for electricity generation. For this book, shale gas raises two questions. First, are these sources a threat to the viability of renewables? Second, can we talk of a renewables transition if other sources can threaten them seemingly out of nowhere?

The first question is still debated. The effect of unconventional gas on renewables depends on the net effect of a series of interlinked events. More gas in the United States could mean less demand for coal, allowing both renewables and natural gas to grow (Lee, Zinaman, and Logan 2012). However, as US demand for coal decreases, global coal prices may go down, too. This development may spur more coal imports by Europe and other regions. This, in turn, could negatively affect renewable energy producers in these areas. A similar view is held by Schrag (2012), who contends that the shale gas boom has reduced prices to the point at which investments in renewables might be postponed because of the low return to electricity. At the same time, he also contends that the gas and renewable energy industries share similar goals: they both would benefit from pushing coal out to shore up prices. Furthermore, from an environmental perspective, gas would be a better companion to renewables than coal. Gas plants could be combined with intermittent renewables to provide a stable and flexible supply of power (Krupnick, Wang, and Wang 2013).

The answer to the second question—whether the growth of unconventional gas overshadows the growth of renewables—is a clear "no." To begin with, shale gas does not have nearly as big a structural effect as renewables. Infrastructures and technologies remain the same. Conversely, renewables have a much deeper effect on the structure of the electricity sector. Furthermore, the extent of the growth of shale gas compared with renewables is actually not that different. A few back-of-the-envelope calculations prove this point. In 2007 (first year with reliable data for shale gas in the United States), shale gas generated about 65,743 million kWh of electricity.[9] In 2013, that number had increased to about 407,948 million kWh, a fivefold increase. Let us now contrast this with the growth of solar photovoltaic and wind power. Together, they generated about 35,062 million kWh in 2007. In 2013, they generated about 144,913 million kWh, which represents a threefold increase. Considering that shale gas was still only 40% of all natural gas produced in the United States in 2013, the relative growth of modern renewables was at least comparable, and in many ways much larger, than that of shale gas. At the same time, the growth of renewable energy is a global phenomenon, whereas commercial shale gas extraction has so far been largely limited to the United States. Globally, the growth of renewable energy outpaces the growth of shale gas by a wide margin.

2.2 Historical Energy Transitions

Despite the challenges noted in the previous section, renewable energy has expanded quite rapidly in recent years. Yet all energy transitions have faced technological challenges. Is the transition to renewables different? To answer this question, let us consider the history of energy transitions.

Before the industrial revolution, the world's energy needs were being met—or often unmet—by traditional biomass. Biomass can be used as a fuel for fire (Price 1995). It can and was also used to feed animals that were then employed for farming, increasing the returns of agriculture. Hydropower has also been used for a long time. Water mills were known by ancient Greeks and Romans. Their importance increased in the Middle Ages, and their use gradually expanded across industries (Cipolla 1994). In the early stages of the industrial revolution, a few regions used renewable technologies to help local farming or even early industrial interests. This was the case in New England, where hydropower was used to power factories, or in the American Midwest and West, where wind power was used to pump water (Carlin, Laxson, and Miljadi 2003; Steinberg 2004). Yet all these technologies were either unavailable or proved unable to unlock enough energy to fuel industries on a grand scale. What was needed was abundant and affordable energy.

Both the cost and quantity required of energy tend to be a function of the conversion efficiency of a given energy source. An energy source is efficiently converted if the ratio of output to input energy is high. For instance, an open fire has a conversion efficiency of 5%, meaning that only 5% of the energy contained in the wood is ultimately used in the cooking process (Smil 2010). Humans were able to develop their societies by replacing inefficient energy sources and conversion technologies with more efficient ones. The industrial revolution was fueled with steam engines, the conversion efficiency of which approached 25% by the year 1900. Steam was soon superseded by diesel engines, and their efficiency ratio grew from 25% to 50% during the 20th century. The unparalleled development of Western economies is thus rooted in the increased efficiency of their energy use. Due to low conversion efficiency, the average American household only used about 10% of its 100 GJ per capita gross consumption profitably in 1860. In other words, 90% of the gross consumption was wasted due to inefficient conversion. Currently, the typical US household can exploit 150 GJ per capita of the gross consumption of 375 GJ per capita, yielding the much better efficiency ratio of about 40% (Smil 2010). Thus, households have 15 times as much energy at their disposal, whereas gross consumption has increased less than fourfold.

How did these transitions occur? Consider the case of the Second Industrial Revolution around 1860–1890. By that time, the power of coal had been harnessed in industrial regions in both North America and Europe. Soon a new way to use and process energy—electricity—would revolutionize factories across the world (Devine 1983). The conversion of power was more efficient and—once power lines had been installed—the provision of electricity more convenient than that of coal. However, the transition was slow. As Atkeson and Kehoe (2007, 68) note, the transition from steam- and water-based energy to electricity took decades: "[if] we measure the diffusion of electricity in manufacturing starting in 1869, we see that it took 50 years for electricity to provide 50 percent of mechanical power." Concretely, electric motors still represented less than 10% of all energy used in manufacturing in the United States in 1899. Fast forward to 1929, when the rate had increased to more than 80%. The reason is that it took years for key players—factory managers—to fully understand how to take advantage of this new energy source. For a long time, relying on proven techniques was superior to trying out something quite new. It is only gradually and through experimentation that electricity started to dominate coal in US factories.

A more recent, if partial, transition is that of nuclear energy. It illustrates both the dynamic logic and the complexities of energy transitions. As Rai, Victor, and Thurber (2010) note, nuclear power was, at least until the late 1960s, a marginal technology. In the 1950s, for example, "[potential nuclear manufacturers] Westinghouse and General Electric were aware of the dismal prospects for economically competitive nuclear power, and the great complexities that remained to be resolved" (Balogh,1991, 304). Nuclear researchers tried to make the technology suitable for commercial use by civilians, but it was only in the second half of the 1960s that electric utilities began investing on a large scale. In the 1970s, the use of nuclear power expanded as plants were constructed based on the orders of the 1960s. Then the growth of nuclear power in electricity generation stopped despite the 1973 oil crisis. Balogh (1991) ascribes the woes of commercial nuclear power to the nuclear experts' loss of authority and, thus, an increasingly skeptical public. According to Walker (2004, 244), the 1978 Three Mile Island accident was decisive in instilling fears about nuclear power among the US population as "[t]he dual legacy of the crisis was, on the one hand, to galvanize regulatory and operational improvements that reduced the risks of another severe accident and, on the other hand, to increase opposition to the expansion of nuclear power."

Energy transitions are processes that generally take decades to materialize. However, the causes of historical energy transitions are different from the causes of the nascent renewable energy transition. The steam-to-electricity transition appears to have been entirely motivated by gains in energy efficiency. Devine (1983, 348) contends that the shift to "electricity [was] important [because] these forms of energy could be used with greater productive efficiency than coal and water power, producing more goods and services per unit of capital, labor, energy, and materials employed." The diffusion of this revolutionary technology occurred in a decentralized and evolutionary manner. The companies that learned to take advantage of electricity gradually crowded out those that did not. The selection of fuels depended only on their affordability and convenience. Even the arrival of nuclear power followed similar rules, although our previous discussion shows that the demise of nuclear power in the United States was not economic. Although initially supported by eager governments (Cowan 1990), its ability to become a major source of electricity originated from an ability to generate abundant power at a relatively low marginal cost.

The key insight for us is that these past transitions—from biomass to steam to electricity to diesel engines and later nuclear power—all emphasized efficiency and affordability at the expense of other considerations. As soon as the price of new technologies decreased enough, they could be used to unleash more power. These transitions took time because decisions in the energy sector were generally decentralized or, when the government was involved, made with the goal of satisfying rapid growth in the demand for energy. Naturally, as we discussed, this take simplifies a complex story. It is not true that nobody cared about the environmental repercussions of fossil fuels. We mentioned the notion of "white coal," that is, hydropower, which some saw as a clean alternative (Landry 2012), and we met figures such as Gifford Pinchot who encouraged the development of clean electricity over coal. Local initiatives against coal-induced air pollution in places like Pittsburgh already generated spirited debates (and lawsuits) in the late 19th century (Tarr and Clay 2014). In the midst of rapid social changes, people such as Henry Thoreau longed for a pristine environment. However, in practice, fossil fuels still achieved a remarkably dominant position.

This logic does not apply to the current renewable energy transition. For the first time, our societies are not solely maximizing quantity and minimizing cost. Instead, the transition is almost entirely driven by efforts to take into account the negative externalities generated by fossil fuels and nuclear power. These can provide abundant power at low cost, but they damage the environment and people's health. The attention to negative externalities radically changes the cost–benefit calculation that has prevailed until now. At the risk of oversimplifying history, the motto used to be, "We want as much energy as possible for as cheap as possible." Now policymakers and voters around the world say, "We want energy that is abundant, cheap, but also clean." The key challenge now is to strike the right balance among these three considerations. Technical constraints limit possible combinations of abundance, cost, and cleanliness. The eventual balance will be the result of a political process that we sketch in this book.

2.3 The Early Years of Renewable Energy

In the early 20th century, investments and the discovery of new oil fields in the Middle East made oil more affordable than ever. A barrel of oil cost $70 (in 2012 prices) in 1870. A hundred years later, the price had dropped to $10 (BP 2013). By 1970, the world seemed awash in oil and free of any constraint that could be imposed by energy sources. Production continuously increased over the same period, fulfilling the energy needs of our societies.

This world of cheap and abundant fossil fuels came to an abrupt end in 1973. The idea that oil was virtually free had to be rethought overnight. In the wake of the West's support for Israel during the Yom Kippur war, the major Arab oil producers decided to impose an embargo on oil exports to the industrialized democracies. The consequences were dramatic and exposed the overreliance of Western economies on oil from the Middle East. The embargo was particularly successful because demand for oil was strong—decades of cheap oil had created an addiction to fossil fuels in industrialized countries—and years of underinvestment into new production meant there was little slack to compensate for a drop in supply (Yergin 1988). The price of oil jumped from $17 per barrel in 1973 to $54 a year later, the highest price at the time (in real terms) since 1877.[10]

These sudden increases had devastating effects on Western economies. A century of low oil prices had deeply transformed the energy system on which industrialized economies relied. Although oil was the main source of fuel for transportation (cars, planes, ships), it also played an important role in electricity generation. In the United States in 1970, for instance, 11% of all electricity generated came from oil.[11] The only major alternatives were hydropower, gas, and coal. In the absence of feasible alternatives, businesses had no choice but to pay the higher bill or stop producing their goods. The world entered a recession for the first time since the Second World War, while inflation crept up in many countries (Blanchard and Gali 2007). Unemployment increased virtually everywhere. In the United States, it almost doubled from 4.9% in 1973 to 9% two years later. In France, the oil shock marked the end of the "Trente Glorieuses," or "Glorious Thirty," the post-Second World War period that was associated with high growth and rapidly increasing wages. This era was referred to as "il miracolo economico" (the economic miracle) in Italy and the "Rekordåren" (record years) in Sweden. This idyllic period was now coming to an end, and "energy security" became the buzzword of the day.[12] Analysts now realized how reliant the United States and other industrialized countries were on foreign oil supplies. Indeed, energy was suspected to be the "Achilles heel of United States power" (de Carmoy 1978). Energy and oil became a core issue of national security, and the second oil crisis in 1979 only reinforced this impression. President Carter referred to the shock as the "moral equivalent of war" (Herring 2006).

National security hawks did not have the monopoly of worries about fossil fuels. The environmental movement also increasingly intervened in the debate. In the United States, the origins of the modern environmental movement can be traced to the 1960s and 1970s. Of course, its roots are older, going at least to the mid-19th century. But the makeup of the movement evolved considerably at the time (Wellock 2007). The emphasis shifted from a desire for the sound management of natural resources to a more radical preservation of ecosystems at risk. This new wave of environmentalism was fueled by a broader desire for social change, tying together various interests across US society (Dunlap and Mertig 1992).

The evolution of the environmental movement is reflected in popular culture. Rachel Carson's *Silent Spring*, a blistering account of the detrimental effect of DDT, a pesticide, was published in 1962. The latent support for environmental conservation began to turn into activism in the aftermath of the 1969 Union Oil Company oil drilling accident off the coast of California (E. Smith 2002). The catastrophe, the third largest spill on record in the United States, triggered a visit by President Nixon and a short-lived ban on drilling in the region. Furthermore, it became a focal point for the burgeoning environmental movement. The 1970s witnessed the gradual expansion of the proenvironmental movement in many Western societies (Van Liere and Dunlap 1981). Donnella Meadows and her colleagues (1972) published *Limits to Growth* in 1972, a controversial study that emphasized the Earth's finite resources. Periodic shocks, such as the Three Mile Island nuclear meltdown in 1979, reinforced suspicions about traditional energy sources and shifted support toward finding alternatives to the status quo (McVeigh et al. 2000). By the early 1980s, a strong prorenewable coalition had crystallized in many countries. Besides the sheer strength of its size, it also had been at the origin of many governmental (e.g., the US Environmental Protection Agency) and nongovernmental institutions (e.g., the World Wildlife Fund; see Dunlap and Mertig 1992).

In this context, policymakers had to respond to the oil shock. Before 1973, there was little "energy policy" to speak of (Sørensen 1991). The US government attempted to regulate energy in the early 20th century but with little success and even less coherence (Clark 1974). Despite occasional alerts (mostly during or after the World Wars), cheap and abundant oil tended to be taken for granted. The successive oil shocks discredited this paradigm. The time had come for frontrunner governments to implement and consolidate energy policy. This led, for instance, to the creation of the US Department of Energy in 1977.

The political response to the oil crisis was twofold. First, governments attempted to reduce their dependence on fossil fuels by improving energy efficiency. If demand could be reduced enough, then prices and consumption would go down. This would alleviate the pressure on growth while keeping people satisfied, although there is some doubt about whether energy efficiency policies can work because reducing prices may lead people to consume more (Greening, Greene, and Difiglio 2000; Herring 2006). Nonetheless, reducing oil consumption appeared to be a rational strategy in the face of the shortages experienced in the West. It also appealed to those who wanted to reduce their ecological footprint and favored a less consumerist society. This argument found support in unexpected quarters. President Carter made a forceful appeal against materialism, claiming that "[w]e have learned that piling up material goods cannot fill the emptiness of lives which have no confidence or purpose" (Herring 2006). In practice, this translated into policies that, for instance, imposed limits to heat loss in buildings or energy efficiency requirements for boilers (Sørensen 1991). Although these policies curbed energy intensity in some countries, such as Denmark, others remained largely unaffected. Indeed, Carter's loss to Ronald Reagan symbolizes the obstacles that this line of reasoning faced in the West.

The second strategy was to invest in new energy technologies. If alternative sources—ideally homegrown—could be found, then the oil issue would disappear. Because many Western countries were relatively poorly endowed with fossil fuels, the solution had to come from somewhere else. Part of the response was to encourage substitution by investing in coal and natural gas. However, these sources created new environmental issues. Renewable energy appealed to environmentalists, who did not wish to exhaust finite resources, and it pleased those who worried about reliance on oil imports from foreign countries. Wind energy received a lot of attention mostly because wind mills had existed for centuries. In the United States, the Wind Energy Program was launched in 1973 with the aim "to accelerate the development of reliable and economically viable wind-energy systems and enable the earliest possible commercialization of wind power" (Thomas and Robbins 1980, 323). Similarly, solar energy was the darling of the first generation of renewable energy enthusiasts. President Carter launched the Solar Energy Research Institute in 1977 and installed solar panels on the rooftop of the White House—the panels would eventually be removed by Ronald Reagan and reinstalled by Barack Obama. Similar to the case of wind, interest in solar energy predates the 1973 crisis. In an article published in *Science* five years before the first oil shock, Glaser (1968) envisioned a future in which satellites would absorb solar energy and beam it to Earth, and then receiving stations would collect it. Viable applications only left the laboratories toward the end of the 1970s, with few large projects in the pipeline. To expand, these new technologies needed local sponsors and supporters.

2.4 Localized Expansion

Although a lot of enthusiasm about renewable energy was generated in the aftermath of the 1973 and 1979 oil shocks, initial investment materialized slowly and was limited to a handful of frontrunner countries. Sometimes, as in the case of wind, the leaders were even more localized. The 1980s marked a period of trials and errors. Although energy output remained fairly low in absolute terms, this period paved the way for the boom we are now experiencing. The rapid development of wind energy technology is a case in point. A modern wind turbine produced 300 kW in 1989. Ten years later, this number was multiplied by a factor of five to 1.5 MW. As of 2017, the largest turbines have a capacity of 9 MW.[13] This dramatic expansion in the size and power of wind turbines illustrates the rapid development of renewable energy technologies during the past four decades, again motivating the question of why renewables have made so much progress in such a short period of time. Although interest in solar and wind energy was high from the beginning in the West, production remained low. In the early years already, the centrality of local policies was recognized.

California illustrates the role of local incentives in the development of new energy technologies. *Windpower Monthly* noted, "[w]ind power did not choose California as its womb; California chose to foster wind power" (Righter 1996, 784). In 1980, California had no wind turbines (Righter 1996), and by the end of the decade, 15,000 turbines had been installed (Ackermann and Söder 2002). Although the number of turbines has regressed slightly since then, the total amount of energy produced has increased massively.[14] California was crowned the "historic leader in wind energy development, both in the United States and internationally" (Bird et al. 2003, 6).

The origins of this unprecedented boom are found in the combination of growing environmental concern among the population and the 1975 election of an environmentalist governor, Jerry Brown. Geographical conditions for wind power are average at best in California. The state only ranks 17th in the United States for wind energy potential (Bird et al. 2003). Instead, benign political conditions were central in making California the beacon of wind energy. Governor Brown implemented a shrewd set of policies to ensure that the wind industry could take hold. First, he took advantage of federal regulations to encourage small energy producers. President Carter's Public Utility and Regulatory Policies Act (PURPA) forced large utilities to buy energy from smaller producers, including renewable energy producers, provided that the cost of their electricity was below a certain price. This price was to be determined by utility regulators at the state level, allowing different states to select different levels of implementation in the federalist spirit. Brown's plan was to staff regulating institutions, such as the California Energy Commission, with officials who were supportive of renewables (Righter 1996). Regulators then set the price at which utilities had to buy electricity from independent power producers at extremely high levels as compared to other states. In fact, the price was up to seven times as high in California as in other states. Thus, Brown had managed to create both supply and demand in a market that did not exist ten years before. This aggressive policy triggered an unparalleled period of growth that lasted until the early 1990s. By 1991, 80% of worldwide wind energy production was located in California (Gipe 1991).

California later faced competition for leadership from an unlikely place, Texas, better known for oil and conservative political ideology. Texan renewable energy production was negligible until the late 1990s, and the share of energy generated by renewables remained below 2% until 2000. Then suddenly renewables began to grow at an astonishing rate. Within ten years, the share of renewable grew to more than 5%. Even more impressive is the actual growth rate of renewable generation. Wind energy increased from 300,000 MWh in 1999 to 16 million MWh in 2008—by a factor larger than 50.[15] In the same year, total renewable energy production in Texas was about 20 million MWh (almost all of which came from wind, biomass being quite negligible), compared with 4 million MWh in 1990 . Even more striking, most of the growth spurt occurred after 2000. As was the case in California, PURPA gave the initial trigger and local leadership proved key. In 1999, then-Governor Bush enacted a renewable portfolio standard (RPS) in the context of power sector deregulation (Langniss and Wiser 2003; Hurlbut 2008). Under an RPS, utilities have to provide some fraction of their electricity from renewables, and Texas showed that such a policy can effectively support the growth of renewable energy (Rader 2000).

Despite these success stories, the 1990s and early 2000s marked the decline of US leadership in the wind industry. The new leaders were European. Although Europe appears to be somewhat of a late bloomer, the roots of its success can also be traced to the 1980s. Two success stories are particularly informative for our argument, those of Germany and Denmark. Denmark was an early leader in the development of wind technology. In fact, Danish expertise was instrumental in California's success. By 1990, half of the existing Californian wind turbines and related equipment were produced in Denmark (Gipe 1991). As Heymann (1998) argues, Denmark's success was largely driven by dynamic, small entrepreneurs who emphasized trial and error in the development of new wind technologies. Currently, Denmark is the sixth largest European country and 11th worldwide by installed capacity, with more than 4 GW (Global Wind Energy Council 2013b). This is quite an achievement for a country that ranks 113th worldwide and 22nd in Europe in terms of population. Perhaps even more impressive, renewable energy provides about 40% of the country's electricity, with wind power representing about three-fourths of this impressive total. By this measure, Denmark has made more progress in its sustainable energy transition than any other country.

Denmark underwent this transition because key policymakers supported it. The initial impetus came from concerns about energy security in the 1970s. Energy was almost entirely provided by imported oil, and there was no nuclear industry to pick up the slack (Heymann 1998). Furthermore, political opposition to nuclear power was formidable, with activists expressing grave concerns about safety and waste disposal (Hadjilambrinos 2000, 1112). In this, Denmark was not so different from many of its Western counterparts, although many had already built or were in the process of building nuclear power plants. The reason that wind energy picked up in Denmark is that the country offered ideal conditions for its development. The origins of the Danish wind success story are threefold. First, Denmark's flat topology means that it has an abundance of wind. Second, the country has historically used wind energy, most notably during the two World Wars (Gipe 1991). This history provided the country the technical expertise and prior experience that facilitated the adoption of new, yet somewhat familiar, technology (Heymann 1998). Finally, a sizable and power-hungry agricultural sector created demand for energy that could be provided by the wind industry. It also ensured local support for new installations, as agricultural cooperatives benefited from the profits that electricity generation created (Toke 2002).

Danish policies were unique in their encouragement of local development of wind turbines (Toke 2002). By enrolling local supporters, policymakers and investors ensured that local opposition to new installations would remain minimal. Although Denmark and the United Kingdom offered over long periods of time similar levels of financial support to wind energy at the country level, only the former encouraged these local initiatives. Overall, the impact of these policies is impressive. When the second oil crisis occurred, wind capacity represented 1 MW.[16] By 1990, it had increased to 326 MW. Wind capacity continued to expand; ten years later, capacity reached 2.4 GW and currently represents more than 4 GW. The Danish case thus offers a picture of the feasibility of the renewable energy revolution.

In many respects, the German case is just as impressive as the Danish one. By 2012, Germany had the largest installed wind capacity in Europe and the third largest worldwide. Wind energy has long been an important energy source for Germans. In 1904, more than a century ago, the country hosted more than 18,000 wind mills (Ackermann and Söder 2002). Although Germany remained at the forefront of new technological development until the 1950s, interest in wind power faded due to the abundance of cheap oil and nuclear energy. However, in Germany as elsewhere, the two oil shocks and the 1986 Chernobyl nuclear catastrophe led to a radical shift in public opinion and elite ideology. Wind energy began to grow. Thanks to favorable policies, capacity increased from 1 GW in 1995 to 8 GW in 2001 and 31 GW in 2012 (Ackermann and Söder 2002; Global Wind Energy Council 2013b). Denmark focused all its efforts on wind, whereas Germany also diversified its renewable portfolio. Solar energy capacity was about as large as wind capacity in 2012, with more than 30 GW installed (EPIA 2013), and all of this capacity has been installed within a decade.

Unlike Denmark, the preexisting conditions in Germany were not particularly favorable to renewables. The geographical conditions for wind or solar power were uneven. Abundant local coal resources—mainly lignite—made the country less vulnerable to oil shocks at a time when oil was a key fuel in the power sector. Both coal and nuclear power had powerful friends on the left and right of the political spectrum. In the country's federal structure, many veto players could obstruct new policy. The transformation of the German energy landscape began under intense political pressure from new social movements. In the 1970s, the antinuclear movement started protesting against new plants (Wüstenhagen and Bilharz 2006). The movement progressively developed its political wing, the Green Party, in the 1980s. Worries about acid rain and the powerful shock from Chernobyl expanded public support for alternatives to nuclear and coal power. Initial attempts to build a renewable energy industry failed miserably (Michaelowa 2005). Projects such as GROWIAN—a wind turbine park—were launched with little political or local support and had disappointing results. The key moment was the politicization of the energy issue. By the end of the 1980s, Germany, led by Hermann Scheer, a Member of the Bundestag, hosted one of the first major lobby groups for solar energy (Scheer 2007). Around the same time, the wind energy lobby organized too and rapidly influenced policymaking (Michaelowa 2005). These lobbies built alliances with other interests, such as farmers, providing them with enough support among both Social Democrats (center-left) and Christian Democrats (center-right) to promote pro-renewable policies. Public opinion soon shifted against traditional energy sources. Although the country could be described as reliably pronuclear by the mid-1980s, both solar and wind power were more popular by the early 2000s. As described earlier, a boom in renewable energy generation followed. Germany showed that a major power could overcome the carbon lock-in.

2.5 Globalizing Renewable Energy

Today, investment in renewable energy is not limited to Europe. This is particularly true of solar energy (Solangi et al. 2011). In the United States, for example, solar energy capacity remained fairly stable over the 1990s, and it was only in the 2000s that the situation changed radically. In 1990, 367 GWh of solar energy was generated in the United States. This number climbed to 493 GWh in 2000, a fairly minor increase. However, solar energy generated more than 4,000 GWh in 2012. This is a tenfold increase over twelve years, a remarkable achievement. Notably, the boom is not solely fueled by governmental subsidies. Firms such as Google or Kohlberg Kravis Roberts, a private equity company, have invested significant amounts of capital into solar power plants.[17]

Even more important, many observers are already looking beyond the West. The attention of policymakers, investors, and scholars alike is now drawn to Asia. Japan has become a major player in solar production (Solangi et al. 2011), and so have China and India. India's current energy situation, for example, is by most accounts one of severe difficulty. The national electricity grid suffers from regular blackouts and fails to reach a vast share of the population. In fact, about 400 million individuals have no access to electricity (Government of India 2011). In turn, these poor infrastructures represent an impediment to the economic growth and poverty alleviation of the country (Ailawadi and Bhattacharyya 2006).

The Indian government has developed ambitious plans to tackle these issues. First, the government deregulated the largely bankrupt electricity market to attract private investors (Sharma, Nair, and Balasubramanian 2005). Second, it invested in alternative energy sources—including renewables—to mitigate the negative consequences of the existing weaknesses of the regular grid. Unsurprisingly, the solar and wind industries have been the government favorites. The wind industry has strongly benefited from both public and private support. Investments have increased rapidly from less than $2 billion in 2009 to about $6 billion in 2011 (Shrimali et al. 2013).

These investments have made the wind industry particularly impressive. Installed wind capacity represented more than 18 GW at the end of 2012, making India the world's fifth largest wind producer (Global Wind Energy Council 2013b).

In recent years, the solar energy industry has also attracted a lot of attention in India. Solar power offers both grid and off-grid alternatives to fossil fuels, and it is economically attractive in India. Under the Jawaharlal Nehru National Solar Mission plan, India hopes to install 20 GW of solar power by 2022 (Shrimali et al. 2013). This target is ambitious given that at the end of 2013, India's capacity from solar energy was 2.1 GW.[18] Despite these modest beginnings, India's solar industry has attracted impressive amounts of capital. In 2011, investments reached nearly $5 billion, only $1 billion less than the wind sector (Shrimali et al. 2013).

If India is a rising major player, then China is already a giant. Electricity generation from renewables had reached about 110 billion kWh in 2011 (excluding hydro, including biomass), about three times the amount generated in India (Shrimali et al. 2013). Installed wind power capacity reached more than 75 GW in China by the end of 2012 (Global Wind Energy Council 2013b). This figure is equivalent to the capacity of Germany, Spain, United Kingdom, Italy, France, and Portugal—the six biggest European producers—combined and represents 26.8% of total worldwide capacity. China's wind market is still growing fast since the country installed a third of worldwide new wind capacity in 2012. Indeed, wind was the main reason that China left its competitors so far behind in renewables (Shrimali et al. 2013).

The success of China's wind program is matched by the country's solar energy industry. Over the last decade, China created more than 8.3 GW of solar energy capacity (EPIA 2013). In 2012, the country installed more than 15% of new capacity in the world, and it is the leader outside Europe. Indeed, only Germany and Italy are larger solar players than China. What makes China's contribution even more meaningful is that it is the world's premier provider of solar panels. For instance, solar exports from China to the European Union climbed to $28 billion, whereas those to the United States reached $2 billion in 2012.[19] China's phenomenal growth has since triggered several trade disputes. Perhaps it is a sign of the industry's maturity that renewable energy is now a key topic for international trade lawyers.

Whither renewables? Predicting energy trajectories is hard. Analysts at the IEA did not expect the recent growth of renewables. This is not surprising because recent energy history seemed to give ample reason for pessimism—the last major change occurred seventy years ago with the arrival of nuclear power. We note two major paths of development that lie ahead. First, renewables will gradually become a mature source of power for the grid. With this, future debates will revolve around the preferred level of public support that will be invested. To some degree, these debates have already started in countries such as Denmark and Germany.[20]

Second, a major development in renewables is the enhanced competitiveness of distributed (off-grid) power generation. This is a key development of renewables because it largely bypasses utilities and thus represents a direct threat to their business models. A study by Satchwell, Mills, and Barbose (2014) contends that if residential solar panels were to reach 10% of the US market, utilities' earnings could drop by 5% to 41%. Such a market share is not unimaginable based on projections from the BNEF (2012).[21] The success of distributed power will necessitate advances in power storage. Both solar and wind energy require favorable conditions (sunny, windy weather) to operate best. The most widespread way to store renewables electricity is to use pumped-storage hydroelectricity, but not all countries can do this. Thus, scientists and private firms have conducted research in recent years to create storage batteries, such as Tesla's "Powerwall."[22] Enhanced batteries would allow the use of renewable energy even when the original source of power (sun, wind) is absent. Similarly, new distribution technologies, such as the smart grid, will begin to remove impediments to the growth of renewable energy (Niemi, Mikkola, and Lund 2012; Baker et al. 2013).

Distributed power generation is also important in the context of development policies in emerging countries. Poorer countries often suffer from deficient electric grids. Decentralized systems are particularly appealing given the obstacles to the expansion of the grid, especially to rural areas (Brass et al. 2012). Indeed, off-grid solutions have made clear progress. For instance, in recent years, three million solar home systems have been installed in Bangladesh (Khandker et al. 2014). The deployment of renewables in such a context is quite different from the logic of industrialized countries, but it underscores the global diffusion of these technologies.

3 Policy Responses to External Shocks: A Theory

The growth of renewable energy is an example of a broader class of political-economic developments that stem from policy responses to external shocks. Drawing on insights from political economy and historical institutional analysis, we develop a theory of how policy responses to external shocks could trigger slow-moving but dynamic changes that produce systemic effects over time. The theory is generally applicable, and we use it to gain analytical leverage on the complex problem of renewable energy.

Our argument rests on the notion that, in the absence of supportive government policy, renewables cannot break the carbon lock-in. Even if the prices of fossil fuels increase due to concerns about scarcity, an immature technological alternative cannot replace them unless the price increase is extreme. The development, demonstration, and commercialization of a new technology are long and uncertain processes (Grübler, Nakićenović, and Victor 1999), and the energy sector has historically not been characterized by rapid breakthroughs of new technologies (Smil 2010). These factors conspire against the kind of technological and commercial breakthroughs that have revolutionized telecommunications and information technology. In the case of renewable energy, the need for government policy is particularly clear because fossil fuels generate negative externalities, including air pollution and carbon dioxide emissions, which are not reflected in energy prices.

In a society characterized by carbon lock-in, government policy is necessary to level the playing field and promote alternatives to fossil fuels (Unruh 2002). The prospects of renewable energy depend on a variety of factors. Our theory focuses on politics and policy in society, but it is clear that techno-economic constraints impose limits to renewable energy development. Even if government policy succeeds in promoting renewables, we do not expect renewables to rapidly replace fossil fuels and produce a wholesale phaseout of coal, oil, and gas. Instead, a renewable energy transition can be said to begin if renewables become a normal part of the energy system and a serious competitor to fossil fuels outside niche applications. Such a transition requires the technical, economic, infrastructural, and institutional development of a social system for renewable energy.

Escape from the carbon lock-in and the transition to renewable energy involves two distinct problems. First, electricity markets present large-scale coordination problems (Stern and Holder 1999; Unruh 2000; Schmidt and Marschinski 2009; Dangerman and Schellnhuber 2013). Their governance requires the coordinated efforts of utilities, regulators, and transmission operators to respond flexibly to consumer demands, possibly across multiple countries. For instance, one cannot simply turn off nuclear plants and replace them with solar panels. The stability of the system depends on how well different actors coordinate their operations in both the short and long run. The need for coordination strengthens the carbon lock-in because a single agent cannot change the status quo. To effect change, a shock must not only shake people's beliefs about the merits of the status quo but also change their expectations about how others will behave in the future. Given the scale of the electricity market, only a major shock is likely to trigger a shift of behavior away from the carbon lock-in.

Second, the transition to renewable energy is dynamic. After a shock, the natural tendency of a system based on large, irreversible investments is to move back to the old equilibrium (Arthur 1989; Cowan 1990; Gersick 1991; Unruh 2000). Agents that prospered under the old carbon-based system of coordination have strong incentives to return to it. As the effect of a shock dissipates, a system characterized by carbon lock-in tends to return to the past equilibrium. With such strong centripetal forces, an external shock on its own is not enough to break a carbon lock-in. The key feature of new energy technologies is that their survival and success require years of support and investment, and the growth of new interest groups is decisive for breaking the new path for the power sector.

In developing the theory, we draw on insights from both positive political economy (Stigler 1971; Wittman 1995; Acemoglu and Robinson 2008; Lake 2009) and historical institutionalism (Thelen 1999; Pierson 2000; Hacker 2004; Thelen 2004; Peters, Pierre, and King 2005). To understand the political history of renewable energy, a synthesis of these two broad approaches is necessary because neither alone has sufficient analytical power. In our synthesis, political economy factors determine how a society responds to an external shock and starts to follow a certain path over long periods of time. Political economy, understood as an approach that emphasizes strategic behavior and material interests, provides useful insights into interactions between interests and institutions in a structural context. Historical institutionalism can shed light on critical junctures and path dependence in the evolution of renewable energy policy. We expect both kinds of factors to play important roles in the development of renewable energy. By drawing on both traditions in an eclectic but purposeful manner, we can, similar to our earlier work (Aklin and Urpelainen 2013b), show how politics modifies and conditions the effects of path dependence on society.

Substantively, our primary inspiration is the literature on sustainable energy transitions (Jacobsson and Bergek 2004; Sandén and Azar 2005; Jacobsson and Lauber 2006; Könnölä, Unruh, and Carrillo-Hermosilla 2006; Atkeson and Kehoe 2007; Grübler 2012; Aklin and Urpelainen 2013b; Dangerman and Schellnhuber 2013). Emphasizing the social context and importance of energy technology, we adopt a broad view of renewable energy transitions. As Wüstenhagen, Wolsink, and Bürer (2007) argue, the choice of energy technology is ultimately a social decision. In addition to technical and economic considerations, the use of renewable energy depends on perceptions, beliefs, and ideology. The transitions literature has grown rapidly in recent years and made major contributions to our understanding of changes in energy policy. A core insight from this literature is that improving the environmental sustainability of the energy sector requires a broad and deep societal transition involving producers, consumers, and policymakers (Unruh 2000; Verbong and Geels 2007). Following Aklin and Urpelainen (2013b) and Torvanger and Meadowcroft (2011), we pay particular attention to the political dynamics of sustainable energy transitions. Decisions to promote renewable energy are made in a setting characterized by economic structures, public opinion, interest groups, political institutions, and partisan ideology. External shocks such as surging oil prices or nuclear accidents perturb the existing sociopolitical equilibrium, creating a window of opportunity for renewable energy.

We also draw on economic studies of technological change and spillovers (Atkeson and Kehoe 2007; Acemoglu et al. 2012; Cheon and Urpelainen 2012; Nemet 2012). These studies recognize the dynamic and endogenous nature of technological change, emphasizing that past policies and investments shape the fortunes of a new technology in the future. Our primary insight and contribution is that these dynamics are not only economic. Previous policies create winners and losers. If the number and clout of winners grow fast enough, these policies create their own future political support and become entrenched, similar to the growing support for trade liberalization on the demise of import-competing industries, such as textile and apparel manufacturers, in the United States (Hathaway 1998). This dynamic setup captures the growth of renewable energy in a more nuanced way than a politically static world, in which constellations of societal interests are fixed, as in traditional theories of pluralist policy formulation (Truman 1951).

The third important source of ideas is the institutional perspective to policy formulation (Ruttan 1997; Aoki 2001; Spiller and Tommasi 2003; T. Moe 2005), including scholarship on the relationship between domestic politics and the international system (Gourevitch 1978; Ikenberry 1986; Putnam 1988). Policy responses to external shocks may be reactions to a major exigency, but they are still made against the backdrop of existing institutions that constrain and shape agency and choice. The political system is based on a collection of interrelated rules that policymakers and their constituents mostly accept as a given. Because these rules determine how agendas are set and authority wielded, they shape the political histories of renewable energy in different countries.

Our theory is most applicable to democratic policymaking. In a democratic system, interest groups compete for policy influence in the shadow of public opinion, and the government's decisions reflect the imperative of winning the next election (Becker 1983; Wittman 1995). The setting of democratic competition for office and policy is ideal for the application of our theory. Empirically, this is also the most relevant setting because the growth of renewable energy began in industrialized democracies.[1] Although some autocratic regimes, with China leading the way, have recently made major progress in renewable energy development, their newcomer status means that the key to solving the empirical puzzle lies with understanding policy and outcomes in industrialized democracies.

The key components of the theory are summarized in Figure 3.1. To begin with, a severe external shock allows renewables to begin their growth. Without some kind of an external perturbation to the system, the existing system for energy production does not allow an alternative to fossil fuels and nuclear power to emerge due to the carbon lock-in. The external shock calls into question the rationality of the fossil infrastructure, and the government has an incentive to respond by implementing policies that promote renewable energy. These policies open a window of opportunity for renewable energy, and the external shock protects the policies from political challenges. Initially, little political opposition to renewables occurs.

Despite initial success in the promotion renewable energy, we expect a political backlash against it over time. The very success of renewables

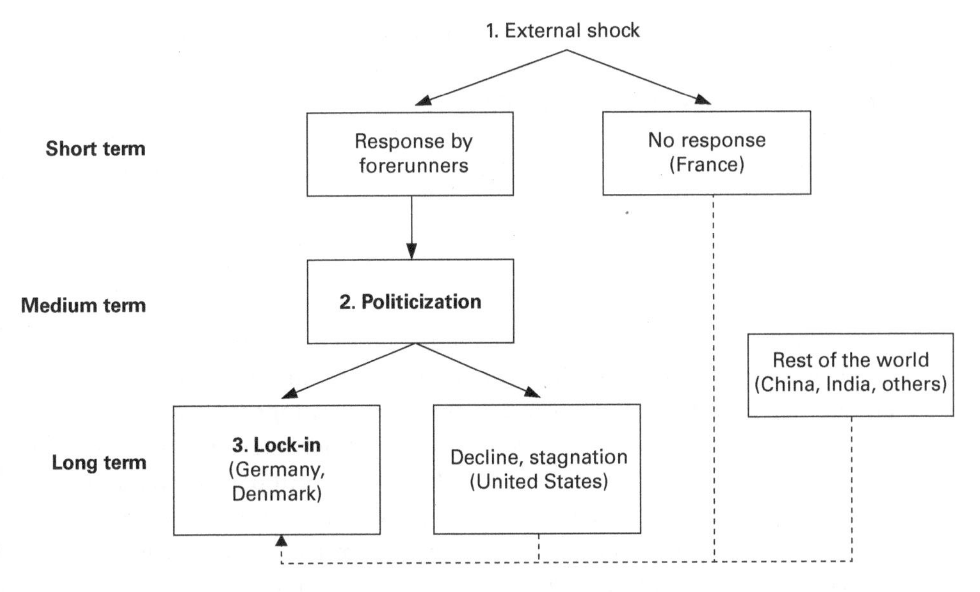

Figure 3.1
Overview of the theoretical argument. An external shock triggers a policy response in frontrunner countries, but the outcome of a politicization process determines whether the renewable energy industry undergoes a lock-in.

transforms them into a threat to incumbent interests. As renewables become increasingly competitive, they begin to threaten a variety of social interests, such as fossil fuel producers. The threatened interests form a coalition to oppose further support for renewable energy, arguing that renewables are an expensive and ineffective policy response to the social problems they are supposed to solve. As a result, renewables become the subject of a political debate and struggle. The balance of power between the advocates and opponents of renewable energy determines the outcome of the political struggle.

In some countries, the advocates of renewables win the political contest and a lock-in of renewable energy ensues. These countries implement aggressive policies that allow renewables to grow at a fast pace, bringing down the cost of renewable energy generation through commercial innovation, learning by doing, and economies of scale. As governments and investors in other countries notice this success, they also start to support and invest in renewables. If the outcome of the political contest is favorable to renewables in many major economies, the initial success of renewable energy is globalized and allows renewables to become a truly international phenomenon. Industrialized countries as well as emerging and least developed economies begin to consider new renewables—and not just more conventional solutions, such as large hydroelectric dams—as a credible response to their political, economic, and social problems. In other words, both renewable energy policies and generation converge across countries over time.

3.1 External Shocks

If modern economies have been under a carbon lock-in for decades (Unruh 2000), what sparks change in the power sector? We begin with the notion of an external shock that opens a window of opportunity for new policies and innovations in the society. As noted earlier, we define an external shock as a major, abrupt event that reveals the weakness of current policy and is not the direct product of a government's own policy.[2] In contrast, an internal shock would be directly caused by a government's own actions, such as misguided policy. The shock can be either international or domestic. According to this formulation, an external shock is a negative event that creates a crisis and calls for a policy response. An external shock may undermine current policy by showing that the economic cost of conventional energy is higher than expected. For example, concerns about the exhaustion of available oil in California in the late 1960s and the 1973 international oil crisis would qualify. An external shock could also be any other rapid increase in the international oil price due to international conflict or a terrorist attack. Such a shock would challenge the fundamental assumption that fossil fuels are readily available to generate economic growth.

Alternatively, the external shock may reveal that conventional energy solutions carry high environmental, health, or other costs. The Chernobyl and Fukushima nuclear accidents, or the environmental damage caused by an oil spill, would be examples of this kind of an external shock. Such shocks would force governments to consider the environmental effects of their current energy policies and strategies. If a nuclear reactor in Chernobyl creates a radioactive cloud and causes a swing in public opinion against nuclear power in Western Europe, this counts as an external shock in Western European countries. Similarly, if a record heat wave in the summer draws attention to the dangers of climate change in the United States and undermines the position of coal in power generation, this counts as an external shock in America. In both cases, the countries under consideration are indirectly responsible for the problem at hand. Public opinion about nuclear power would be less important in Western Europe if there were no nuclear power plants in the region; the burning of coal in the United States generates carbon dioxide emissions that contribute to global warming. However, neither the Chernobyl accident nor summer heat waves in the United States were directly caused by governments.

In studies of energy and climate policy, external shocks are sometimes considered "critical junctures" or "focusing events." In his study of Eurasian geopolitics, Calder (2012, 53) defines a critical juncture as "a historical decision point at which there are distinct alternative paths to the future. . . . Crisis *breeds stimulus for change*" (emphasis original). For Karapin (2016, 58), "external *focusing events* can rapidly shift the content of the public agenda and encourage the simultaneous mobilization of diverse actors." The common denominator is that a critical juncture or focusing event prompts a shift in the public debate and mobilizes different actors to demand change. We apply this general notion to understand the development of global energy.

In the case of renewable energy, the effect of an external shock such as a nuclear accident depends on the constellation of domestic interests and institutions. In some countries, such as Germany, the Chernobyl nuclear accident left an abiding impression on public opinion (Ruedig 2000). In other equally exposed countries, such as Finland, the accident has not deterred subsequent governments from investing in nuclear power (Litmanen and Kojo 2011). More generally, our approach draws heavily on the "second image reversed" tradition in international relations (Gourevitch 1978; Putnam 1988; Milner 1997; Ward 2006; Aklin and Urpelainen 2013b). This body of literature highlights the relationship between domestic politics and the international system, arguing that changes in the system can have profound effects on domestic political outcomes. We conceptualize external shocks as the primary channel of international influence on energy policy. Domestic energy policies are formulated in the shadow of international energy markets (Levi 2013) and their expected local and global environmental effects (Barrett 2003; Urpelainen 2009). Therefore, external shocks can disrupt the prevailing equilibrium characterized by carbon lock-in.

External shocks are abrupt in nature. They are potent drivers of political and economic change exactly because they are difficult to anticipate and require swift responses. Solow (1974, 7), focusing on the overuse of exhaustible resources, noted that

> resource markets may be rather vulnerable to surprises. They may respond to shocks about the volume of reserves, or about competition from new materials, or about the

costs of competing technologies, or even about near-term political events, by drastic movements of current price and production. It may be quite a while before the transvaluation of values—I never thought I could quote Nietzsche in an economics paper—settles down under the control of sober future prospects. In between, it may be a cold winter.

Recent scholarship on political processes has shown that policymakers tend to pay attention to highly salient problems and crises. As Jones and Baumgartner (2005, 5) put it, "[d]isproportionate information-processing leads to a pattern of extreme stability and occasional punctuations, rather than either smooth adjustment processes or endless gridlock." External shocks focus the attention of policymakers to an urgent problem that requires immediate action, and decisions are made under pressing time constraints. The abrupt nature of external shocks means that the initial responses to these shocks have a random and perhaps even chaotic characteristic. This scenario contrasts with the slow intensification of external pressure, as in the case of the slow mounting of government debt until a country defaults. Similar to the argument that economic crises engender reforms that would otherwise be political unpalatable (Drazen and Grilli 1993), we expect external shocks to draw their power from a rapid change in the external environment that creates uncertainty about the future. Without a shock, the gradual addition of pressure, such as slowly increasing oil prices, would not force government policymakers to consider experimental policies, such as alternative sources of energy. External shocks are thus important because they open a window of opportunity for new policies that otherwise would not have any chance of succeeding. As Karapin (2016, 62) puts it, a window of opportunity is important because during it "policy makers are intensely focused on [the problem], and significant policy change is possible, though not inevitable. If the advocates of policy change fail to take advantage of this opportunity, they will have missed their chance until the next window opens."

The key problem that the carbon lock-in proposed by Unruh (2000) presents to societies is that socioeconomic structures that have been based on abundant fossil fuels for decades, if not centuries, are remarkably resilient to change. This argument builds on Arthur's (1989) more general insight that in a dynamic environment, even an inferior technology or practice can become dominant if it has some initial success and there are increasing returns to scale or positive network externalities. In such a setting, an external shock may be necessary for engendering change and breaking the entrenched socioeconomic patterns that maintain and renew the dominance of existing practices. This idea is neatly captured by Walter Lippmann in the aftermath of the Great Depression (Hiltzik 2011, 53):

There are good crises and there are bad crises. Every crisis . . . is either a disaster or an opportunity. A bad crisis is one in which no one has the power to make good use of the opportunity and therefore it ends in disaster. A good crisis is one in which the power and the will to seize the opportunity are in being. Out of such a crisis come solutions.

The effects of the external shock vary. An oil price spike would alert consumers to the dangers of fossil fuel dependence and constrain economic growth (Kilian 2008). Both public and elite opinion would become more favorable to exploring alternatives when the previous expectations about abundant, inexpensive fossil fuels would be challenged by new information and circumstances. A nuclear accident would raise safety concerns among the public, turning the public opinion against nuclear power.

This logic may not apply to countries that are completely dependent on fossil fuel extraction. Karl (1997) documents the effects of oil price surges on Venezuela, which experienced a rapid increase in export revenue. Given the country's abundant oil resources, the oil shocks deepened the society's dependence on oil for everything. Similarly, the global dangers of climate change would not induce energy policy reforms in a country that is completely dependent on oil export revenue. Although many oil exporters, such as Saudi Arabia, have slowly started to invest in renewables, and solar in particular, we see little potential for breaking these countries' extreme carbon lock-in.

It is also important to recall that renewable energy is not the only response to an external shock. Whether in addition or instead of renewables, a government could try to promote a certain fossil fuel, such as coal as a substitute for oil, or decide to go nuclear. The government could also invest in energy conservation and the exploration of fossil fuels. Our goal is to explain renewable energy responses, but we do investigate alternative policies and their relationship to renewables in the case studies. To illustrate, the French nuclear drive turns out to be a key reason that the French government initially eschewed renewable energy.

To understand policy responses to external shocks, we next identify the government as the central actor of the model. In formulating policies, it considers both their effects on political survival and the payoffs to core constituencies. Following Grossman and Helpman (1994) and Aidt (1998), the government is assumed to maximize a political support function that comprises the payoffs to both the mass public and special interests. In the case of energy policy, the mass public is concerned about energy prices and the negative environmental externalities of energy production, such as air pollution or nuclear waste. Special interests consider a wider range of factors.

Energy producers are primarily interested in their profits, whereas environmental organizations primarily worry about the environmental effects of energy policy. Industrial consumers of energy also consider prices, whereas producers of energy technology are interested in expanding their markets (Cheon and Urpelainen 2013).

In a world characterized by carbon lock-in, the policymaker's strategy without a forceful external shock is straightforward. Because the carbon lock-in puts alternative sources of energy at a clear disadvantage, the goal of energy policy is to ensure that fossil fuels are inexpensive and abundant. In terms of the government's political support function, the most important components are the energy prices that consumers pay and the profits to energy producers. Environmental or resource scarcity considerations do not yet play a major role, given a history of low energy prices and growing consumption that fuels economic expansion. To the extent that energy policy exists, it is geared toward regulating consumer prices and supporting efforts to expand the extraction and distribution of fossil fuels. In this environment, any major policy initiative aimed to promote a new, immature source of renewable energy would face long odds. In the absence of an external shock, the government's efforts to promulgate a renewable energy policy would face resistance. Without major environmental, scarcity, or security concerns, it is difficult to mobilize an "advocacy coalition" (Sabatier and Jenkins-Smith 1993) for renewables. The carbon lock-in creates an uneven playing field that strongly favors fossil fuel technologies.

Recent scholarship in historical institutionalism has challenged the primacy of external shocks as the cause of change by emphasizing endogenous sources of perturbation (Hacker 2004; Thelen 2004; Beland 2007; Hacker and Pierson 2010). These scholars emphasize endogenous change in the form of "conversion" (existing policies and institutions are modified for new purposes), "layering" (new elements are gradually added to an otherwise stable system), and "policy drift" (changing circumstances modify the effects of existing policy) (Beland 2007, 22). In our case, this line of inquiry faces limitations, however. Because no renewable energy policies existed before the 1973 energy crisis, and virtually all existing policies and institutions were designed to sustain the carbon lock-in in the absence of societal interests advocating for change, there is little reason to believe that conversion, layering, or policy drift would have brought about major change without an exogenous shock. This is counterfactual reasoning, of course, but the carbon lock-in raises doubts about alternative histories and futures featuring gradual change without exogenous shocks. Certainly few indications of such gradual change were evident before 1973.

An external shock can challenge the stability of the carbon lock-in. If the dangers of carbon lock-in were made apparent by a strong external shock, the fundamental assumptions underpinning the broad contours of past energy, or lack thereof, would be questioned by a variety of societal interests. Changing circumstances would elevate levels of uncertainty about the future and force policymakers to face a complex, confusing reality (Laird 2001). If ideology is defined as an integrated collection of "assertions about goals and values . . . categories for defining situations or problems . . . and causal theories or empirical hypotheses" (Snyder, Shapiro, and Bloch-Elkon 2009, 163), then the external shock casts doubt on the ideology of the carbon society. For example, a rapid surge in oil and coal prices would call into question the fundamental assumption that the most important fossil fuels are abundant and readily available to fuel economic growth at a low cost. A nuclear accident would raise concerns about the safety of nuclear electricity.

To illustrate the effects of the external shock on the political environment, consider the debate on energy policy in the United States in the wake of the 1973 oil crisis. This abrupt crisis is an ideal illustration of the idea that external shocks have profound effects on the political environment. The United States had occasionally experienced concern about access to oil earlier. In fact, American scientists worried about scarcity in 1908 already (R. Stern 2016). However, this situation was different. Earlier shortages would have unpleasant consequences, but the country was by no means as dependent on oil as it was in 1972. Wood was the dominant fuel until 1880, and coal dominated all other sources until the 1950s. To illustrate the gap between coal and oil, consider that in 1920, the United States consumed 15.5 quadrillion BTU from coal compared with 2.7 from oil.[3] Access to oil mattered for strategic reasons, but it did not threaten the industrial structure of the economy.

The oil shock called into question the premise that energy consumption would grow rapidly in the United States at least until the year 2000. One of the most vocal opponents of an energy policy based on fossil fuels, Amory Lovins (1976, 217), summarized the argument against the conventional approach as follows:

> Some people think we can use oil and gas to bridge to a coal and fission economy, then use that later, if we wish, to bridge to similarly costly technologies in the hazy future. But what if the bridge we are now on is the last one? Our past major transitions

in energy supply were smooth because we subsidized them with cheap fossil fuels. Now our new energy supplies are ten or a hundred times more capital-intensive and will stay that way.

What is notable about this paragraph is the emphasis on the difference between past and future. If fossil fuels are now expensive across the board and nuclear is not a feasible solution, Lovins (1976) argues, policy must be built on the fundamental principle of a "soft energy path." Laird (2003, 29) further argues that the advocates of a soft energy path believed that this path could bring about a more profound social transformation: "convinced that the technologies that made up the energy system would powerfully influence society and politics, they believed—much like nuclear advocates—that creating new technological systems would eventually bring about the society they wanted." This vision was different from the "large-scale, bureaucratic, hierarchical institutions" of the time:

They thought that a decentralized society, relying on solar power, would be a more just society, promoting self-reliance and reducing the income gap. Some argued that the solar movement should align itself with social justice movements as well as with the environmental movement. The new society would need to be based on a very explicit understanding of ecology, as desirable social goals would follow from the creation of an ecologically viable community. (Laird 2003, 44)

The external shock is a particularly potent force for change if it shapes expectations about the future. Epiphenomenal shocks, such as temporary oil price increases due to an unusually cold winter, give no immediate reason to challenge the fundamental assumptions of the carbon society. The key difference between epiphenomenal and permanent shocks is whether policymakers and the public have reasons to believe that the effects are temporary. A cold winter shapes oil prices for a few months, and the sudden depletion of a major oil field could do so for years to come. However, if oil prices increase because of alarming news about dwindling reserves, this external shock raises difficult questions about the need to change energy policies and find alternatives to oil. The expectation of permanent effects means that the society will have to live with the consequences of today's misplaced energy policies for a long time, which is a powerful political argument for change and mobilizes the advocacy coalition for change.

When an external shock befalls the society, policymakers become less averse to formulating new policies that they would have previously considered too risky. In a new environment, policymakers face the difficult challenge of implementing untried policies. However, the lack of experience with such policies does not deter them from trying to react to the external shock. Because the external shock renders existing policies ineffective and potentially counterproductive, some kind of a policy response is necessary. Although policymakers cannot generally fall back on previous policy responses to external shocks due to the novelty of the challenge, they nonetheless have to react somehow to the shock. In the language of Hirschman's (1970) "exit, voice, and loyalty," the external shock reduces the attractiveness of loyalty to business as usual, and in the absence of plausible exit options—an industrialized country cannot function without power—voice remains the only plausible option. Thus, policymakers begin to experiment with new ideas and approaches to solve the problem at hand.

Weyland (2002) studies the psychological foundations of such a situation in the context of market reforms in the "fragile democracies" of Latin America. He argues that as policymakers and governments consider policies, their response depends on whether the status quo promises to produce significant losses with a high level of certainty. Building on Kahneman and Tversky's (1979) prospect theory, Weyland (2002) argues that risky economic reforms arise out of necessity. As long as policymakers expect to survive without major changes in their policies, they want to avoid risks, even if the current policies can hardly be considered optimal. But if policymakers expect clear losses from the status quo, avoiding those losses becomes their top priority. We argue that this logic can be applied to renewable energy. If an external shock casts doubt on the rationality of a strategy based on fossil fuels, then renewable energy is an alternative strategy, as explained in the previous chapter on renewable sources of energy. Because they are good for the environment and technically abundant, they have intuitive appeal as a response to external shocks, undermining the case for fossil fuels. They also do not carry the kinds of risks associated with nuclear electricity.

Although the government holds the initiative in responding to the external shock, policy formulation is a complex process with multiple stakeholders (Chubb 1983; Grossman and Helpman 1994; Aidt 1998; Cheon and Urpelainen 2013). Various interest groups attempt to influence the content and nature of the government's renewable energy policy, with important consequences for both their effectiveness and distributive implications. For example, energy producers could lobby to secure support for the research and development (R&D) of clean energy technologies. As interest groups try to exert influence on the policies, their primary goals generally are different from the policy payoff to the mass public. Realizing that the government has a strong incentive to enact renewable energy policies, interest groups try to secure rents from the new policies.

Chubb (1983, 181–212) provides an interesting illustration of this process during the Ford and Carter administrations in the United States. In the years 1976–1979, the federal government annually allocated between $1 and $4 billion to energy R&D. The main interest groups active in this sector focused on increasing their share of the available resources. Many key players in the oil and gas sector had "significant interests in coal, uranium, solar, or other energy technologies" (Chenoweth and Feitelson 2005; Chubb 1983, 186). Consequently, they also had an interest in all kinds of energy funding, including renewables. According to Chubb (1983) and McFarland (1984), a key feature of energy R&D lobbying in the years of the oil crisis was the lack of conflict among interest groups. Perhaps a type of collusion, each interest group requested funding for their preferred form of R&D without challenging the positions and arguments of other interest groups. Writing from a critical perspective, Reece (1979) also notes that much of the solar funding was actually allocated to traditional energy interests.

For these companies, renewables created an opportunity for R&D of new technologies. In a time characterized by high levels of uncertainty, this opportunity was valuable as a diversification strategy. This is precisely why external shocks have the potential to break the carbon lock-in that prevents renewable energy from blossoming. The government's willingness to implement risky and daring policies increases in a highly uncertain environment that requires a policy response. At the same time, interest groups realize that the new policies create both an opportunity for securing rents and a potential threat depending on their design. Therefore, interest groups become active players in the formulation of renewable energy policies. Instead of categorically refusing to consider said policies, as most interest groups would have done before the external shock, the political process of policy formulation is based on the assumption that there will be a policy response. The question is what this policy response is and how it is designed.

Socioeconomic transitions are long and complex processes (Smil 2010), and it would be naïve to expect that even an abrupt external shock would have immediate effects on the society. A government cannot expect to dismantle the existing structures that maintain and perpetuate the carbon lock-in simply by implementing a series of policies. Instead, the government's policy responses to the external shock create a window of opportunity for the growth of renewable energy. They are not enough to overcome the carbon lock-in immediately, but they are the beginning of a process that holds promise for change over time. One reason for this opportunity is the existence of "niche markets" for new technologies, including renewables

(Lopolito, Morone, and Taylor 2013). Even if renewable energy is not initially fully competitive with fossil fuels on a national scale, it may be able to compete in some settings. For example, an exemplary site for wind energy may allow pioneers to test and expand new technologies. Similarly, solar technologies may begin their development in remote rural locations that do not have access to the national electricity grid. The importance of these niche markets lies in their ability to provide renewables with a space for growth at the early stages of technological development. In this regard, the theory we propose is significantly different from standard accounts of the punctuated equilibrium because we do not expect a "revolutionary upheaval" (Gersick 1991).

For the case studies, we now formulate a series of expectations. As noted in the introduction, we analyze the comparative political history of renewable energy across countries and over time. The statement of expectations is intended to facilitate this process. By clearly specifying the empirical implications of our argument, we can structure the case studies and evaluate the explanatory power of our logic. Some of our expectations are descriptive statements about the common effects of shocks and other changes across countries. Others are closer to classical hypotheses, emphasizing the covariation of explanatory and dependent variables, and yet others focus on the processes we expect to see in the cases.

First, when an external shock befalls a society that is heavily dependent on fossil fuels, or perhaps atomic power, as in the case of a nuclear accident, a temporary opportunity exists for a government to initiate renewable energy policies that previously would have been out of question. This window of opportunity is temporary, however, and it closes as soon as the worst effects of the shock wane. For example, an oil price shock would last only until oil prices decrease again or the society adjusts to a reality of higher energy prices. Therefore, our expectation has a strong temporal aspect. Even if the external shock has permanent structural consequences for the economy, the window of opportunity for a novel policy response is not open forever. The novelty and salience of the external shock, combined with the fear of permanent effects, open and close the window.

Shock Expectation 1 (effect of external shocks). *External shocks create a window of opportunity for government policies that support renewable energy deployment in the long run.*

Here, it suffices to offer a hypothetical illustration. Consider a rapid increase in energy prices across the board due to the sudden depletion of a major oil field. Because oil is still used as a substitute for fuels such as coal and natural gas in the power sector, this kind of an oil shock would force governments to consider policies that reduce the society's dependence on oil. The government would form various scientific and technical committees while also soliciting comments from business and civil society interests. In due course, proponents of renewable energy would propose the formulation of policies that support the R&D of solar, wind, and modern biofuel resources. Only a few years earlier, such proposals probably would have been met with ridicule among the broader policy community. Due to the oil shock, however, the government would consider the possibility of such a policy. Interest groups would recognize an opportunity for profitable subsidies, and so they would also support these policies. Even if the central thrust of the government's policy response were elsewhere, such as in the deployment of nuclear power, renewables would nonetheless receive more support than was possible before. The government's initial reaction to the shock is to search for alternatives in a confusing situation characterized by tremendous political pressure to act. The search for these alternatives creates opportunities for different societal interests to propose their preferred solutions, and some of these focus on renewables. The salience and potential permanence of the shock mean that the government has strong political incentives to consider the proposed renewable energy policies, which encourages interest groups to channel their energies into securing gains from said policies.

The example also highlights the importance of the social, economic, and political settings. There is nothing automatic about the effect of the external shock on renewable energy policy. As the government ponders policy responses, it can select from a wide range of possible responses. Renewable energy is not the only available solution. The government may well choose to invest in nuclear energy instead or expand domestic oil and gas exploration depending on factors such as the preferences and behavior of the organized interests that participate in the policy process. At the same time, the government's ability to formulate policies also depends on institutional capacity and constraints (Ikenberry 1986). The government's successful efforts to promote renewable energy in new circumstances would depend on the careful crafting of policies that support the growth of an immature industry and an unproven technology (Grübler, Nakićenović, and Victor 1999; Loiter and Norberg-Bohm 1999; Norberg-Bohm 2000). The formulation of appropriate policies in a complex environment requires a competent bureaucracy, and widespread variation exists in the ability of different governments to formulate energy policies. As Ikenberry (1986, 137) puts it, "[b]ecause the methodology of state action differs, so too does state capacity to influence political outcomes."

The state's willingness and ability to promote renewable energy in response to an external shock depends, in addition to the severity of the shock, on a variety of modifying factors. There are a number of such factors, and we face here the difficult decision of selecting a manageable number that may be expected to be particularly important. For the case of renewable energy, we argue for the central importance of the following factors: energy security and elite ideology. In discussing these factors, it is important to remember that at the time of the shock, there is no strong and organized advocacy coalition for renewable energy yet. The external shock befalls a society characterized by carbon lock-in, and the interest group support for renewable energy must come from either a motley crew of enthusiasts for alternative technologies or industries with at best shallow interest in the rents that the government's policies would create.

We start with structural factors that could increase the appeal of renewable energy to the government. Because these factors are always relevant to energy policy decisions, they play a role in decisions about renewables even if the issue does not yet have a high political profile. We argue that the foremost structural factor to consider is the country's dependence on foreign imports of fossil fuels. Energy security has always been a core consideration in the formulation of energy policy, and concerns about the scarcity of fossil fuels raise the specter of dependence and vulnerability (Yergin 1988; Li 2005; Chalvatzis and Hooper 2009; Müller-Kraenner 2010). Moreover, the mirror image of imports is domestic production. If domestic energy production is high, then producers such as oil, nuclear, and coal companies have significant political and economic clout (Stenzel and Frenzel 2008; Aklin and Urpelainen 2013b). Policies that reduce the profitability and competitiveness of traditional energy sources are not in the interest of said producers, and so the government's ability to promote renewable energy is diminished.

The second factor that we emphasize is elite ideology. We define it as the government elite's positive and normative beliefs about the functioning of the society and its political institutions. Because there is no strong advocacy lobby for renewables yet because they are a new idea not in widespread use, the decision to formulate renewable energy policies requires a policy-making elite that is aware of this option and at least somewhat favorable to considering renewables: leaders "who are interested in or committed to [change] can help to prioritize it on both the public agenda and the political decision-making agenda, and to assemble a broad coalition in favor of it" (Karapin 2016, 58). By the elite, we refer to actors within a society that participate in policy formulation, either directly within the government or indirectly outside it, but with access to the policymaking process. Therefore, elite ideology depends on both the views of key elected officials and their allies, such as technocrats, academics, the media, and influential nongovernmental organizations. The history of environmental policy suggests that "epistemic communities," which share a common belief in the presence of a problem and appropriate solutions to it, can shape elite ideology to a significant extent (P. Haas 1989, 1992). Elite ideology may also depend on the personal experiences of key policymakers, the historical context, and even the electoral incentives of political parties, whose leaders may choose their ideology on instrumental grounds.

The choice between a "soft" and a "hard" energy path, which Lovins (1976) discussed in his article, is a great illustration of how an external shock can change thinking in a society as well as an example of the role of elite ideology. Unlike many of the more recent critics of fossil fuels, Lovins (1976) emphasized the scarcity of fossil fuels. He argued that a soft energy path emphasizing conservation and renewables, instead of heavy reliance on fossil fuels and nuclear power, would provide a more solid foundation for economic prosperity and environmental sustainability in the United States. Although this idea did not fully penetrate the American elite beliefs, it became influential enough that the Carter administration was willing to invest billions of dollars in solar power and other renewables.

The 1973 oil shock illustrates the notion that elite ideology was a critical determinant of whether policymakers accepted the notion that fossil fuels are scarce and that shock was the harbinger of an era of unprecedented scarcity. This notion, in turn, was influential in shaping whether policymakers saw a window of opportunity for renewable energy as a solution to an impending scarcity crisis. In the literature on resource scarcity, the belief in scarcity is the decisive difference between the Neo-Malthusian and Cornucopian views of resource management (Chenoweth and Feitelson 2005). The Neo-Malthusian school argues that natural resources are scarce and on the verge of exhaustion, whereas the Cornucopian view emphasizes that scarcity increases prices and encourages new resource exploration and technological development. Because we cannot objectively verify the validity of the two views, elite ideology is ultimately based on subjective perceptions of the resource scarcity issue.[4] Cornucopians would dismiss the Lovins (1976) argument against fossil fuels as naïve and economically destructive, whereas Neo-Malthusians would consider it necessary to avoid a socioeconomic collapse on an impending surge in the prices of fossil fuels.

According to Laird (2001, 7), ideological considerations are particularly important in the case of immature technologies because the uncertainty related to their future undermines rational cost–benefit calculations:

> Under such immense technical uncertainly, people's ideas about what constitutes a good political and social order, and which institutional and technological arrangements they think will further that order, come to dominate policy-making debates, since long-term interests are hard to identify and predict and institutions may be embryonic or nonexistent.

Shock Expectation 2 (modifying role of energy imports and elite ideology). *The window of opportunity will be exploited most effectively by governments in countries that depend on energy imports and have an elite ideology that favors renewable energy.*

Systematic measurement of elite ideology is difficult. Although many of our explanatory variables can be measured consistently across countries, there is no easy way to measure elite ideology in multiple cases. For evaluating the role of elite ideology, a qualitative approach is ideal. We examine primary and secondary sources that indicate how government decision makers, as well as other actors participating in the policy formulation process, initially thought about the role of renewable energy among possible responses to the 1973 oil crisis and other external shocks. We consider elite ideology favorable to renewables if there is, by and large, a consensus about the possible benefits of renewables and their importance as a component of the policy response to external shocks.

Various expectations can be formulated about the interactions between these variables. One possibility is that energy dependence and supportive elite ideology are complementary forces. In this case, we would expect a policy response especially when both are present. Another possibility is that the two factors are substitutes, which would imply that the presence of one would be enough to prompt renewable energy policy. In the empirical analysis, we can consider the strength of the evidence for each possibility.

Because there is no advocacy coalition for renewable energy yet at the time of the external shock, some other plausible explanatory factors lose importance.[5] For one, we would not expect a clean technology industry to play an important role in pushing for renewable energy policy. Although a clean technology industry would benefit from such a policy, at this stage, said industry does not yet exist on a scale large enough to shape policy.

Our final expectation is that the preliminary renewable energy policies should not initially provoke vocal political opposition.[6] One reason is that the stakes are simply not high enough yet. Given the past record of carbon lock-in, renewable energy has yet to emerge as a serious challenger to fossil fuels. Although renewables could eat into the market share of fossil fuels on the margin, the effect would be too small to become the core concern of energy utilities. Another reason is that, instead of mobilizing against renewable energy, interest groups can reap large gains from focusing their efforts on capturing rents from the government's new policies. In an admittedly ideological book, for example, Reece (1979) documents the "corporate seizure" of solar policies in the United States during President Carter's tenure. Finally, the height of the shock does not provide an ideal environment to voice opposition against the urgency of new energy policies. Although major energy corporations ensured that renewables would not threaten their emphasis on domestic exploration and nuclear power, they had little reason to be concerned given how far renewables were from becoming commercially viable. At the same time, billions of dollars of federal funding were made available for research, development, demonstration, and deployment (Chubb 1983). The same energy industry that advocated the continuation of a strategy based on fossil fuels also benefited from subsidies given to the slowly emerging solar industry.

Shock Expectation 3 (lack of political opposition to the policy response). *During an external shock, there is little political opposition to renewable energy policy.*

3.2 Politicization

Although the external shock does create the political demand for renewable energy, it does not yet present an immediate, salient threat to influential political constituencies that reap gains from the country's reliance on fossil fuels. In the short run, experimental policies that allow the growth of renewable energy are simply not threatening enough to the incumbent energy producers and their key customers, such as heavy industry and homeowners. At the same time, producers and consumers of fossil fuels maintain an interest in directing renewable energy policy toward their own goals. As discussed briefly earlier, corporate interests in the United States tried to secure, with considerable success, rents and resources from the various solar policies that the Carter administration implemented during his tenure, which was characterized by high energy prices and coincided with the 1979 Iranian revolution that further increased said prices. This corporate strategy underlines the idea that, instead of trying to undermine the relatively harmless solar policies of the federal government, business interests focused on capturing the rents at the expense of, say, the idiosyncratic crew of "appropriate technology" (Schumacher 1973) enthusiasts who sought state support for decentralized, distributed energy generation.

After the initial external shock wanes, renewables face new challenges. As time goes by, however, it is no longer clear that generous investments in renewables are still warranted. At this stage, political conflict about renewable energy emerges, and the process of policy formulation depends critically on the politicization of renewable energy. By politicization, we refer to the process whereby renewables become contested and the object of a divide between opponents and supporters. In our model, the conflict between opponents and supporters is determined by partisan politics, interest groups, and public opinion. These political economy factors shape the long-term consequences of the initial external shock. The outcome of interest is how much opposition to renewables exists. Because our focus is on the political backlash against renewables, we discuss politicization in terms of opposition as opposed to support. This approach makes sense because we focus on frontrunner countries that responded to an external shock by experimenting with renewable energy.

The beginning of the politicization stage can be identified by looking at the price differential between renewable energy sources and fossil fuels. When the early investments in response to the external shock cause the price differential to decrease sufficiently to make the large-scale deployment of renewables a possibility with the support of government policy, political conflicts over renewable energy erupt. At this time, potential losers from increased use of renewable energy realize the stakes and begin their mobilization against government policy in support of renewable energy policy.

Under low levels of opposition, the core arguments for renewables remain widely accepted, and the debate focuses on specific questions, such as the cost of certain policies. Under high levels of opposition, the foundations of renewable energy policy are challenged. Opponents of renewable energy launch a frontal attack against the policies, trying to stop the growth of renewable energy. As noted in the introduction, our goal is to understand and explain renewables, not to comment on their virtues.

Over time, the favorable environment for renewables created by the shock disappears. As renewables grow, they become a credible challenger to fossil fuels. Moreover, proponents of renewable energy begin to demand policies that allow renewables to move from experimentation toward commercialization and rapid large-scale growth. Constituencies who expect to lose from renewable energy mobilize in opposition. Although the external shock initially allowed renewable energy to grow, over time the prorenewable consensus—or, at the least, the lack of opposition—breaks down, and opponents challenge the rationality of investment in renewables. Widespread opposition to renewable energy is only expected in countries that made initial investments in response to the external shock. Where no external shock was felt or such a shock did not prompt policies, opponents see no need for politicizing the issue. Because renewable energy is not growing in the first place, its potential detractors do not consider it a threat to their interests. Therefore, opposition should be limited to countries that made investments in response to the external shock.

Renewable energy has several different opponents. Heavy industry worries about high electricity prices, fossil fuel producers about markets for their product, and conservatives about heavy state intervention. Depending on how policies are formulated, for example, the cost of installing renewable electricity capacity may fall in varying degrees on ratepayers and taxpayers. Although policy design influences the distribution of both costs and benefits, we can identify political constituencies that are likely supporters or opponents depending on their dependence on inexpensive energy and fossil fuels in particular.

Advocates of renewable energy argue that it is a rational response to the problems that energy systems based on fossil fuels cause. The response of the opponents is the contestation of this claim. The opponents challenge the idea that renewables are a rational, benevolent solution to the problems of energy security and pollution or risks associated with nuclear energy. They argue that renewables are too expensive to be a practical alternative to fossil fuels. They argue that solar and wind power are intermittent sources that depend on a baseload of nuclear or other fossil fuels, and they contest the technical potential of biofuels and biomass. The future of renewable energy policy, and therefore the deployment of renewables, depends on how vocal and persuasive the skeptical arguments of opponents of renewable energy are and how much political influence these detractors of alternatives to fossil fuels have. This process of politicization is the litmus test of renewable energy. If renewable energy enthusiasts win the political conflict that follows, so that the government has continued political support to formulate supportive policies, renewables grow rapidly over time. If they lose, then the carbon lock-in continues until the next external shock shakes the society.

In the United States, the conservative movement has opposed renewable energy policies for ideological reasons, arguing against the state intervention needed to support renewables (McCright and Dunlap 2003). In 2007, the US House of Representatives adopted an energy bill that would have required electric utilities to derive 15% of their power from renewable energy sources. Although the renewable electricity requirement did not become a law due to opposition in the Senate, the arguments made against it by conservative legislators are illustrative of political opposition to renewables. Republican Representative of Alaska, Don Young, argued that the bill does nothing to solve America's energy problems, saying that, "it tells us to turn the lights out, that's what this bill does. . . . There is no energy in this bill at all."[7] For better or worse, this representative questioned the plausibility of generating sufficient energy from renewable sources of energy, instead demanding policies that support domestic oil and gas exploration.

A clear, if perhaps symbolic, example of the politics of renewable energy is the installation of solar panels on the US White House. This practice was initiated by Jimmy Carter for water heating, but in 1986, the Republican Ronald Reagan made a public statement against renewables by removing the solar panels from the rooftop. In August 2013, Democrat Barack Obama again brought back the solar panels as a symbol of his interest in renewable sources of energy.[8] Although the presidential solar panels themselves are unimportant, they provide a powerful illustration of the political significance of renewable energy. Carter's initial decision to install them was a public symbol of his interest in using solar energy to solve the energy crisis in the United States, whereas Reagan's decision to remove the panels symbolized his lack of enthusiasm for renewables.

The novel element of our argument is the situation of renewable energy in the years that follow the external shock. We do not argue for a rapid transformation of the energy sector as a result of the critical juncture created by the shock. Instead, we expect renewable energy to make modest progress while remaining dependent on the support of the state for growth and expansion. Politicization follows the government's initial response to the external shock exactly because renewable energy is neither irrelevant nor dominant. Renewables are at the cusp of becoming a major component of the energy sector, but their growth can still be stopped if the government decides to stop the implementation of supportive policies. Recognizing this situation, the opponents of renewable energy have strong incentives to challenge the government's strategy and stop the formulation of new renewable energy policies.

Another key feature of the politicization stage is that the growth of renewables is no longer dependent on external factors, such as international oil prices. Indeed, quantitative research shows that oil prices play a less significant role in explaining the growth of renewable energy in countries with the potential for positive reinforcement (Aklin and Urpelainen 2013b). Our theory can explain this loss of importance. By the politicization stage, renewables have advocates who lobby for supportive policies for various reasons, ranging from climate mitigation to green jobs, and the success of this lobby is enough to contribute to renewable energy growth through policy and lower generation costs. Thus, our approach offers an explanation for why renewable energy was able to grow rapidly even under low oil prices in key countries such as Denmark and Germany.

The process of politicization has multiple forms. The opponents of renewable energy can use the media to cast doubt on the rationality of investments in renewable energy, cautioning against increasing energy prices and loss of industrial competitiveness (McCright and Dunlap 2003). They can also lobby politicians directly, trying to convince them of the need to abandon renewable energy policies and perhaps support their message with campaign contributions or promises to mobilize voters (Chubb 1983; Markussen and Svendsen 2005). The totality of these different forms is the key component of the politicization that threatens the future of growth in renewable energy.

The intensity of opposition to renewables varies across countries and over time. In some countries, renewables have become the subject of intense political debate and conflict from the early years. In others, opposition has been delayed and remains at low levels. To capture this variation, our first expectation seeks to predict the degree of opposition based on the strength of the government's initial response. Because the challenge to renewable energy is driven by concerns about its success, we expect the degree of opposition to depend on the success of the initial renewable energy policies. If these policies produce large reductions in the cost of renewable energy and create a conducive environment for investment, then renewables begin to pose a real threat to fossil fuel producers, heavy industry, and ideological opponents of renewable energy.

The German case illustrates this idea. Until the enactment of the 1991 *Stromeinspeisungsgesetz* that prescribed a feed-in tariff to support renewables, the opposition of renewable energy was limited because renewables did not yet pose any kind of a threat to energy utilities. When the economic competitiveness and political support for renewables grew, however, the conservative Christian Democrats started to shift back to conservative energy policies that emphasized coal and nuclear power with support from the industry (Jacobsson and Lauber 2006). It was the growing importance and promise of renewables that provoked the opposition. Until the 1991 feed-in tariff, the high electricity prices that renewables spurred had not yet materialized, and therefore opposition remained at low levels.

In previous work, we have proposed that as the strength of the renewable energy constituency grows due to supportive policies and decreasing economic costs, governments become more receptive to this constituency's demands (Aklin and Urpelainen 2013b). While building on this analytical framework, we argue that the initial growth of renewable energy in the aftermath of the external shock tends to endanger the growth of renewable energy because of a political valley of death. Although the growth of renewable energy creates positive reinforcement in the long run, the first consequence of the emergence of renewable energy as a serious contender to fossil fuels is a backlash.

If anything, the expectation of positive reinforcement, path dependence, and lock-in strengthens the political backlash against initially successful renewable energy policies. As the opponents of renewable energy update their beliefs about the severity of the threat that renewables pose to their interests based on the track record of renewable energy policy, their concern is informed by the expected growth of renewable energy. Under rational expectations, the opponents of renewables understand that they have a lot to lose if they let renewables enter the stage of rapid and endogenous growth. In the United States, for example, electric utilities have become increasingly hostile to rooftop solar, which presents a threat to the traditional electric utility model by reducing electricity demand, as the cost of solar home systems and their installation has decreased.

Because variation in the politicization of renewable energy across countries has many causes, we face the challenge of focusing our attention on the most important factors. Indeed, these factors are not the same as the factors that determine the government's initial response to the external shock. Because the politicization process takes place in an environment in which the earlier external shock is less relevant than before, we need to construct another analytical model to explain the covariates of intense opposition. The main difference between the politics of renewable energy during the initial external shock and the politicization episode lies with the structure and organization of competing interests. During the external shock, the possible losers from renewable energy do not yet consider it a top priority. At the same time, however, there is no organized lobby for renewables. Because neither side is well organized or capable of effective political action, the government's initial response to the external shock is not the subject of intense political conflict. During the politicization stage, both sides grow stronger.

To understand the determinants of opposition is to understand when and how the opponents of renewable energy manage to cast doubt on the wisdom of investing in renewables. Therefore, we train our analytical lens on factors that shape a government's incentive to continue support renewables amid political conflicts. These factors are public opinion, partisan politics, and interest group influence. To understand the politicization process, we deliberately focus on these incentives, as opposed to factors such as political institutions or the timing of elections. In the case studies, we also consider other factors and evaluate both competing and complementary explanations for the outcomes, but here we prefer to maintain a sharp focus on a small number of carefully chosen factors. Based on these factors, we again derive a series of expectations about the level of opposition to renewable energy in a society. In the case studies, we also consider the possibility of interactive effects of various configurations.

The first factor of importance is public opinion. Because renewable energy has now become a contentious topic, with the two sides arguing about proper policies, public perceptions of the issue modify the strategies of policymakers. If the public is strongly in favor of renewable energy policies, then the government's political incentives are in alignment with the preferences of the renewable energy advocacy coalition. Conversely, a suspicious and hostile public can turn the government against renewable energy because supportive policies would undermine the government's electoral prospects. As Wüstenhagen, Wolsink, and Bürer (2007) explain, public acceptance of renewable energy is particularly critical for those forms of renewable energy that could be perceived to have negative effects, such as wind energy.

Recall that the public's overall perception of renewable energy is highly positive in most countries. Opinion polls from Germany suggest that people's preferences shifted from traditional energy sources to renewables in the early 1990s (Wüstenhagen and Bilharz 2006). This change has been persistent. In a survey conducted by Gallup in 2013 in the United States, three-quarters of respondents wanted more solar and wind power. In contrast, two-thirds desired less nuclear power and coal.[9] Similar patterns are observed in Europe. According to a survey by Eurobarometer (2013) held in the same year, 75% of respondents thought that increasing the share of renewable energy by 20% by 2020 was about right or even too modest.

Given this positive perception, the real issue for understanding the public's response is the willingness to pay for new policies. Because renewable energy policies either consume tax revenue or increase energy prices, they are costly to the public, at least in the short run. Even if the public's perception of renewable energy is generally positive, the level of public support for renewable energy policies decreases as the cost of these policies increases. Therefore, a theory of the relationship between public opinion and the politicization of renewable energy should focus on factors that shape the public's willingness to pay. We argue that the public's willingness to pay depends to a large extent on how concerned people are about the problems that renewables could solve. On the one hand, renewable energy is a possible solution to the problems caused by nuclear power. On the other hand, renewables can also address climate change. If the public has a high willingness to pay for renewable energy, then the opponents of renewable energy cannot easily sway the government's views or activities. The public's strong support for renewable energy leaves the renewables advocacy coalition fighting from a position of strength, and therefore opposition to renewable energy remains low.

Politicization Expectation 1 (public opinion and opposition). *The less concerned the public is about nuclear power and climate change, the higher the degree of opposition to renewables.*

We do not expect energy security to be important for understanding politicization because dependence on foreign energy imports is worrying to a large majority of the general public. Different from nuclear power and climate change, energy security is not an issue that divides the public. Although actual energy security has historically varied a lot both across and within countries, public concern about it tends to be persistently high across countries. US public opinion is divided on many issues, but it agrees that energy security is a problem. A survey conducted in 2006 found that 82% of respondents were concerned or very concerned by energy shortages and their effect on the economy (Council on Foreign Relations 2012). Citizens from other countries are worried, too, because the global average of people who expressed similar levels of concern was 77%. A poll from the German Marshall Fund showed in 2008 that 87% of Americans expect energy dependence to affect them personally within ten years. Again, this fear was shared by the respondents from twelve European countries, where 81% shared these beliefs. These views were captured in a quote from Bloomberg about the Keystone pipeline project: "[the interviewee] says she backs the pipeline for its economic benefits and sees it as an alternative to oil imported from unfriendly countries. 'We get along with them better,' she says, referring to Canada."[10]

Although public opinion plays an important role in shaping the politicization of renewable energy, it is far from the only decisive factor. Because renewable energy and environmental issues more generally constitute but one component of the public's overall evaluation of the government (List and Sturm 2006) and many voters pay little attention to policy in any case (Grossman and Helpman 1996), there is plenty of room for elites and interest groups to influence policy formulation. The next factor to consider is the political clout of prorenewables interest groups. First, strong environmental movements are in general conducive to the growth of renewables (Kim and Urpelainen 2013b; Sine and Lee 2009; Vasi 2011). Their interest in renewable energy stems from its positive environmental effects at local, national, and international levels. Although some conservationists may oppose certain forms of renewable energy due to their perceived negative local effects, such as windmills killing birds, virtually every major environmental group today is in favor of replacing the extraction of fossil fuels with renewable energy generation. For example, Vasi (2011) shows that environmental groups have played an important role in promoting the use of wind energy in Denmark, Germany, the United States, and Spain. These groups were motivated by concerns about climate change and other environmental problems associated with the use of fossil fuels.

In practice, environmentalists can promote wind energy in several different ways. Kim and Urpelainen (2013b) present a game-theoretic model of how an advocacy group can influence the use of a clean technology, such as renewable energy, by both lobbying the government for new policy and campaigning among end users to increase their interest. For example, an environmental group could first lobby for the government to implement a renewable energy feed-in tariff and, having done that, encourage customers to subscribe to green electricity services. Indeed, Kim and Urpelainen (2013b) find a certain strategic complementarity between the two activities. If environmental groups only campaigned among end users, then the government might not feel any pressure to enact renewable energy policy. If environmental groups only focused on policy advocacy, then they would forgo an important opportunity to promote renewable energy. Empirically, the patterns of campaigning reported by Vasi (2011) are consistent with the general predictions of the Kim and Urpelainen (2013b) model.

The key economic interest group that has an interest in more ambitious renewable energy policy is the clean technology industry. This group comprises firms that manufacture and install windmills and solar panels, equipment for biodiesel production, wood pellets for combustion, and so on. The clean technology industry profits from renewable energy policies because said policies create demand for the industry's products. Increased demand raises market prices while allowing the industry to capitalize on economies of scale. Consequently, the clean technology industry has strong incentives to support renewable energy policies. The available empirical evidence for this claim is strong. Both Cheon and Urpelainen (2013) and Aklin and Urpelainen (2013b) find that strong clean technology groups have played an important role in promoting renewable electricity generation in industrialized countries over time. According to Michaelowa (2005), German wind technology producers have achieved high levels of success in influencing German renewable energy policies. Lewis and Wiser (2007) find that, in general, governments have a strong interest in promoting the commercial interests of the wind energy industry. All these studies are consistent with the notion that the clean technology industry has become an important player in the formulation of renewable energy policy and has access to key policymakers at the highest levels of government. The potential of the clean technology industry can also be measured by investigating patenting activity in the clean technology sector in the early years of renewable energy before policy begins to endogenously create new interest groups.

A particularly important prorenewables group is small producers. Bayer and Urpelainen (2016) analyze the determinants of FIT formulation and find that democratic countries with electoral rules that favor rural areas are particularly likely to adopt an FIT—a policy that enables small producers to reap profits from renewable electricity generation at the expense of utilities. If the number of such small producers grows sufficiently large, then their political clout can prove decisive for prorenewable energy policies. Although traditional energy interests remain powerful, a large number of small producers in politically important areas can exert influence over energy and climate policy. Indeed, in our earlier work (Aklin and Urpelainen 2013b), we have found that governments of industrialized countries consider the future implications of investing in renewable energy policies that create new, prorenewable interest groups.

Politicization Expectation 2 (prorenewable groups and opposition). *Politically powerful clean technology industries and environmental groups reduce opposition to renewables.*

We now turn our attention to economic opponents. Because renewable energy policies eradicate negative externalities, generate profits, and impose costs, different interest groups have a direct interest in renewable energy policy. Those expecting to secure profits and rents from renewable energy policies support them, and those expecting to lose mobilize opposition against said policies (e.g., E. Moe 2015).

One key cleavage is between the heavy industry, which expects to lose from higher energy prices, and the clean technology industry, which expects higher profits due to more demand for its products (Cheon and Urpelainen 2013). For heavy industry—aluminum, iron, steel, paper, chemicals, paper, pulp, and the like—incentives to oppose renewables stem from the higher electricity prices that reliance on expensive renewable energy creates. Because renewable energy sources are not economically competitive with fossil fuels in a society characterized by carbon lock-in, policies that mandate the use of renewables increase the cost of electricity. This increase is a grave threat to heavy industry, given that its profits depend on access to affordable electricity. Where electricity is a key input in the production process, even modest increases in electricity prices can greatly reduce the industry's profitability. Under international competition, domestic producers may also lose market share because their foreign competitors continue to enjoy inexpensive electricity.

Unless the government is ready to fully subsidize the higher cost of renewable electricity to compensate heavy industry for its losses at a high fiscal cost, this interest group has a clear interest in undermining renewable energy policies. Typical renewable energy policies, such as feed-in tariffs and portfolio standards, raise the cost of power by forcing utilities to purchase expensive renewable energy, reducing the profitability of heavy industry. As a result, heavy industry has an incentive to oppose renewable energy policies by politicizing the issue. This industry prefers to minimize the cost of electricity and avoid any policies that focus on expensive renewable energy instead of less expensive fossil fuels. The dependence of heavy industry on stable access to large loads of power also reduces their interest in renewables because of the technical issue of intermittency. According to several interviews with staff of major Finnish paper manufacturers conducted by Luukkanen (2003), Finnish industry engineers considered modern wind and solar power a largely trivial source of power. These engineers argued that the techno-economic potential of new renewable generation in Finland was so limited that it would play no role in the short run, and even in the long run the ability of renewables to truly replace conventional power sources was subject to doubt.

As we have already explained, fossil fuel producers also have much to lose from renewable energy policy. As E. Moe (2015, 21) puts it, "in so many countries, there is a strong institutional bias in favor of the present energy structure, based on fossil fuels (and sometimes nuclear) and on big, centralized energy utilities distributing electric power to a vast number of industries and households." If renewables begin to displace fossil fuels, then the market price of fossil fuels decreases. For example, if renewable portfolio standards create demand for wind and solar electric capacity by forcing utilities to meet a certain proportion of their total production from renewable sources, then the demand for coal for electricity generation decreases. The decrease in demand causes a decrease in the market price for coal and, with production costs remaining unchanged, dampens the profitability of coal mining. Although all industrialized countries have a power sector in common, only some of them also have upstream production of fossil fuels. Australia, Canada, Poland, and the United States are only some examples of the group of countries with abundant fossil fuel endowments. In such countries, we would expect the fossil fuel industry to oppose renewable energy policy. Major fossil fuel producers have a direct economic incentive in retaining their market share.

Fossil fuel producers should become more politically active when climate change becomes a salient concern for the public. The importance of fossil fuel producers should be significantly amplified by concerns about climate change. Although major oil companies, unlike coal producers, may not have a direct interest in stopping renewable power in electricity generation, they may oppose climate policies that would enhance the competitiveness of renewable power (D. Levy and Kolk 2002; E. Moe 2015). If major emitter countries chose to significantly reduce their carbon dioxide emissions, perhaps under a global climate treaty such as the Paris Agreement, then large fossil fuel producers would not be able to continue selling their primary products on the world market (Carbon Tracker Initiative 2014). In a world of ambitious climate policy, huge amounts of fossil fuel deposits would be "stranded assets" and lose their value. Therefore, we expect the fossil fuel industry as a whole, and not just in the power sector, to play a major role in determining the outcome of political conflicts over climate and energy policies that favor renewables.

Politicization Expectation 3 (antirenewable groups and opposition). *Politically powerful fossil fuel producers and heavy industries increase opposition to renewables.*

Because political interests are almost without exception represented by political parties in industrialized democracies (Benoit and Laver 2006), we finally consider partisan ideology. Given how much attention the left–right divide has received in the political economy literature (Boix 1998; Garrett 1998), this salient cleavage is a natural point of departure for the analysis. We argue that leftist parties have stronger incentives to enact renewable energy policies and less interest in allowing the opponents of renewables to politicize the issue. Although the right wing has both ideological and material reasons to oppose renewable energy, for the left wing, renewables are an attractive proposition. The left–right divide stems mostly not from disagreement about the benefits of renewables but from differing valuations of the costs. Both the left and right value energy security and environmental quality, but their willingness to implement policies and create regulations is different. Members of right-wing parties believe that the costs of state intervention into the economy are high, whereas most left-wing activists and leaders worry about market failures in the absence of public policies (Boix 1998; Garrett 1998).

Even if both sides of the aisle agree on the need to promote renewable energy, their differing expectations regarding the costs of interfering with the market allocation of resources drives a wedge between them. The role of partisan ideology and economic factors in public opinion about renewable energy can be seen in the Republican–Democrat difference in public support for various energy policies in the United States. According to an opinion poll by Leiserowitz et al. (2013), in April 2013, as many as 69% of all Democrats supported a 20% renewable electricity requirement for energy utilities, even if the annual cost was $100 to every American household, whereas only 43% of Republicans supported such a policy. At the same time, the partisan difference in support for tax rebates for fuel-efficient vehicles and solar panels was much less pronounced: 81% of Democrats and 66% of Republicans supported such a policy. The more costly portfolio standard for renewables was more divisive.

Based on this logic, we expect partisan ideology to play an important role in the politicization of renewable energy. Left-wing governments should be supportive of renewable energy and not receptive to the arguments of the opponents of renewable energy, whereas right-wing governments should be more willing to politicize renewable energy and dismantle or abandon policies that could help renewables grow. Therefore, we expect levels of opposition to increase with partisan shifts to the right within the government of an industrialized country. If a leftist government is replaced by a cabinet on the right, then politicization levels increase because opponents of renewable energy have better access to the most important decision makers in the country.

However, right-wing parties are not opposed to renewable energy in all circumstances. Sometimes the constituencies of right-wing parties derive considerable benefits from renewable energy policies. In Germany, for example, the generous feed-in tariff for solar power has allowed some farmers to generate most of their annual income from the installation of solar panels.[11] Because farmers have traditionally supported conservative parties in Germany, even a conservative government coalition may face a political backlash for trying to reduce the feed-in tariff. Although there should be a left-right difference on average, in any particular case, it is important to investigate the specific constellation of interests and consider possible reasons for deviations from the general rule. By analyzing the left-right cleavage in terms of government shifts within a country, we go some way toward avoiding comparisons between apples and oranges.

Another key dimension of the party system is the role of green parties. While left- and right-wing governments have different incentives and hold different ideological positions with respect to state intervention, green parties are characterized by their heavy emphasis on environmental protection. Regardless of their position on the left-right axis, the supporters and leaders of green parties have a special interest in policies that protect the environment. In the energy sector, this interest manifests itself in high levels of interest in renewables. These sources of energy reduce environmental pollution, mitigate climate change, and replace nuclear power. The power sector is among the most important sources of environmental destruction that industrialized societies cause, and green parties are determined to protect the environment. Even if the green parties recognize the economic costs of renewable energy, it is such a priority topic for them that their willingness to pay is high.

Consider, for example, the preferred energy policy of the German Green Party (Jacobsson and Lauber 2006). One of the party's defining goals was the shutdown of all remaining nuclear plants in Germany, and the main alternative to atomic power was renewable energy. The Greens were impassionate advocates of a federal feed-in tariff that would allow renewables to compete with nuclear power and coal on a level playing field. Given their principled opposition to nuclear power and concern for climate change, both the party leadership and their potential voters agreed that aggressive policies to support renewable energy were needed.

Based on this logic, we expect strong Green Parties to reduce opposition to renewables. When Greens have political clout, it is hard for opponents of renewable energy to mobilize effectively against renewables. A strong Green Party is such an important player in the political system that mainstream parties in the government hesitate to dismantle or abandon renewable energy policies. If the Green Party is in the government, then other coalition members depend on its votes for enacting their preferred policies.

Opponents of renewable energy would have to pay a high price to achieve their goals, which reduces their incentive to mobilize in order to politicize the issue of renewable energy.

The electoral success and political clout of Green Parties depend on several factors. Most important of these are electoral institutions and public opinion. When elections are based on proportional representation, small parties with specialized agendas generally have good electoral prospects, and this benefits the Greens (Müller-Rommel 1998). In majoritarian systems, such as the United States, electoral victory requires securing the majority of votes, and this is hard for a niche party with a highly specialized agenda. In proportional systems, in contrast, small parties have a good chance of securing enough votes in at least some districts. This is where public opinion becomes important. Although a proportional electoral system enables the success of Green Parties, they still need to secure enough votes to win seats. A Green public opinion may, in addition to shaping the policies of mainstream parties, promote change by strengthening the position of Green Parties. Even in this case, it is possible to evaluate the effects of Green Parties on renewable energy by investigating variation in Green Party strength while holding the electoral system and public opinion constant.

Politicization Expectation 4 (partisan ideology and opposition). *Compared with their left-wing counterparts, right-wing governments are more opposed to renewables. Furthermore, strong Green Parties reduce opposition to renewables.*

The role of political institutions, such as the difference between proportional and majoritarian electoral systems, is an important question for understanding the fortunes of renewable energy. For explaining variation in the dynamic development of renewable energy in a politicized environment, however, an essentially static approach has its limitations. Even in a proportional system, it is possible that Green Parties fail to play a pivotal role in the formation of governing coalitions; similarly, right-wing parties could win several elections over time. For this reason, we believe that emphasizing partisan politics, as opposed to electoral institutions, is the more fruitful approach.[12]

To summarize, the nature and extent of politicization of renewable energy depends on several key factors. Public opinion shapes the government's incentive to grant access and consider the complaints of constituencies and interest groups that expect to lose from increased use of renewable energy. Among political parties, the Left and the Greens are sympathetic to renewable energy due to their ideological commitments, whereas environmentalists provide support outside the legislative institutions. Heavy industry and fossil fuel producers expect heavy losses from renewable energy and mobilize to politicize and criticize renewable energy, but the clean technology industry counters their opposition with demands for more aggressive renewable energy policy.

3.3 National Lock-In and Global Convergence

If renewables prove triumphant at the politicization stage, then they move toward a lock-in. Stated differently, renewable energy begins to establish itself as a standard option in energy planning. The improved performance and growing political support for renewables allow the society to "escape" the earlier carbon lock-in (Unruh 2002). Although renewables will not completely replace other energy sources any time soon, their status as an important source of energy is consolidated. Renewables achieve lock-in for the same reasons that allowed carbon lock-in earlier. Path dependence is fundamental to both economics (Arthur 1989) and politics (Pierson 2000). As technologies become mature, their ability to capture the entire market increases. As policymakers implement policies, they understand that their choices have effects on both present and future outcomes.

The lock-in that characterizes renewable energy is, however, quite different from the traditional carbon lock-in. In the case of carbon lock-in, fossil fuels and nuclear power achieved a virtually unassailable position in the energy structure.[13] This is not the case for renewables, as they continue to grow in the shadow of the carbon lock-in. Instead, the renewable lock-in ensures that renewables are no longer under an immediate danger of losing their support. Their reduced cost, environmental benefits, positive public image, and growing advocacy coalition ensure that the government continues to support renewables and promote their growth. The lock-in of renewable energy is a combination of two distinct yet related changes. First, the techno-economic performance of renewable energy increases due to lower costs and technological advances. Second, the social acceptance and public image of renewable energy improve as a result of improved performance and familiarity. Although analytically distinct, these two factors are mutually reinforcing.

On the economic side, the two key mechanisms that promote the lock-in of renewables are learning and economies of scale (Arrow 1962; Hisnanick and Kymn 1999; Nemet 2012). Manufacturers, installers, and users of renewable energy equipment learn over time how to deploy renewables with improved performance and at a lower cost. These learning effects are a key reason that the cost of renewable energy has decreased rapidly over time, with reduced costs strongly associated with the accumulation of experience and total installed capacity at the global and national levels. Additionally, economies of scale can directly contribute to the competitiveness of renewables. As demand for renewable energy grows, larger units such as massive wind and solar farms can be installed. The large size of these installations brings down the cost of electricity generation even further, although the cost reductions are realized slowly over decades.

In addition to economic lock-in, renewables benefit from changes in the political power balance. As renewable energy generation increases, so does the number of economic actors who benefit from it (Aklin and Urpelainen 2013b; Michaelowa 2005). To illustrate this notion, consider an expansion in the wind energy capacity of a country due to generous renewable energy policies. The producers of wind energy will demand that these policies be continued so they can continue to benefit from them in the long run. If some nontrivial proportion of the wind power equipment is manufactured within the country, then the owners and employees of the manufacturing companies will also demand that the policies be continued. Moreover, these manufacturing companies have clients that benefit from the technology and would rather purchase it with government support. The creation of these new renewable energy constituencies endows the government with new incentives to continue and expand policies that support renewable energy. For empirical evidence, Cheon and Urpelainen (2013) show that as the share of renewables in a country's national electricity mix grows, the ability of heavy industry to prevent further growth of renewables is diminished.

The feedback loop between policy and cost is characterized by positive reinforcement. If a policymaker wants to achieve a certain level of renewable energy growth, it can do so at a lower cost over time as the competitiveness of renewables increases. Even modest subsidies, tax incentives, and regulations can secure large gains in the growth of renewables in the energy mix if the cost differential between the most competitive renewable and traditional energy sources is diminished. In turn, policies promote further learning and economies of scale, allowing even greater gains in the competitiveness of renewable energy.

In a globalized economy, renewable energy investment and policies are not strictly national processes. If renewable energy gains ground in one economy, then the resulting policy learning and cost reductions, as well as positive publicity, encourage the growth of renewables in other countries. The success of renewables in one country increases their economic competitiveness in other contexts as well, and improved policies allow other governments to replicate successful approaches while avoiding pitfalls. The main economic driver behind the process of diffusion is learning. Although economies of scale may not influence outcomes across national boundaries, learning does. New technologies and best practices are not geographically restrained. For example, if a Chinese manufacturer of solar panels manages to reduce the cost of said panels through product innovation or simply thanks to generous subsidies, then businesses and consumers in other countries can invest in solar electricity generation at a lower cost than in the past. One country's ability to reduce the cost of renewable energy creates opportunities in the global marketplace, causing demand to increase and allowing the use of renewables to spread across borders (Lewis and Wiser 2007).

The same logic applies to renewable energy policies. If a certain policy, such as the feed-in tariff, proves to be successful in one country, then other countries expect gains from replicating the success. Governments tend to be risk-averse, and seeing a successful policy model being implemented in another country is a reliable signal that the policy produces tangible benefits. For example, Smith and Urpelainen (2014) show that, among industrialized countries, the enactment of feed-in tariffs by neighbors is a strong predictor of a feed-in tariff. This pattern is consistent with the idea that governments follow the lead of other countries in similar circumstances.

The spread of renewable energy is also promoted by powerful forces of policy diffusion. The spread of new policies across countries has drawn many a scholar's attention in recent years, and the evidence suggests that policy innovations diffuse across borders relatively easily (Dobbin, Simmons, and Garrett 2007; Simmons, Dobbin, and Garrett 2008). Governments not only learn from their neighbors and partners but also mimic other governments in a less rational and conscious fashion, the result being "isomorphism" of institutions and policy formulation (J. Meyer et al. 1997). As a result of conscious learning and unconscious emulation, policies and institutions diffuse across borders, and evidence suggests that this is the case for environmental and energy policy as well (Busch and Jörgens 2005; Aklin and Urpelainen 2014).

Given the powerful forces of diffusion, we expect a global convergence of renewable energy policies and outcomes. Although renewable energy has long been considered an expensive luxury good, our theory predicts that wealthy as well as poor countries find renewables worthwhile. Over time, the decreasing cost of renewable energy encourages more and more countries, including those that are the least developed, to invest in solar, wind, and related power sources. Although renewable energy may remain more expensive than competitors, such as coal and natural gas for electricity generation, the political and economic virtues of renewable energy promote its use across the world.

The first reason that renewables become increasingly popular relates to the simple economics of fossil fuel scarcity. Many countries, ranging from Japan to India, have only limited fossil fuel resources. Their economic well-being is highly dependent on the prices of fossil fuels. They do not reap any benefits from high fossil fuel prices, and they pay a high price for their fossil fuel imports. Because the total cost of electricity from renewables is much less sensitive to fuel prices than the total cost of electricity from fossil fuels, they are a potential insurance against a future characterized by high and volatile world market prices for fossil fuels.

Reduced dependence on fossil fuels also enhances a country's energy security (Asif and Muneer 2007; Müller-Kraenner 2010). Because renewable energy can be produced locally even in countries without fossil fuel reserves, its supply is not vulnerable to extortion or blackmail by a foreign adversary. Even countries that are not worried about the economic risks associated with fluctuating world market prices for fossil fuels may value the security benefits of renewable energy enough to make investments, provided that the cost of renewable energy has reached acceptable levels. New fossil fuel resources such as shale gas undermine the incentive to invest in renewable energy, but uncertainties about their abundance and concerns about their environmental effects continue to provoke interest in renewable energy.

In addition to these benefits, governments may pursue the strategy of "green job" creation or seek to build a renewable energy industry in the hopes of increased exports (Lewis and Wiser 2007; Cheon and Urpelainen 2012). Investment in renewable energy is a visible political strategy that allows a government to claim credit for the creation of new manufacturing jobs and successful export businesses while protecting national and global environments.

At low costs, renewable energy is also an increasingly attractive response to environmental problems. The health and environmental costs of generating electricity from coal, the dominant fossil fuel in the power sector, have proven to be high (Muller, Mendelsohn, and Nordhaus 2011). As the cost of renewable energy decreases, it becomes an increasingly lucrative strategy for dealing with the environmental and health problems associated with the power sector.

Lock-In Expectation 1 (global convergence of renewables). *As the cost of renewable energy decreases, supportive policies spread among countries, promoting global convergence in both policy and outcomes.*

As the cost of renewable energy decreases over time and it becomes an attractive solution to different societal problems, political controversies surrounding it also diminish. The decreased cost of renewable energy reduces the incentives of previous opponents to politicize renewable energy. Lower production costs mean that heavy industry has less interest in opposing renewables, whereas enhanced economic competitiveness of renewables also allows the power sector to profit from them. Due to the reduced need for state support, ideological opposition to state intervention is also reduced, and fiscal conservatives worry less about the cost of renewables. The lock-in of renewable energy reduces the degree of opposition to renewable energy.

The decrease in the degree of opposition to renewables, resulting from a political-economic lock-in, is important. If renewable energy improved only in terms of reduced cost, then it would not start to decarbonize the energy sector until it became fully competitive with fossil fuels. Due to a long history of carbon lock-in, this scenario is unrealistic on a national scale. Wind power, solar photovoltaics, sustainable biomass, and small hydroelectric plants may prove competitive in isolated settings and narrow geographic locations, but this does not mean they can grow on a large scale without supportive public policies.

To illustrate this pattern of politicization over time, consider the case of rooftop solar. Initially, solar home systems were a niche product that did not provoke much passion in the political arena, symbolism notwithstanding. As their economic credentials have improved, however, electric utilities have begun to express concern about the effects of rooftop solar on the profitability of the traditional utility model. Over time, we would expect these conflicts and disputes to decrease because rooftop solar becomes standard equipment for the modern home, and electric utilities can no longer expect success in preventing or reducing the use of rooftop solar through policy.

Lock-In Expectation 2 (global growth of political support for renewables). *As the cost of renewable energy decreases and the number of people producing or using it increases, the political controversies surrounding it are mitigated.*

It is useful to consider a typology of various cases. In original frontrunners, where renewables survived the politicization, the new lock-in allows renewables to continue their growth at a lower cost and with less political opposition. In other industrialized countries, government and society experimented with renewables during earlier shocks, but renewables did not survive the political backlash against them after the shock waned. In these cases, we expect renewables to return with force. In yet other industrialized countries, renewable energy was never on the table. Here we expect renewable energy to grow despite the lack of a solid foundation. Finally, in emerging economies and the least developed countries, renewables enter the game for the first time. Despite differing histories, we expect a convergence of policies and outcomes over time, due to both better economics and less conflictual politics.

Although we foresee convergence of outcomes over time, an important difference exists in the process of renewable energy development between industrialized and developing countries. In industrialized countries with a sophisticated infrastructure for exploiting fossil fuels, the lock-in of renewables should require more time. With the exception of the original frontrunners, which have already made extensive investments in renewables, other industrialized countries are faced with the issue of replacing fossil fuels, and possibly nuclear power, with renewable energy. In developing countries, however, renewable energy is but one investment in a rapidly expending energy infrastructure. Although developing countries are also subject to a carbon lock-in due to the globalization of the energy industry (Unruh and Carrillo-Hermosilla 2006), they also face the pressing need of energy infrastructure development. Under such pressure, investment in renewables is not tantamount to replacing fossil fuels. The degree of carbon lock-in before the advent of renewables is, therefore, a decisive factor in the pace at which renewables can grow in different countries.

Lock-In Expectation 3 (reduced opposition to renewables in different countries). *Opposition to renewable energy decreases more rapidly in countries that have previously made large capital investments into an energy infrastructure based on fossil fuels.*

To summarize, the lock-in of renewables is both economic and political. Reduced costs increase the economic attractiveness of renewables as a solution to various problems in the society, and these cost reductions mitigate opposition to renewables among political constituencies. The joint effect of these political and economic changes is to allow the lock-in of renewables in frontrunner and other countries. The long path from the initial external shock to lock-in and global expansion through the difficult process of politicization is complete.

3.4 Research Strategy

Our theory generates a number of empirical predictions. Some focus on variations within countries, and others are about cross-national variation. Therefore, we must consider the appropriate research strategy carefully. The problem is compounded by the fact that the initial part of our story concerns a fairly small set of countries. We thus use comparative case studies as our primary empirical method (King, Keohane, and Verba 1994; Gerring 2004; J. Levy 2008). The first important question concerns sampling (Achen and Snidal 1989; Seawright and Gerring 2008). To some extent, the nature of the population in question is defined by our theory. The first step of our argument contrasts the responses of countries to the energy shocks of the 1970s and 1980s. This presupposes that we need to draw countries that were in a position to answer to such a shock. In 1973, investing in renewables was never a realistic policy option for a poor country. Thus, we select cases from the set of industrialized countries with high levels of income. These countries plausibly had control over their energy policy.

For the case studies, we adopt a two-pronged approach. First, we select countries that exhibited variation on the main independent variables of interest: vulnerability to imported oil and elite ideology. This strategy ensures that we can assign differences in outcomes to these factors. Second, to reduce the risk that our findings are driven by other variables, we match selected cases based on their resemblance on other dimensions. As such, we follow the "most similar" approach of Seawright and Gerring (2008), which is based on Mill's method of difference (Sekhon 2004). We select a total of six country case studies for the analysis, ensuring that each frontrunner country is paired with a similar country for a structured, focused comparison along the lines of George (1979). Specifically, we compare and contrast the following groups of countries: the United States and Germany with France and the United Kingdom, as well as Denmark with Finland. These two sets of countries all contain well-functioning democracies with similar income levels. Their population had a similar cultural makeup. Furthermore, they had access to similar technologies. They did differ, however, in terms of their elite ideology and exposure to the successive energy shocks. In chapter 4, we show that the United States and Germany (unlike France and the United Kingdom), as well as Denmark (unlike Finland), responded to the oil and nuclear shocks by planting the seeds for the renewable energy industry.

Before beginning the case studies, we demonstrate that our theory applies to the group of industrialized countries in general. In Table 3.1, we compare our three frontrunners with other countries before the 1973 oil shock. First, we examine their gross domestic product (GDP) per capita and population. Two of the three frontrunners are large countries, whereas Denmark is small; all three are relatively wealthy, but only Denmark is among the three wealthiest countries on the list. On the other hand, all three frontrunners are quite dependent on oil imports. Denmark, in particular, is in a position

Table 3.1

Cross-national comparisons among the three frontrunners and other industrialized countries before the 1973 oil crisis. The gross domestic product (GDP) per capita numbers are given as constant 2005 prices in US dollars ($). For Switzerland (1980) and New Zealand (1977), the GDP per capita numbers are more recent. Oil imports are from the International Energy Agency (IEA) World Energy Balances and Statistics; all other data are from WDI (2013).

Country	GDP per capita (1971, $)	Population (1971, in millions)	Oil imports (1971, ktoe[1])	Oil imports per million people (1971, ktoe)	Oil in electricity generation (1970, %)
Denmark	25,359	5	10,813	2163	69
Germany	17,431	62	31,214	508	12
United States	20,991	210	101,851	485	12
Australia	18,129	13	2,748	211	5
Austria	16,948	8	3,196	426	7
Belgium	17,668	10	6,054	624	52
Canada	18,373	22	10,000	455	3
Finland	15,886	5	3,181	692	28
France	18,007	52	6,401	123	22
Greece	11,368	9	1,706	194	35
Iceland	23,261	0.2	531	2576	3
Ireland	12,748	3	2,135	712	54
Italy	14,837	54	3,188	59	49
Japan	15,671	110	33,769	307	59
Luxembourg	27,681	0.3	1,367	3992	18
Netherlands	20,426	13	9,673	744	33
New Zealand	17,915	3	895	309	3
Norway	25,314	4	4,102	1052	1
Portugal	7748	9	1,490	173	15
Spain	11,732	34	1,271	37	27
Sweden	21,803	8	20,392	2518	31
Switzerland	42,657	6	7,912	1276	6
United Kingdom	17,812	56	20,486	366	19

[1]Kilotons of oil equivalent

of vulnerability: only Luxembourg, Iceland, and Sweden imported more oil per capita at the time. These numbers suggest that even a simple indicator of dependence on oil imports predicts the three frontrunners well. Finally, all three frontrunners, similar to the majority of industrialized countries, continued to use large amounts of oil for electricity generation. Again, Denmark is an outlier, with more than two-thirds of their electricity generated from oil.

To increase confidence in our findings, in the case studies, we examine a range of relevant outcomes. We begin by demonstrating that all countries under consideration implemented similar policies before the 1973 oil crisis. At this time, their policy trajectories began to diverge, and we provide direct evidence for the relevance of the external shock and the contingent national factors in generating the differential policy trajectories. We examine the presence or absence of direct and indirect renewable energy policies. In the absence of actual output—little renewable power was generated in the 1970s—we focus on the kinds of tools that were instrumental in the future deployment of this technology. An example is the level of R&D investments. In addition to secondary sources on the cases, we use newspaper articles, official documents, and a small number of interviews with policymakers and other stakeholders. Given our focus on the comparison of multiple cases, we do not claim to shed new insights into policy formulation in any particular case, with the partial exception of the understudied cases of the United Kingdom and Finland. Instead, our main focus is on summarizing and synthesizing the available research to derive new insights from structured comparisons under a unified analytical framework.

The next step of our argument focuses on variations among and within frontrunners during the politicization stage. This point is important: the predictions about the degree of politicization (and its effects) are conditional on having responded to the shock. That is, politicization is only relevant for countries in which renewable energy has become policy issue. Therefore, in chapter 5, we continue to follow the countries we selected in chapter 4 but with a greater focus on those countries that have responded to the oil shock by investing in renewables: the United States, Germany, and Denmark. We briefly examine the absence of renewables over this period in France, the United Kingdom, and Finland, but our focus is on the countries that have gone on a path initially favorable to renewables. Methodologically, we continue to draw on the "most similar" approach. Germany is as close of a comparison case to the United States as is reasonable to expect, except on our key independent variables.

These case studies can also be illustrated with the help of a table showing cross-national comparisons. This is done in Table 3.2, which focuses on the early years of the politicization period to avoid conflating causes and consequences of renewable energy policy. We compare public opinion against nuclear power, renewable energy patents, the size of the industrial sector, Green Party shares, as well as fossil (oil, gas, coal) and nuclear energy production.[14] As the table shows, these basic indicators highlight possible reasons behind the divergence of the United States versus the more resilient frontrunners, Denmark and Germany. Denmark has relatively high levels of opposition to nuclear power, a capable clean technology industry, a small industrial sector, and no fossil fuel production at all. Germany does not initially seem to be a good candidate for political success of the pro-renewables coalition, but nuclear opposition is relatively high already in early years, and, critically, the German Green Party did better in elections in the 1980s than any other Green Party in the world. The United States has low opposition to nuclear power, a small clean technology industry, lots of industrial activity, and a large fossil fuel industry.

These cross-national comparisons set the stage for case studies over time. Given that we have three aggregate explanatory variables (parties, interest groups, and voters), we cannot trace their effects with three static country cases. Fortunately, we can do so by examining variations within these countries. We can use variation in governments, lobbies, and people's preferences to validate the effects of these three factors. We are thus not comparing three countries but three countries over a number of years. Our approach here is related to process tracing to the extent that we explore historical processes (D. Collier 2011). However, we also take advantage of exogenous changes inside countries, such as the election of a rightist government, to evaluate their effect on domestic energy policies.

Finally, in chapter 6, we explore the lock-in period. Overall, we predict a large increase in the deployment of renewables, driven by decreasing costs and higher investments. The third phase consists of two related predictions. For one, we anticipate that frontrunners—mostly Denmark and Germany—continue to experience sustained growth. In these countries, renewables have survived the pushback from the politicization phase. Renewables become increasingly competitive while still benefiting from the general support of political elites. If opposition arises it does not last long because decision makers have political and economic incentives to maintain high growth rates for renewables. The predictability of the policymaking process encourages investors to fund the addition of new capacity. Investments and

Table 3.2

Cross-national comparisons among the three frontrunners and other industrialized countries at the beginning of the politicization stage, around 1980. Nuclear opposition is the mean percentage of respondents opposing nuclear power in 1978 and 1982 Eurobarometer and Public Attitudes Toward Nuclear Power polls; renewable energy patents are from the OECD patent statistics database; industry is from WDI (2013); Green Party shares are from the Manifesto Project Database and extended by the authors; fossil fuel (oil, gas, coal) and nuclear energy production is from IEA World Energy Balances and Statistics. Normalization by population is based on the 1971 numbers from the previous table.

Country	Nuclear opposition (1978–1982, %)	Renewables patents per million people (1976–1984)	Industry, value added (% of GDP)	Average Green seat share (1980–1990)	Fossil, nuclear production per 1,000 people (ktoe,[1] 1980)
Denmark	42	3.20	27	0.00	0.00
Germany	36	0.54	41	4.29	2.83
United States	28	0.65	34	0.00	4.63
Finland		0.87	38	1.61	0.40
France	36	0.62	32	0.00	0.69
United Kingdom	31	0.38	41	0.00	2.05
Australia		1.75	38	0.00	4.57
Austria		0.93	36	1.75	0.33
Belgium	38	0.52	35	2.08	2.43
Canada		0.00	37	0.00	4.29
Greece	49	0.11		0.00	0.34
Iceland		0.00	36	0.00	0.00
Ireland	41	0.00	35	0.00	0.26
Italy	36	0.02	38	0.00	0.21
Japan		0.15	39	0.00	0.31
Luxembourg	40	0.00	30	0.67	0.00
Netherlands	51	0.31	33	0.37	5.38
New Zealand		0.00	31	0.00	0.67
Norway		1.67	39	0.00	5.89
Portugal		0.00	30	0.26	0.01
Spain		0.00	37	0.00	0.33
Sweden		5.74	32	1.01	0.85
Switzerland		2.98	35	1.99	0.60

[1]Kilotons of oil equivalent

innovation, in turn, reduces the cost of renewables, making them more powerful. This constitutes our first hypothesis.

On the other hand, we claim that the rest of the world benefits from the progress made by the frontrunners. Innovations cut the cost of deploying renewables. This enables policymakers that previously could not afford, politically speaking, to promote renewables to finally do so. Denmark, Germany, and, to a lesser extent, the United States have operated since the 1970s as political entrepreneurs. Years of testing various policies and investing directly or indirectly in the various renewable energy industries have paid off. Newcomers, shielded from the costs of these trials, can now reap the renewable energy premium.

We show how the diffusion of renewables occurred across industrialized and developing countries. The major change during the past ten years is that developing countries have embraced renewables as a solution to their own issues. China has sought to use renewables to quench its thirst for energy and solve its pollution issues (Y. Chen et al. 2013). Countries such as India perceive renewable energy to be helpful in tackling rural electrification. Brazil, which already had a biofuel industry at the time, has continued to invest vast amounts to reduce its dependence on foreign fossil fuels and develop its own industrial capacity.[15] Thus, across all continents, governments have increasingly decided to support renewables. To test our third lock-in expectation, we investigate differences in the degree of political opposition to renewables between industrialized and developing countries, finding that renewables are much less controversial in the developing world. This finding is consistent with the notion that the higher degree of carbon lock-in in the industrialized world makes policy shifts toward renewable energy politically more controversial than in the developing world, where any kind of new energy infrastructure is considered a priority.

II A Political History of Renewable Energy

4 External Shocks: Destabilizing the Carbon Economy

This chapter explains how the oil crises of the 1970s destabilized the energy economies of several industrialized countries. We begin with a broad overview of the relationship between the major oil shocks, new concerns over the safety of nuclear energy, and renewable energy policy in leading industrialized countries. Next, we investigate the cases of individual countries to gain a better understanding of exactly how policymakers responded to external shocks.

The key to understanding the initial growth of renewable energy lies with the massive oil shocks that prompted policymakers to make ambitious changes to their energy policies. Given that renewables had virtually no commercial applications in 1973, it was only due to the rapid change in the overall energy economy that policymakers were willing to invest in renewable energy. From the United States to Germany and Denmark, government officials were operating in an environment characterized by strong societal demands for action, high levels of uncertainty about the future, and the availability of a policy alternative that seemed to offer a way out of a difficult situation. The common factor that the frontrunners in renewable energy shared was the vulnerability to the external shock under permissive structural and institutional conditions.

The spirit of the time is nicely illustrated by President Jimmy Carter's televised speech on April 18, 1977, which called America's energy crisis, entering its fifth year, "the moral equivalent of war":

> Tonight I want to have an unpleasant talk with you about a problem unprecedented in our history. With the exception of preventing war, this is the greatest challenge our country will face during our lifetimes. The energy crisis has not yet overwhelmed us, but it will if we do not act quickly. . . . Because we are now running out of gas and oil, we must prepare quickly for a third change, to strict conservation and to the use of coal and permanent renewable energy sources, like solar power.[1]

This passage is remarkable for two reasons. First, the rhetoric clearly betrays the severity of the energy crisis as perceived by the US political elite. Second, renewable energy is considered but one solution to the problem of oil crisis, next to conservation and even increased reliance on domestic coal resources.

The next section provides an overview of policy responses to the 1973 and 1979 oil crises. We collect and present the available data to demonstrate the strong association with the key external shocks and the renewable energy policy responses of many governments. We demonstrate the severity of the external shocks and the vulnerability of many industrialized countries to them, and then we characterize the policy responses of different governments. The evidence presented in this section sets the stage for a more detailed analysis of specific country case studies. Before conducting the individual case studies, we present and discuss a summary table that foreshadows the key results from the case studies. The cases are grouped under three sections. The first section evaluates the factors that led both Germany and the United States, despite their differences, to implement experimental renewable energy policies as a response to the oil shocks and problems associated with nuclear energy. The next section then solves the puzzle of why France and the United Kingdom chose not to invest in renewable energy at the time despite many similarities with Germany and the United States. Finally, to highlight the contrast between forerunners and laggards even more sharply, we explain why the successful Danish strategy of aggressive wind power deployment was not replicated in another small Nordic country, Finland, despite similar vulnerabilities and problems.

4.1 The 1973 and 1979 Oil Crises: Of Shocks and Policy Responses

In 1973, the Western world abruptly moved from a situation in which oil was an infinite resource to one in which oil appeared scarce. Previous episodes of concerns about declining oil supplies—and there had been a few (R. Stern 2016)—paled in comparison to the disruption generated by the oil shock.[2] Cheap and abundant oil could no longer be taken for granted. Oil prices are important because oil shocks trigger economic and financial crises (Hamilton 1983; Sadorsky 1999). Oil is an input to production; if prices go up, then inflation increases. In the absence of domestic producers, or when these producers are owned by a small and concentrated elite, aggregate welfare likely decreases. Furthermore, increases in oil prices are felt by citizens through higher gasoline and heating costs.

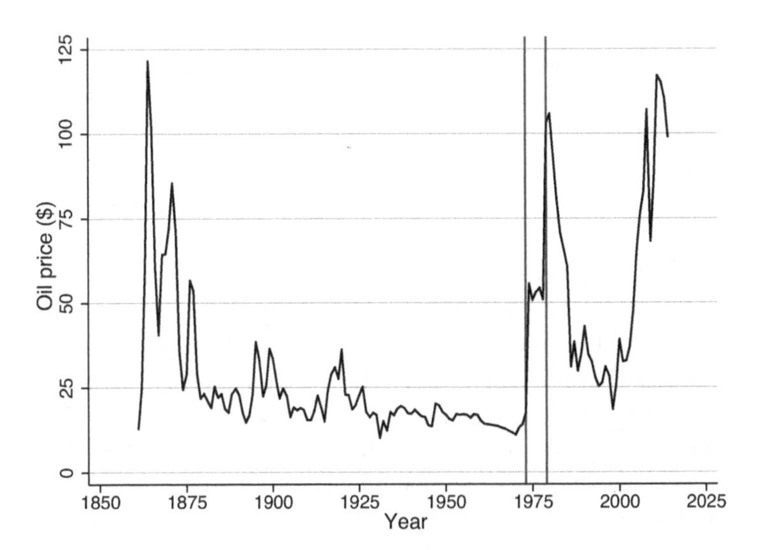

Figure 4.1
Oil prices, 1861–2014. The prices are measured per barrel in constant 2012 dollars. The vertical bars indicate the years of the oil crises, 1973 and 1979.
Source: BP (2015).

As we show in Figure 4.1, oil prices remained well under $50 per barrel for the first three quarters of the 20th century. Prices exploded in 1973 to about $50, remained flat for a few years, and then doubled to more than $100 per barrel in 1979. The two oil shocks shaped prices as well as oil consumption. Although the general pattern of global oil consumption is one of largely uninterrupted growth, this description masks important changes during this period. Between 1965 and 1973, world consumption increased by 157 million tons of oil per year (BP 2013). During the period of volatility between the two oil crises, 1973 to 1979, consumption increased by 58 million tons per year only, which represents a drop of 99 million tons per year. From 1979 to 2012, consumption increased worldwide by an average of 31 million tons per year.

The change may even be understated because the slower growth in oil consumption occurred while major countries, such as China and India, experienced an economic boom in the 2000s. Looking at Organisation for Economic Co-operation and Development (OECD) countries only, we observe the same striking pattern. Between 1965 and 1973, the yearly increase in oil consumption was about 115 million tons. The postcrisis drop was radical for these countries, as consumption increased by only eight million tons per year between 1973 and 1979. From 1979 to 2012, consumption has actually decreased by about a million tons per year. Therefore, the amount of oil consumed in OECD countries is almost the same today as it was in 1973, despite population and economic growth.[3]

The slowdown cannot be explained by high oil prices. By 1986, the price of a barrel had gone back down to $30 (in 2012 prices). The oil shock was truly temporary. What persisted was a change—in some countries—in government policy. The fright engendered by the crisis led various energy policy reforms, most of which aimed at improving energy efficiency. The idea was to reduce demand to match the reduction in supply and so policies aimed at regulating energy consumption. The United States, for instance, implemented a series of aggressive measures to tackle these issues, with many pieces of new legislation enacted between 1973 and 1979 (Kirby 1995). One such reform concerned cars. The government sought to increase mileage from 14 miles per gallon to 27.5 by the mid-1980s.[4] Governments also encouraged energy-saving investments. In Sweden, for instance, the government offered low-interest loans to reduce the energy consumed to heat houses (Geller et al. 2006). Finally, many OECD countries attempted to stem the growth of energy consumption by investing directly and indirectly in R&D (Geller et al. 2006). Japan invested vast sums to the "Moonlight Project," which sought to create new and more efficient gas turbines (Richards and Fullerton 1994). By 1978, the powerful Ministry of International Trade and Industry (MITI) devoted 14% of its R&D budget to this project.

Besides energy efficiency, governments planted the first seeds of the future growth of renewable energy. Of course, output initially remained limited, large hydroelectric facilities and some biomass notwithstanding. However, research and innovation picked up. According to Bettencourt, Trancik, and Kaur (2013), by the mid-1970s, the number of patents for renewable energy exceeded new patents for both fossil fuels and nuclear energy. As we detail below, public R&D in renewable energy increased significantly in a number of countries. To give but one example, in Germany, R&D in renewable energy jumped from $1 million in 1974 to $137 million at the time of the second oil crisis (almost all of which went to solar and, to a lesser extent, wind).

The private sector also acted. In the 1970s, industrial actors began to file patents, especially on wind but also on solar technologies. Between 1976 and 1980, the United States filed 62 patents on renewable energy under the Patent Cooperation Treaty (OECD 2014a). In fact, by the late 1970s, almost 1% of all patents were related to wind or solar technology. Figure 4.2 shows the early boom in research during this decade.[5] This level of activity would

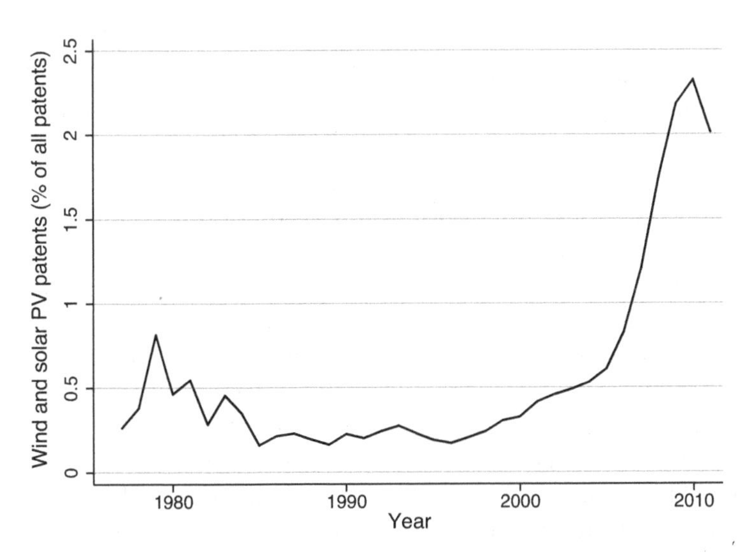

Figure 4.2
Wind and solar patents as a share of all patents filed by year under the Patent Cooperation Treaty (PCT), 1976–2011.
Source: OECD (2014a).

be unmatched until the mid-2000s. Overall, these first steps paved the way for a major push for renewables observed in some countries.

In summary, the oil shocks triggered a set of policy reactions that had real repercussions on fossil fuel consumption. At the same time, troubles with nuclear power also played a role. Nuclear energy policy has been particularly vulnerable to radical shifts in public support. Despite its military origins, nuclear power has been used for civilian purposes in a number of countries (Cowan 1990). France's nuclear industry, for instance, stems from its military program and defense policy (Finnis, Grisez, and Boyle 1988). Essentially used for the production of electricity, nuclear power has historically been a contentious source of energy. Its development for civilian purposes in the postwar period was celebrated as the triumph of humankind over nature and a sign of progress (Gamson and Modigliani 1989). The reason was that, although the fixed capital cost of reactors was high, nuclear energy promised abundant electricity at low marginal cost.

This idyllic view was shattered in many countries by a succession of accidents. In the United States, a series of mechanical and human failures led to the Three Mile Island accident in March 1979. Although the catastrophe on its own turned out to be less serious than initially feared (Walker 2004), subsequent hearings revealed the lack of oversight and accountability of the power plants (Fischer 1997). The construction of new nuclear power plants froze for decades. In 1986, the Chernobyl disaster further strengthened the antinuclear movement, especially in Europe. Chernobyl accelerated the demise of nuclear energy in many countries. Although more than 150 GW of new nuclear power electricity were under construction worldwide in 1980, only 50 GW were scheduled in 1990, and the number has been declining further since then (Fischer 1997).

Despite its appeal, nuclear energy thus suffered a significant blow. The blow was not fatal everywhere, as nuclear power continued to flourish in places such as France, but in many countries the nuclear option was off the (political) table. Instead, over the next decades, investing in renewable energy became a major policy alternative. As an immature technology, renewables certainly faced long odds. It is otherwise hard to explain why the IEA ignored the issue for so long. In the mid-1970s, renewables could not yet offer a realistic replacement for fossil fuels and nuclear power. Forty years later, wind and solar energy have become respectable, rapidly growing energy sources. The path from 1973 to today, however, has not been linear. Some countries have been consistently supportive (Denmark) or dismissive (France) of renewable energy. However, in many others, renewable energy has experienced both successes and failures. To understand the current state of affairs, the rest of this chapter documents and explains the initial short-term responses to the energy shocks of the 1970s and early 1980s, before the time of intense politicization.

4.2 Summary of Comparative Case Studies

Having described the broad contours of the 1970s oil shocks, new concerns over nuclear power, and policy responses, we now turn to the more detailed analysis of six country cases. Given that the external shock is largely common to the countries under analysis, the analysis is based on the comparison of differential policy responses in different types of countries. We begin by demonstrating that all countries under consideration implemented similar policies prior to the 1973 oil crisis. At this time, their policy trajectories began to diverge, and we provide direct evidence for the relevance of the external shock and contingent national factors in generating the differential policy trajectories. Our case studies combine comparative analysis with process tracing. Given our focus on the comparison of multiple cases, we do not claim to shed new insights into policy formulation in any particular case, with the partial exception of the understudied cases of the United

Kingdom and Finland. Instead, our main focus is on summarizing and synthesizing the available research to derive new insights from structured comparisons under a unified analytical framework.

To evaluate the theory, we begin by examining two major industrialized countries that can be considered forerunners with regard to the initial responses to policy responses during the first decade after the 1973 oil shock: the United States and Germany. We choose these two countries for analysis because they are conspicuous frontrunners. Both have a history of aggressive exploitation of fossil fuels and interest in nuclear power, and yet they also made significant investments in renewable energy in the 1970s and early 1980s. These cases are also important because these two countries are the largest economies among industrialized countries, and their energy choices are of major significance for the development of energy policy more generally.

In the United States and Germany, the rapid and unexpected increase in the prices of oil and other fossil fuels cast a dark shadow of doubt over the government's energy projections and planned policies. Among the wide range of policies that the government implemented, renewable energy played an important role, and the mechanics and details of policy formulation closely accord with the basic logic of our theory. Dependence on foreign fossil fuels, growing worries about the abundance of domestic energy resources, new social movements, and elite ideology guided renewable energy policy.

We next contrast the US and German cases to those of France and the United Kingdom. The latter two industrialized countries suffered from the oil shocks at least as much as the United States and Germany, yet their responses were different. The French government committed even more strongly to nuclear power, completely ignoring the renewable energy sector. The United Kingdom did virtually nothing to change energy policies, instead continuing on the previous track and mostly emphasizing the development of fossil fuels. These outcomes require explanation because they were so different from the US and German policy responses, and our preferred explanation emphasizes the centrality of elite ideology in the French case and, for the British, the expectation of abundant fossil fuels in the future due to the new finds in the North Sea.

From the perspective of research design, the French and British cases have many advantages. Similar to the United States and Germany, both France and the United Kingdom had been exploiting coal, oil, and natural gas to fuel their economic growth. Their political systems and levels of economic development were sufficiently similar for a comparative analysis, and the large size of their energy markets also highlights their similarity to the German and US case histories. However, the outcomes were quite different: neither France nor the United Kingdom chose to emphasize renewables. This contrast approximates the ideal setting for applying Mill's method of difference, although we are careful to also investigate other relevant differences among the cases.

The admittedly tiny elephant in the room is Denmark. Because no other country made so much progress in the development of modern renewables as early as Denmark (Gipe 1991; Heymann 1998; Hadjilambrinos 2000; Agnolucci 2007; Lipp 2007; Vasi 2011), any political history of renewable energy is incomplete without an analysis of the Danish case. The Danish success story of wind energy is so remarkable that it must be given special consideration in any political history of renewable energy. The Danish case allows us to evaluate our expectations and lay the foundation for downstream studies of the politicization and lock-in of renewable energy in different countries, given Denmark's singular contributions to the technological development of wind energy.

The analytical challenge is to find a matched pair for Denmark. We propose that the case of Finnish energy policy at that time is ideal for such a comparative analysis. Relative to Denmark, Finland has historically benefited from abundant biomass resources, but most of these resources have been used by the paper and wood industry. By the early 1970s, Finland's phenomenal economic growth created an energy deficiency that was hard to cover because Finland had no domestic fossil fuel resources. Unlike Denmark, however, Finland invested little in developing renewable sources of energy. In particular, the Finnish government did not attempt to exploit the potential for modern bioenergy. We trace these conscious policy decisions back to elite ideology, especially with regard to the close connections among heavy industry, the government, and the foreign policy elite. Indeed, one of Finland's decisive energy policy choices was to order Soviet nuclear power plants.

These cases are summarized in Table 4.1. The table summarizes the expectations on the external shock and renewables from the previous chapter, along with a summary of findings. As the table shows, we find evidence for the posited frontrunner mechanisms. We also find that in the laggard countries, national countries prevented renewables from even appearing on the political agenda as a serious alternative.

Table 4.1
Summary of expectations and cases.

Expectations	
Expectation 1	External shocks create a window of opportunity for government policies that support renewable energy deployment in the long run.
Expectation 2	The window of opportunity will be exploited most effectively by governments in countries that depend on energy imports and have an elite ideology that favors renewable energy.
Expectation 3	During an external shock, there will be little political opposition to renewable energy policy.
Cases	
USA, GER, DEN	Policy response to oil crisis allows renewables to grow (E1). Domestic conditions shape the nature of the response (E2). Political opposition is initially limited (E3).
FRA, UK, FIN	Although the oil crisis hit these countries, domestic conditions did not allow an effective policy response (E2).

4.3 Frontrunners in Key Industrialized Countries: United States and Germany

"Worries about the Middle East [oil situation] seem a little overdone."[6] This October 1973 statement by Frank Woodfin, an analyst at Standard & Poor's, proved to be more than a little optimistic. As the conflict between Israel and its neighbors intensified, Western support to the former triggered an unprecedented response on the part of Middle Eastern oil producers. By the end of October, Saudi Arabia and seven other Arab countries stopped all oil exports to the United States.[7] The implications of these decisions were not lost on contemporary commentators. A *New York Times* article referred to the "Arab oil weapon" and predicted that "some factories could shut, homes go cold and rationing become a necessity."[8] What was needed was a decisive response, but decision makers had no obvious blueprint. Another commentator in the same article noted, "We are doing a double crosstic blindfolded in an unfamiliar language." The United States and Germany would have to deal with the oil crisis. For the United States in particular, the challenge was amplified by a series of other challenges, such as the weak performance of the military and the political backlash against the Vietnam War, the collapse of the gold standard, and major environmental problems.

More generally, Western industrialized countries found themselves in uncharted territory. In some respects, they did not have a choice, given their dependence on oil. The embargo and record prices forced consumers to cut back on their appetite for oil. On the eve of the first oil crisis, the United States consumed 834 million tons of oil per year, or about a third of worldwide consumption. By 1982, US consumption had decreased to 695 million tons. Total oil consumption remained below the 1972 level until the mid-1990s and the Clinton-era economic boom. Oil consumption grew again but never exceeded 120% of 1972 levels. By the end of the first decade of the 2000s, the United States had returned to the same consumption levels as in 1973, despite a 50% increase in population.

The United States emerged as a frontrunner in the post-oil crisis world. Throughout the 1970s, the politically most salient and controversial policies were related to the regulation of fossil fuel production, consumption, and pricing (Ikenberry 1986; Tugwell 1988; M. Jacobs 2016). Conservatives argued for liberalization and lax regulations to encourage domestic production, whereas liberals advocated direct state intervention and efforts to regulate the energy industry. These measures, however, did not solve the crisis. As M. Jacobs (2016, 312) puts it, "Americans' unwillingness to pay more for their energy presented policy makers with a fundamental obstacle to constructing long-term policy . . . the political power of consumers led to holding down prices even as environmentalists called for conservation."

Perhaps a more effective long-term response, then, was to improve energy efficiency. For instance, in 1975, the federal government began the Corporate Average Fuel Economy (CAFE) program (S. Anderson et al. 2011). This program required carmakers to improve the miles per gallon ratio for new cars. The program celebrated early successes. The gap in fuel efficiency between the United States and Europe decreased considerably, and the number of liters consumed by an average US car per 100 kilometers went from about 18 in 1975 to less than 12 in 1985 (Zachariadis 2006). However, the program stalled in the 1980s. Instead of progressively increasing standards, the requirements remained flat. As a result, the liters per 100 kilometers ratio remained flat at about 12 from the mid-1980s until today. In the meanwhile, Europe managed to reduce its average consumption from 10 liters per 100 kilometers in 1975 to about 8 in 1985 and about 7 in 2000.

The third response—and the one directly relevant to renewable energy—consisted of increased energy research and development budgets. Initially, public investments aimed at improving energy efficiency. Worries about oil scarcity encouraged projects that would reduce demand for fossil fuels. At the time of the first oil crisis, public funding in efficiency research amounted to a paltry $5 million (2012 prices) (IEA 2014c). By the end of President Carter's administration, R&D spending had increased to $707 million. In total, the United States had spent more than $2 billion by the time Ronald Reagan entered the White House. These investments, however, were dwarfed by investments in alternative energy sources. Nuclear was initially an option. R&D in nuclear energy doubled between 1974 (about $2 billion) and 1979 ($4.5 billion). The Three Mile Island nuclear incident of 1979, however, cut these hopes—and investments—short. By 2000, spending had receded to $355 million. The big winners of the nuclear debacle were renewable energy sources. Total R&D spending on renewables boomed from $38 million in 1974 to $1.6 billion in 1980. The primary beneficiary of this spending spree was solar energy. During Carter's presidency, research for solar power captured about 60% of renewables R&D spending. Geothermal was second, with $360 million. Wind and biofuels generally lagged behind. At the same time, R&D on fossil fuels increased only from 286 million (1974) to 2 billion (1980), meaning that the renewables–fossil fuel ratio went from about 1:8 to 4:5.[9] These patterns are consistent with our first expectation: an external shock is required for governments to push for renewable energy policy.

Increased domestic coal exploitation was on the table throughout. The United States has the world's largest coal reserves, and both the government and energy sector pushed for additional use. Most important for our purposes, however, coal was never considered a competitor for renewable energy. Coal and solar were considered replacements for oil, and even the Carter administration did not see any contradiction between promoting both at the same time. After all, alternatives to oil were motivated by an energy crunch not environmental considerations. At the same time, the popularity of nuclear power plummeted because the slowdown in the growth of electricity consumption made the huge capital investments required for nuclear power plants economically unattractive, the costs of construction grew because of tightened safety standards and other issues, and environmental opposition intensified. As a result, "all orders [of nuclear power reactors] placed after 1973 have been canceled."[10] Indeed, Baumgartner (1989, 62–63) argues that the process of nuclear policy formation has been open to various antinuclear groups outside the scientific community, meaning that "[a]nti-nuclear activists in the United States have successfully exploited the opportunities which the complicated policy making process gives them to shift the debate into those arenas where they can do best."

There was no public R&D for wind to speak of when the first oil crisis occurred. Perhaps surprisingly, wind energy remained a distant third in the renewable pecking order, with $147 million in 1980. Yet wind energy was the first renewable energy source to become a serious provider of electricity. In fact, the early success of wind underscores the second part of the US response to the oil crisis, namely, the role of local renewable energy advocates in California. State policy in California is a prime example of how subnational leadership played an important role in the American renewable energy response (van Est 1999; Karapin 2016). In fact, state authorities had already shown environmental leadership before 1973. In 1970, for example, Governor Ronald Reagan signed the California Environmental Quality Act (CEQA), a series of laws that required state officials to evaluate the environmental impact of their activities. According to Karapin (2016, 123), "[a]t this time in California, longstanding environmental organizations saw a surge in membership . . . [s]everal political actors took the opportunity presented by this upsurge in attention to press for a strong environmental review law in the state." Although passing the law was not difficult, "the law quickly proved to be a powerful tool to slow and call into question both private and public construction projects . . . environmental groups had gained leverage to influence large construction projects, including new power plants" (Karapin 2016, 124).

The 1973 oil crisis gave renewable energy its decisive momentum in California. In 1973, a workshop that included scientists, entrepreneurs, and industry representatives paved the way for the development of wind energy (van Est 1999). Already in 1974, as an early response to the oil crisis and the growing demand for environmental protection, the California legislature enacted the Warren-Alquist Act, which became a law in 1975 and established the California Energy Commission (CEC),[11] a government agency that was to play a central role in state energy policy.

The role of the oil crisis in the passing of the Act is easy to see, as "just a few weeks after Governor Reagan vetoed [the original] bill in October 1973, the five-month-long Arab oil embargo began" (Karapin 2016, 124). As Wellock (1998, 143–144) notes, Reagan soon realized that the situation had changed dramatically: "[c]orrectly reading public sentiment, Reagan called for decisive action and began encouraging conservation measures and even implored Californians not to return to their 'wasteful ways.'" As Sawin (2001, 164, 166) recounts, the CEC was specifically established for the purpose of "weaning the state off of fossil fuels." According to van Est (1999, 36), the CEC initiated a wind energy program already in 1976, seeing "[w]ind technology . . . as environmentally benign, and capable in the short term of reducing the severity of oil price escalation and ultimate scarcity." Notably, these developments came well before Brown's 1977 wave of appointments of soft energy advocates in 1977 to both the CEC and another influential body, the California Public Utilities Commission, which built additional elite support for renewables in the administration.

The major challenge remained the provision of cheap energy. In California, nuclear was overall a popular option to increase electricity production, but it faced the opposition of a determined fraction of the population (Wellock 1998, 4). The other alternative, reducing consumption, was unpopular among labor and industries alike (van Est 1999). Over time, the pressure to deal with California's energy problems intensified, with the combination of the Three Mile Island nuclear accident and the second oil crisis in early 1979 (Karapin 2016).

To deal with these problems, Governor Jerry Brown became interested in wind energy as an alternative. California was by no means a leader in wind energy before Brown's governorship, but the CEC's interest grew in the late 1970s. Brown had "won the 1974 gubernatorial election by a 50% to 47% margin over his Republican opponent" and, over time, became increasingly enthusiastic about renewable energy (Karapin 2016, 128). The combination of external shocks, political difficulties related to the conventional alternative of nuclear power, and a narrow electoral victory that brought a proenvironmentalist into power was enough to turn California into a renewable energy leader.

Disagreements between Governor Brown and President Carter encouraged the CEC to develop its own programs and thus gave California's renewable energy policy an original twist. Taking advantage of the vaguely formulated federal legislation, especially the 1978 PURPA, Brown encouraged the development of small-scale electricity generation across the state with attractive tax incentives.

The result was a spectacular wind energy boom. Large corporations, such as Boeing, were eager to enter a potentially lucrative market thanks to their technical expertise. California went on, in the words of a wind energy expert, "to dominate worldwide wind development in the early 1980s" (Gipe 1995, 30). Although wind energy capacity grew slowly around the world, in California it went from 0 MW in 1980 to 1,235 MW in 1986 (Sawin 2001). By 1986, virtually all wind energy capacity in the United States was found within California—a state of affairs that continued into the late 1990s (Sawin 2001).

To summarize, the US response was a combination of R&D policy at the federal and deployment support in California. At the federal level, R&D was important because it reduced the cost of renewable energy technologies, solar technology in particular.[12] California's emphasis on deployment was important because it created an opening for actual wind energy installation, thus enabling learning by doing in the industry. As we shall soon see, the Danish wind industry—another key forerunner in renewable energy—capitalized on this opportunity and sold the world's best wind energy technologies to the dynamic Californian market. As Hale and Urpelainen (2015) show, California's demand supported the growth of the Danish wind technology industry, with major effects on the global competitiveness of wind as a source of electricity.

How can we explain US renewable energy policy of the time? Both at the federal level and in California, the modifying factors identified in our second expectation—energy imports and elite ideology—played a role. The 1973 oil crisis and its 1979 successor brought the rapid growth of oil and energy consumption in the United States to an abrupt end, calling into questions many decades of energy predictions and expertise. As Graetz (2011, 21) puts it, the mirage that cheap energy "would last forever was shattered" with the oil shocks and the vulnerabilities of the US economy that they revealed. M. Jacobs (2016, 4) notes that President Nixon's top advisers called the Arab oil embargo an "energy Pearl Harbor." Among the public, the "sense of crisis, whether people believe it was real or manufactured, quickly triggered a panic" (M. Jacobs 2016, 5).[13] In California, where renewable electricity grew faster than anywhere else thanks to a conducive policy environment, oil production had already peaked, and there was little hope of recovery (Nehring 1975). The perception of an energy crisis in California was further compounded by the 1969 Union Oil offshore platform spill, which "made the linkage between energy and environmental issues baldly apparent" (Laird 2003, 40). These problems with the supply of oil coincided with growing environmentalist opposition to nuclear power and financial problems in Californian electric utilities (Karapin 2016).

Without the sense of a crisis, it is improbable that increases in oil prices would have generated the policy responses we saw at the federal level and in California. As we noted earlier, this was not the first time US elites and the public were worried about oil scarcity, but it was the first time the US economy had become totally dependent on oil and concerns about oil scarcity surfaced. Without the rapid, surprising, and uncontrollable increase in oil prices, economic sectors such as electricity generation and transportation would have had more time to adjust to the changing circumstances. Looking into the future, we can also see that the 1973 and 1979 oil crises were unique in the concern and policy responses that they generated: when oil prices increased again after 2000, no such energy policy response was to be seen.

Elite ideology played an important role as well. For instance, California's energy policy was largely driven by Governor Brown's unusual and highly ideological approach to policy formulation and the preexistence of capable regulatory and scientific institutions. As van Est (1999, 34–35) notes, Brown initially "had few hard and fast notions about energy policy," but, having read the 1974 book, *Energy for Survival* (Clark 1974), became convinced that the United States would have to transition toward a renewable energy economy. In 1977, the "soft energy path" advocate Amory Lovins personally briefed Brown about energy policy and left an abiding impression on Brown's thinking. As Karapin (2016, 129) puts it, Lovins had spent 1977 in California "lecturing and debating. . . . Brown liked the potential for renewable energy to help obviate the need for nuclear power, reduce dependence on oil, and keep down energy prices."

More generally, Brown's approach to energy policy formulation emphasized the importance of bold departures and new initiatives, creating "an atmosphere of positive support for renewables, and led to a chaos of experimentation and a flood of legislation and state programs" (van Est 1999, 35). Without Brown's beliefs that the energy problem needed original and creative solutions, regulatory institutions would probably not have been used to exploit PURPA for renewable electricity generation any more aggressively than the other 49 states did (van Est 1999). Indeed, Sawin (2001, 167) notes that when George Deukmejian, "who had no interest in fostering renewable technologies," replaced Brown as the state governor in 1982, "it was more a matter of benign neglect of renewable technologies than one of intentional destruction."

However, it would be a mistake to attribute California's renewable energy policy to Brown's political persona alone. In fact, the state government's ideology was more broadly favorable to bold energy policy overtures, renewable energy in particular (Sawin 2001). Because of California's rapid electricity consumption growth and the threat of depleting oil resources, in 1972, the Rand Corporation published an influential report on measures to deal with the rapid growth of electricity demand in the state. The significance of this report lies in its timing, as it shows California's openness to measures to reduce electricity consumption before the 1973 oil crisis.

In the absence of regulatory institutions, scientific expertise, and a supportive legislature, Brown might not have achieved impressive results with PURPA and the state and federal tax incentives available for renewable energy. Sawin (2001, 167), for example, notes that the California state legislature was at the time "favorable [to] renewable energy technologies."

In the critical 1977–1980 period, the Democrat-controlled state legislature passed mandates for the use of solar in state buildings, facilitated financing for solar, and set targets for renewable electricity generation in the future (Karapin 2016). These legislative achievements did not depend on Governor Brown's support, but they did reflect the work of environmental groups such as the Environmental Defense Fund and the Sierra Club; in their implementation, the soft-path advocates at the CEC played a central role (Karapin 2016). It was not the case that prorenewables elite ideology was limited to Governor Brown—the entire policy establishment became increasingly interested in this option after the 1973 and 1979 energy crises.

At the federal level, elite ideology was conductive to renewable energy R&D, especially with regard to support for the activities of large energy corporations (Chubb 1983; Reece 1979). Because the evidence for the importance of elite ideology is the clearest in the case of the Carter administration, we initially set aside the Republican administrations of Nixon and Ford. According to Laird (2001, 90, 113), President Carter "spoke out for the creation of a Department of Energy (DOE) . . . before he was elected, during the 1976 campaign," and on the day of his inauguration, he promised a National Energy Plan within 90 days. Although this interest does not yet demonstrate Carter's interest in renewables, it does show that he was in favor of an aggressive approach to dealing with the oil crisis. The 1978 National Energy Plan offers useful insights into the ideology of the Carter administration:

> The diagnosis of the U.S. energy crisis is quite simple: demand for energy is increasing, while supplies of oil and natural gas are diminishing. Unless the U.S. makes a timely adjustment before world oil becomes very scarce and very expensive in the 1980's, the nation's economic security and the American way of life will be gravely endangered. The steps the U.S. must take now are small compared to the drastic measures that will be needed if the U.S. does nothing until it is too late. (Laird 2001, 113)

In October 1977, President Carter and other members of his administration, including Secretary of Energy James Schlesinger and Science Advisor Frank Press, met with Amory Lovins. According to Laird (2001, 138), Lovins "remarked later that Carter had read [Lovins's book] *Soft Energy Paths* and seemed quite knowledgeable about its concepts." In April 1978, Secretary Schlesinger then made a formal request for a Domestic Policy Review (DPR) of solar energy. According to Laird (2001, 139), "[t]he administration conducted DPRs on only a small number of issues, and getting such a review meant that high-level officials, as well as outside constituencies, would devote substantial attention to that issue."

The Carter administration was not at all interested in abandoning fossil fuels or investing in decentralized energy generation to replace the current infrastructure. Regardless, it invested billions of dollars in supporting solar R&D, drastically bringing down the cost of solar technologies, especially by supporting Exxon's research program on solar photovoltaics (Madrigal 2011). The goal of the solar R&D policy was to experiment with renewable energy sources in the context of broader, politically more difficult energy policy choices. Thus, according to Laird (2003, 44–48), "[e]cological solar advocates, who distrusted both big business and big government," ultimately failed to impress their idea of a good energy future on the society. President Carter's consideration of the energy crisis as "the moral equivalent of war" did not push the executive wing to abandon the traditional energy paradigm; the form and nature of US solar energy policy was still based on increased energy abundance through centralized power generation and reliance on major corporations. As Carter put it in a November 1977 address to the nation, "There are three things that we must do to avoid this danger: first, cut back on consumption; second, shift away from oil and gas to other sources of energy; and third, encourage production of energy here in the United States. These are the purposes of the new energy legislation."[14]

It is thus notable that Carter's ideology was in many respects similar to those of his Republican predecessors, Nixon and Ford. Although these two administrations did not act on solar energy as forcefully as Carter, and they certainly did not formulate any policies comparable to Carter's PURPA, the Nixon and Ford views of energy policy are informative about US elite ideology in general. Encouraged by Herbert Stein, the Chair of the Council of Economic Advisors, in November 1973 Nixon announced his "Project Independence" plan, saying in a televised address to the nation that "[w]e are running out of energy today because our economy has grown enormously and because in prosperity what were once considered luxuries are now considered necessities" (Laird 2001, 105). The televised address did not shy away from calling for behavioral change to conserve energy:

> In order to minimize disruptions in our economy, I asked on November 7 that all Americans adopt certain energy conservation measures to help meet the challenge of reduced energy supplies. These steps include reductions in home heating, reductions in driving speeds, elimination of unnecessary lighting. And the American people, all of you, you have responded to this challenge with that spirit of sacrifice which has made this such a great nation.[15]

In a similar vein, President Ford's administration, which delegated much of the preparatory work on energy issues to the Energy Resources Council, "articulated a very similar conceptualization of the energy problem . . . [f]rom their perspective, the energy crisis had the same origins and the same solutions as discussed by Nixon's advisors, some of whom carried over to the Ford administration" (Laird 2001, 108). These plans suggest that Carter's concerns about fossil fuels and the need to find alternatives were not the exclusive property of the Democratic Party but rather a general ideological inclination of US government elites of the time, regardless of their partisanship. On October 11, 1974, President Ford's Executive Order 11814 created an Energy Resources Council.[16] On January 13, 1975, Ford gave a televised address from the White House, telling "Americans that it was time to restore the price mechanism to energy" (M. Jacobs 2016, 135). In the address, Ford said that:

> Americans are no longer in full control of their own national destiny when that destiny depends on uncertain foreign fuel at high prices fixed by others. Higher energy costs compound both inflation and recession, and dependence on others for future energy supplies is intolerable to our national security. Therefore, we must wage a simultaneous three-front campaign against recession, inflation, and energy dependence. We have no choice. We need, within 90 days, the strongest and most far-reaching energy conservation program we have ever had.[17]

These brief illustrations show that the difference between Carter and the two previous Republican presidents was one of degree not kind. In general, the voice of renewable energy advocates, especially Amory Lovins of the Rocky Mountain Institute, in the public debate was stronger than ever before (Lovins 1976). As Laird (2003) shows, the Lovins (1976) approach emphasizing the need for a soft energy path was highly controversial and divided solar advocates into two camps: one continuing to rely on a traditional energy paradigm, with solar being one path to energy abundance, and the other embracing a decentralized, ecological energy future. However, elite ideology clearly sided with the traditional approach. Although federal investment in solar R&D reached record heights during Carter's tenure, Laird (2001) shows in his detailed case study that the traditional assumptions were never really abandoned. Only weeks after Carter's installation of solar water heaters on the roof of the White House, the solar advocacy coalition's "influence in the executive branch eroded severely."

Thus, although solar is traditionally regarded as the major winner in US renewable energy policy during the energy crisis, elite ideology prevented a more profound paradigm shift and ensured that initial solar policy did not challenge traditional energy interests. Nixon, Ford, and Carter all agreed on the importance of increasing domestic use of fossil fuels. Although they disagreed on specific policy measures, such as the importance of price control, they shared a core ideological commitment: oil scarcity is fundamental to US society, and decisive policy measures are necessary to address the problem.

At the federal level, elite ideology also played a role in allowing renewable energy advocates some access to key policymakers. As various authors (Chubb 1983; McFarland 1984; Laird 2001) have noted, the executive branch played a decisive role in renewable energy policy during President Carter's administration in particular. Indeed, during the Carter era, Congress's role was more passive and largely limited to budget appropriations because the executive was able to set the agenda and control policy implementation. Although the Congress did debate and modify Carter's proposals, the key efforts to raise the profile of solar advocates were focused on the White House. Disappointed with Carter's initial response, solar advocates organized a National Sun Day on May 3, 1978, with the result that Carter announced a "$100 million supplemental budget request for the [Department of Energy] solar programs for [financial year] 1979 and the formation of a Domestic Policy Review for Solar Energy" (Frankel 1986, 75–76).

Far from a passive aggregator of interests, the president of the United States, his cabinet, and the entire executive wing were thus in a good position to formulate their preferred policies—continued reliance on the traditional energy paradigm and the use of solar energy to relieve energy scarcity—despite the difficult political situation and the president's decreasing approval ratings. If anything, solar energy was sufficiently powerful for Carter to use it as a way to create political support for his energy policy more generally. As our review of Carter's activities in the field of renewable energy shows, the most important initiative was the major increase in federal R&D support for renewables in general and solar in particular. These activities were initiated by the executive and then approved (and in some cases modified) by the Congress.

The evidence also supports the expectation concerning the lack of political opposition to renewables in general. Although proponents of the radical decentralization of energy through, say, off-grid solar power were sorely disappointed by the government's response (Lovins 1976; Reece 1979; Laird 2001, 2003), even the incumbent energy interests supported renewable energy R&D and coveted it (Chubb 1983; Madrigal 2011). Although it is true that the conservative right wing had reservations about a possible capture of the Carter administration by "solar cultists" (Frankel 1986, 79), this faction was small and politically weak until the inauguration of President Reagan.

The reason for this approach, we posit, was that they simply did not perceive the experimental renewable energy programs as a real threat to their fossil fuel business. As both Chubb (1983) and McFarland (1984) have argued, the special interest politics of R&D during the energy crisis did not pit various groups against each other. Rather, each interest group focused on convincing the executive to increase budgetary allotments to its preferred projects. For incumbent energy corporations, this meant an "all of the above" approach with investment in fossil fuels, nuclear power, renewables, and energy conservation. They did not object to renewable R&D spending as long as such spending did not threaten their core interests in the fossil fuel business.

To illustrate the lack of politicization, consider the 1978 Domestic Policy Review of solar energy (Laird 1990). The goal of the review was to estimate the ability of the United States to use solar power in the year 2000. According to Laird (1990, 53–54), the review "demonstrates clearly the disempowerment of citizens, based on a highly technical definition of the issues." Although the federal bureaucracy explicitly promoted citizen participation and invited commentary by groups outside the government, in practice the technical definition of the issue resulted in depoliticization: government officials "saw the core of the problem in highly technical terms, and the lay public simply had nothing to say that could help them solve that problem" (Laird 1990, 56). Although technical experts debated solar power and the prosolar side won, the issue never became, in a dramatic contrast to the Reagan era that began only three years later, explicitly politicized at higher levels. A lot of debate took place on energy policy more generally, but political history provides little evidence that renewable energy would have been at the center of these debates (e.g., M. Jacobs 2016).

Carter's solar policy illustrates the role of public opinion in the aftermath of the external shock as well. Although little public opposition to solar power existed at the time, the interest of the public in the issue did fluctuate, consistent with the idea of windows of opportunity. In June 1979, Carter's adviser Stuart E. Eizenstat told the president, "[a] strong and vigorous commitment to solar can help us in getting the Congress and the public to swallow the more bitter pills of [price] decontrol and generally increasing prices" (M. Jacobs 2016, 212). At a time when the second oil crisis had brought record oil prices back, the Carter administration saw solar as a popular approach to the nation's energy problems, and Carter's speech on June 20, 1979, captured this idea:

There is no longer any question that solar energy is both feasible and also cost-effective. In those homes now using electricity, a typical solar hot water heating system, such as the one behind me, can pay for itself in 7 to 10 years. As energy costs increase, which is an almost inevitable prospect, that period for paying for this investment will be substantially reduced. Solar energy will not pollute our air or water. We will not run short of it. No one can ever embargo the Sun or interrupt its delivery to us. But we must work together to turn our vision and our dream into a solar reality.[18]

This analysis shows that elite ideology played a role, but it also suggests that the role of this factor cannot be approached from a traditional ideological perspective. There was no clear left-right divide on renewable energy in the United States at this time. Elite ideology promoted renewables because decision makers in California and in the Carter administration were willing to experiment with new approaches to mitigating energy shortages. The Brown and Carter proposals for investment in wind and solar, respectively, did not became major left-right wedge issues at this time. Years later, the standard left-right conflict began to take its toll on renewables. To summarize, the case of the United States shows the power of an external shock. The energy crisis sparked an interest in renewable energy in California and at the federal level. California's leadership in wind energy reflected an unusual combination of Governor Brown's environmental ideology, belief in bold action, and the institutional capacity of the state apparatus. At the federal level, solar power was one of the great winners of the energy crisis, yet elite ideology within the Carter administration prevented a more radical challenge to traditional energy interests. As a result, renewables were never fully politicized in US politics during President Carter's tenure. Only in the electoral debates between Carter and Reagan did a real political debate emerge. In the meantime, however, the federal and California state governments had made important contributions to the early development of renewable energy—developments that were to play a key role in the evolution of renewable energy policy and use in the European context.

We now turn our attention to the case of (West) Germany. The 1973 oil embargo shook the United States as well as Western Europe. West Germany's situation was particularly precarious. By the time of the crisis, Germany imported about 72% of its oil from Arabic countries, or about 74 million tons.[19] To give a sense of Germany's reliance on Middle Eastern oil, at the time, the country consumed about 77 million tons of oil solely for heating, and 149 million tons overall. To situate these numbers, the United States imported about half of its oil from OPEC countries (although less for heating than Germany and more for gasoline and electricity).[20] In many ways, then, Germany's case resembled the US situation. Yet there were differences too. For instance, Germany had made an agreement with the Soviet Union and Algeria that ensured access to natural gas; the Soviet deal, conveniently, entered into force in October 1973 (Victor, Jaffe, and Hayes 2006).[21]

However, oil remained an important weakness for Germany. The prospect of a cold winter worried the Germans, and the newspaper *Die Zeit* wrote (disparagingly) about the fears of a "dark and cold winter."[22] The oil shock made it clear to Germans that oil had finally become a scarce resource.

Germany's initial response was twofold. First, the government and commentators encouraged cutting down on oil consumption. Even ADAC, the powerful German automobile club, encouraged its members to use public transportation.[23] This approach, assisted by the booming oil prices, saw some success. For example, oil consumption in 2014 was about 80% of what it was in 1972. The data also show that Germany diverged from the US trajectory in the 1990s: the United States experienced a rebound of oil consumption in the mid-1990s, whereas Germany's consumption has been declining since the end of the 20th century. This divergence is partly due to differences in infrastructures, such as a dense rail network. However, it was obvious even at the time that curbing the then-growing oil appetite in Germany would be difficult.[24] Improving energy efficiency would also have required investment and cuts in energy consumption; both were seen as impractical and undesirable by the industrial sector.

The second response was to invest in alternative energy sources. Coal appeared particularly appealing because Germany had abundant reserves. Nuclear energy had its supporters, too, and a well-organized lobby. Because it provided fairly cheap electricity, nuclear had the powerful backing of the main (energy-intensive) industry conglomerates.[25] Policymakers initially invested in both research and production of these two resources (Jacobsson and Lauber 2006). Despite the strength of these industries, they turned out politically, economically, and technically only to offer limited relief. In fact, in the 1950s, the coal industry began a long period of decline due to low investments and fierce competition from oil, although coal was still a major energy source in Germany (Prodi and Clô 1975). Furthermore, public opinion soon shifted against both nuclear and coal. Nuclear power was never popular in Germany. Even in 1974, the share of pronuclear respondents was less than 50% (Jahn 1992). From there, public support for nuclear power went mostly downward. After a last high above 40% in 1981, support collapsed to less than 20% by 1988. Nuclear R&D investments, originally among the highest in the world, crashed. Although the federal government spent $1.17 billion in 1974 and $2.1 billion in 1982, R&D dropped to $818 million in 1986. By 2000, spending bottomed at $210 million.

Instead, a significant fraction of the population supported a third approach. Renewable energy already began to attract the attention of policymakers in the early 1970s. To be sure, the initial effort was marginal and represented only a small part of the policy response to the two oil crises (Sawin 2001). Two audiences were particularly receptive to this approach and inclined to lobby for it: environmentalists and researchers. By the early 1970s, the Green movement was rapidly growing. In 1980, the Öko-Institut published a book on *Energiewende*, that is, the transition away from traditional energy sources (Krause, Bossel, and Müller-Reißmann 1980). Meanwhile, this concept has become the basis for the current German energy policy. In 1983, the Green Party stunned observers by entering the Bundestag. The same year, the Wyhl nuclear plant project was indefinitely shelved after continuous pressure from environmentalists and local opposition (Morris and Pehnt 2012). By then support for renewables was not limited to a fringe of Green activists. Instead, renewables were gathering mainstream support. In 1979, Helmut Schmidt, who otherwise was not a supporter of renewable energy, announced the need to overcome his country's addiction to oil through investments in renewables such as solar power.[26] By the end of the 1980s, policymakers on both the left and right agreed that the energy transition was needed. Under the impulse of key figures such as Hermann Scheer (SPD), Hans-Josef Fell (Green), Wolfgang Daniels (Green), or Matthias Engelsberger (CSU) and later Peter Ahmels and Franz Tacke (CDU), prorenewable regulations began getting support across parties at the Bundestag (Yergin 2011b; Hirschi 2008).[27] Tacke, a wind energy businessman and CDU member, neatly summarized the situation: "in this matter we collaborate with both the Greens and the Communists" (quoted in Jacobsson and Lauber 2006, 265). Although the center-right coalition government under Chancellor Helmut Kohl initially was tempted to follow a path similar to Reagan's (i.e., decreasing support for renewables), the succession of shocks in 1973 and 1979 (oil) as well as 1979 and 1986 (nuclear) created a consensus that would last long enough for the renewable energy industry to reach a critical mass.

The second audience was the research community, which included universities, research institutes, and the research branches of some corporations. Public R&D spending increased rapidly in the decade following the first oil crisis. Total R&D in renewables went from $1.5 million in 1974 to $228 million in 1982. The main recipient was solar power, which went from $1.3 million in 1974 to $169 million in 1982. Research institutes and the government also developed test centers, such as the one on Pellworm Island, in the north of the country (Sawin 2001). The role of the research community was particularly important because its members played a key role in developing institutional capacity. This point is underlined by Jacobsson and Lauber (2006, 262): "institutional changes occurred which began to open up a space for wind and solar power; a space which proved to be of critical importance for the future diffusion of these renewables."

The literature on technological innovation emphasizes the role of "niche space" (A. Smith et al. 2014). A niche space is an environment in which a new technology can be developed, protected from immediate market pressure. In these spaces, customers who may prefer new technologies are willing to pay above-market rate prices for the new technology, whereas scientists, entrepreneurs, and firms work to improve and adapt this technology. Market niches are instrumental for overcoming the fierce competition from established technologies, but the development of these niches required the participation of researchers, renewable energy entrepreneurs, and the government (A. Smith et al. 2014). The German government developed institutions and designed policies that would, in the long run, be key for renewable energy. The first institutional reform occurred in 1974, with the creation of the Federal Environmental Agency (Morris and Pehnt 2012) In terms of policies, the *Bundesministerium für Forschung und Bildung* (Ministry of Research and Education), whose focus had mainly laid so far on nuclear, opened up lines of funding for R&D of solar and wind energy. Although the funding for renewables paled in comparison to nuclear research, it was enough to fund a range of projects, including solar energy (Jacobsson, Sandén, and Bångens 2004). Universities and industries enthusiastically stepped in to fill the gap. Together these actors all participated in the early development of renewable energy demonstrations. Some projects, such as the GROWIAN wind park, failed. Others, on a smaller scale, turned out to be much more promising (Bergek and Jacobsson 2003).

These projects laid the foundation for the rise of a renewable energy industry (Jacobsson and Lauber 2006). Encouraged by the emerging demand for Green energy and incentivized by public R&D, a number of firms decided to enter this business. In many cases, they had the know-how and a need for new business ventures. This was particularly true for the wind sector, where both farmers and industrial interests entered the market. Although output remained small in the 1970s, a number of new entrants attempted to find out whether there was any future in this business. Between 1977 and 1991, nineteen industrial firms and universities received public support to fund more than 40 projects. These projects were truly experimental, with turbines varying greatly in terms of size or technical features (Bergek and Jacobsson 2003).

Despite their small output, these projects paved the way for the development of a renewable and wind market. In the 1970s, countries were not defined by their production of renewable energy but by their ability to nurture and protect this niche market. In this respect, Germany, just like the United States and Denmark, was clearly a frontrunner compared with France or the United Kingdom. The Ministry of Research planted the seed for future growth. Germany had more than ten firms that were active in the wind business by the early 1980s. In the mid- and late 1980s, they were ready to capitalize on the rapid change of preferences among CDU and SPD leaders.

In many respects, Germany and the United States appeared to be in a similar position by the early 1980s. Both countries hosted a number of small entrepreneurs who entered the renewable energy market. By tentatively increasing R&D and supporting private investments, officials in these two countries made similar bets. Interestingly, both countries were toying with investing more into nuclear power, yet these efforts were cut short because of two additional shocks, Three Mile Island and Chernobyl. On some dimensions, the United States was in fact ahead of Germany. For instance, the United States issued more patents and spent more money on R&D (between $5.4 billion between 1976 and 1980 compared with $400 million in Germany). Commentators in 1992 argued that "the impact of the USA on renewable energy has been dominant" (Twidell and Brice 1992, 464). As we show in the next chapter, this would not last.

To summarize, both the US and German cases support our expectations. Neither country had much of an appetite for renewable energy until 1973. However, the fortuitous combination of an external shock, perceived dependencies on foreign energy, and tolerant if not enthusiastic elite ideology allowed a policy response that promoted renewables. Due to the limited danger that the renewable energy policies posed to incumbent energy interests, no opposition to them occurred at the time.

4.4 Why Did France and the United Kingdom Not Respond?

Despite variation in the choice of policy instruments and the effectiveness of their overall approach, the two countries described earlier show how renewables found their way onto the policy agenda. Yet not all industrialized countries mobilized as effectively. Indeed, for several reasons, many countries either did not respond at all or chose some other response than renewables. To further scrutinize the validity of our model, we now look at two dogs that did not bark: France and the United Kingdom. Neither country responded to the oil shock by investing in renewables. France had another alternative: nuclear power. The United Kingdom was isolated from the shock because it had abundant coal, oil, and gas reserves.

By 1973, France had undergone three decades of rapid economic development. During this period, referred to as the *Trente Glorieuses* (The Glorious Thirty), France had restored its position as a major economic and political power. Its status as an economic power was cemented by being the third most populated European country (53 million inhabitants), having a higher average income than Germany and the United Kingdom (more than $18,000 in 2005), and a low unemployment rate (about 2.5% in 1970).[28] France's political power was just as remarkable. The country was a permanent member of the United Nations Security Council and, despite the fact that its colonial empire was crumbling, remained a predominant player in vast parts of Africa. In the relatively short period of thirty years, France had experienced a remarkable recovery from a country in ruins to a worldwide power.

On the eve of the first oil crisis, this success came under threat. France's economy heavily relied on imported energy: 72% of the energy that the country consumed came from abroad (WDI 2013). In contrast, Germany's exposure was much lower, with 45% of its energy coming from abroad. A large share of the French imports was used to generate electricity, and the trend was worrying: in 1970, about 22% of electricity was derived from oil; in 1973, that number had increased to 34%. Coincidentally, Électricité de France was still relying on hydropower only ten years earlier, with 71% of its electricity coming from hydroelectricity in 1960. Yet France had already made a commitment to nuclear power. Although interest in nuclear power existed in the 1950s, one had to wait for the return of Général de Gaulle to witness the firm establishment of France as a nuclear power. To satisfy his desire for independence, de Gaulle had supported an ambitious nuclear program that led to the construction of the first civilian nuclear plant in 1962. The ambition of this program was to reduce France's reliance on the US nuclear umbrella (Finnis, Grisez, and Boyle 1988). The influence of this initial decision was significant in terms of its effect on actual energy policies and the beliefs held by the elites. Valéry Giscard d'Estaing, the French president from 1974 to 1981, said that the nuclear "choice has been debated . . . [and] validated by scientists and politicians, including the Communist Party. . . . Since then, it has never been challenged by new governments . . . because there is no alternative."[29]

To finish this project in such a short period of time, France had to rely on the development of powerful energy conglomerates. EDF would become the country's nuclear leader (Hecht 2009). Of course, the contribution of electricity from nuclear power initially remained modest, providing only 8% of EDF's output in 1973. If, however, any hesitation remained about the appropriate path for future energy policy, it disappeared in the fall of 1973. The political response to the crisis was swift. In 1974, Pierre Messmer, then prime minister, announced on television:

> France has not been favored by nature in energy resources. There is almost no petrol on our territory, we have less coal than England and Germany and much less gas than Holland. . . . Our great chance is electrical energy of nuclear origin because we have had good experience with it since the end of World War II. . . . In this effort that we will make to acquire a certain independence, or at least reduced dependence in energy, we will give priority to electricity and in electricity to nuclear electricity.

The so-called Messmer Plan called for a massive increase of nuclear capability. The plan was based on an EDF report which suggested that as many as 200 reactors might be needed by 2000 (Surrey and Huggett 1976).[30] Overall, the Messmer Plan was a tremendous success. Nuclear electricity grew at an unprecedented rate: from 8% of EDF's output in 1973, it jumped to 20% in 1980, 49% in 1983, and 75% in 1990. As nuclear production increased, France's reliance on foreign energy decreased. Although it was still importing 75% of its energy by 1980, France had reduced its share of imports to about 50% in 1990 (WDI 2013). This number has remained stable since then but remains considerably lower than other countries such as Germany (60%).

Perhaps just as important, over the same period of time, EDF consolidated political support for nuclear energy. Although atomic power benefited from the reliable backing of the successive rightist presidents (de Gaulle, Pompidou, Giscard), a period of uncertainty loomed after François Mittérand, a socialist, won the presidential elections in 1981. Leading members of the socialist party disagreed about the energy policy that the country should follow (Collingridge 1984). However, the debate was short-lived. Once in power, Mittérand sided with the pronuclear lobby and approved a nuclear program that was only slightly more modest than initially planned (de Carmoy 1982). Nuclear energy had survived its first political challenge and would strengthen its position as the uncontested cornerstone of French energy policy. One notable strategy pursued by EDF was to fund CGT, one of the biggest trade unions. In parallel, the energy boards advising the government were staffed almost entirely with pronuclear experts. Similarly, abundant money was directed toward the country's most prestigious research institutes and schools (Topçu 2007).

An indication of the nuclear lobby's power is the lack of reaction from the elites to the 1986 Chernobyl catastrophe. Whereas Chernobyl triggered, as we saw, a radical shift in German policy, the French pronuclear consensus remained as strong as ever. Scientists appeared on TV announcing that the country had nothing to fear, which led to the derisive (and apocryphal) statement by opponents to nuclear power that "the nuclear cloud had stopped at the border."[31] This is not to say that there was no opposition to nuclear energy. The *Appel des 400* (Call of the 400), a group of scientists who worried about the detrimental effects of nuclear energy, was founded in 1975. Similarly, popular antinuclear movements popped up in various regions, and Brice Lalonde, the Green candidate to the 1981 presidential election, obtained a positive result with a vote share of almost 4% (Surrey and Huggett 1976; de Carmoy 1982). However, overall, the French response to the oil challenge was both clear ("Tout électricité, tout nucléaire")[32] and popular (65% of support for nuclear in 1982; de Carmoy 1982). Why did this response so starkly contrast with Germany's or Denmark's?

In the short run, France possessed the technological knowledge to push for a solution that would solve the main issue, that is, the lack of natural resources. In that sense, France's response mirrors Denmark's: Denmark had the know-how to develop wind technology, whereas France had the human capital to invest in nuclear energy. A burgeoning nuclear program supported by an army of engineers offered an easy response to the oil shock. Furthermore, by cross-subsidizing the military nuclear program, civilian nuclear energy catered to the international ambitions of de Gaulle and Pompidou. However, this alone does not explain why the nuclear option became such an entrenched policy. In the medium term, the explanation is rooted in domestic politics. Pronuclear policymakers controlled power throughout the 1970s. Every French president since the beginning of the Fifth Republic in 1958 supported the development of military and civilian nuclear capabilities. By the time Mittérand came to power, the political might of EDF was unassailable. The French bureaucratic structure gave pronuclear government officials "virtually complete control . . . controversy has been weak in France compared not only to the United States, but also to other Western European countries. Nuclear power in France is largely considered to be a technical question best left to the experts" (Baumgartner 1989, 50–51). Furthermore, by the early 1980s, the economy relied heavily on cheap and abundant nuclear electricity. Mittérand had little incentive to bow to what remained a fringe movement. Although he did make some concessions to the Greens during the presidential campaign, he had no reason to impose policies that would make France's industries less competitive in the short term. As unemployment increased in the late 1970s, a politically savvy decision maker such as Mittérand knew that he had little to gain from changing course.

Thus, renewable energy remained the "parent pauvre" of the French energy family. Renewable R&D remained pitiful: by 1993, France spent $8 million per year compared with Germany's $113 million (Taylor, Probert, and Carmo 1998). Production remained negligible, too, with wind and solar energy being virtually nonexistent by the early 1990s. Even the share of hydroelectricity was dwindling. France had chosen its own uranium-paved path. This would be reflected in the soaring number of patents issued for nuclear technology from France. The number of patents went from about 1 in 1977 (0.5% of all French patents) to more than 60 in 1980 (2.6%), with a peak of 70 in 1988 (1.5%; OECD 2014a). However, France did not issue more than ten patents for renewable energy technology in any single year before 1998.

At first sight, the United Kingdom's energy policy seemed to share many similarities with that of France by 1973. Following a wave of nationalization under the Attlee government in the aftermath of the Second World War, the energy landscape was dominated by large public corporations. These corporations, organized by energy source, were officially under governmental control but operated independently as monopolies over their respective markets, not unlike EDF. As in France, these corporations and their employees were politically influential. The coal miners, for instance, formed a loyal block of votes for the Labour Party. Furthermore, just like France, the United Kingdom had developed nuclear energy capabilities in the 1950s and 1960s.

However, the situation in the two countries was quite different. The United Kingdom only imported about half of the energy it consumed, compared with two thirds for France (WDI 2013). Additionally, British energy corporations were suffering from tremendous political uncertainty throughout the 1970s (A. Taylor 1996). First, coal interests regularly threatened the government. Even the Labour authorities distrusted them, and with good reason, as the Winter of Discontent (1978–1979), which consisted of a series of strikes against Labour policies, buried the Labour Party. In contrast, EDF (and Gaz de France, its sister company) remained loyal to the French government throughout. Second, these corporations did not benefit from the support of the main rightist party. In fact, Margaret Thatcher and John Major's Conservative governments would eventually privatize Britain's energy sector. Gas was the first conglomerate to become private, leading to the creation of British Gas in 1985. It was followed by electricity (1989), coal (which became UK Coal in 1994), and nuclear (which initially became British Energy in 1996 and has belonged to EDF since 2009). Thus, the government consciously decided to abandon its direct control over energy output. In contrast, EDF's largest shareholder remained the French government, which appreciated the influence that came with the ownership of these large corporations regardless of the president's ideology.

Yet both countries failed to invest in renewables to respond to the challenges of the 1970s. The United Kingdom did not implement a renewable energy policy until the 1990s (Lipp 2007). R&D remained extremely low in the 1980s. Total public renewable energy R&D expenditure was about $50 million in 1980; Germany's was $143 million and growing (IEA 2014c). Most of these meager resources went into hydroelectric R&D. Solar R&D would not crack the $10 million barrier until 2001, whereas wind would get less than $20 million per year throughout the 1980s. In contrast, nuclear R&D would receive between $500 million and $1 billion on a yearly basis in the same period. Even the emergence of renewable energy policies in the early 1990s was accidental. The Thatcher government, which wanted to subsidize nuclear power, had to obtain the European Commission's approval (Mitchell and Connor 2004). To avoid applying for a nuclear power permit in particular, it applied for the right to support non-fossil fuels more generally. Thus, the Non-Fossil Fuel Obligation program, launched in 1990, accidentally gave birth to Britain's renewable industry. However, by then, the United Kingdom was already lagging behind its continental competitors. By 2004, only 4% of its electricity came from renewables (Lipp 2007).

Elite ideology in the United Kingdom was not favorable to renewable energy. Before 1980, many policymakers favored nuclear power, which had been developed for national security purposes (Teräväinen, Lehtonen, and Martiskainen 2011). Thatcher had political reasons to support nuclear power (Wolfram 1999). Coal miners had brought down the previous Conservative government, and she was therefore particularly supportive of substitutes that could make her own government less likely to follow that path (Newbery 1998). Until Chernobyl, nuclear did not raise much opposition. Furthermore, since the late 1960s, the British knew they had large gas reserves located in the North Sea (J. Stern 2004). These gas reserves provided a welcome endowment to fuel the country's thermal power stations. Thus, successive governments from Edward Heath to Margaret Thatcher had a domestic solution to the international energy problem. In fact, they would heavily draw on these reserves starting in the mid-1970s, right after the first oil crisis (Watt 1976). Both Labour and Conservative governments happily collaborated with private oil companies to extract these resources.

This is not to say that the United Kingdom ignored renewable energy altogether. The Ministry of Energy commissioned studies in the late 1970s to evaluate the potential of solar power (Stainforth et al. 1996). Most of these efforts were directed at simple solar designs, generally for warming. Similarly, the private sector and civil society began to mobilize in the same period. The International Solar Energy Society, which brought together solar enthusiasts and scientists, founded a British section in 1974 (A. Smith et al. 2014). Four years later, the Solar Trade Association was founded, offering a platform for businesses that wished to penetrate the energy market. However, these efforts remained confined to a niche market, to borrow the words of A. Smith et al. (2014). Governmental support remained small, especially under Thatcher. As Twidell and Brice (1992, 472) note, the British strategy was exclusively focused on research, and "there was no strategy for seeding a competitive market for products and services."

To summarize, both France and the United Kingdom became renewable energy laggards, although for different reasons. In the aftermath of the first oil shock, France chose to reduce its vulnerability to oil by choosing an alternative path based on nuclear energy. The roots of Britain's nonresponse lie elsewhere, however. Unlike France, the United Kingdom was well endowed with energy sources. It had abundant coal reserves that could be used to produce electricity (A. Taylor 1996). France responded to the oil shock with its *tout nucléaire*, whereas the United Kingdom announced in 1975 that it was the time for "coal first." Britain also had vast amounts of gas and oil reserves located in the North Sea (Forsyth and Kay 1980). These reserves were too costly to exploit before the oil shock, but they became profitable under high oil prices. They would prove immensely helpful: by 1981, the United Kingdom was a net exporter of energy (WDI 2013). This situation would last until 2006, when the United Kingdom started to be a net importer again. Thus, the United Kingdom did not invest in renewables because it was not necessary. Its resources were sufficient to absorb the shock from the two oil crises. In fact, the crises made this resource economically valuable, making the entry of new competitors such as renewables even less likely. In summary, the effect of the shocks in the United Kingdom was to strengthen the grip of fossil fuels, along with some investments in nuclear power.

4.5 Contrasting the Danish and Finnish Cases

No political history of renewable energy is complete without an account of Denmark's wind energy boom. As in the case of the United States and Germany, this saga begins with the 1973 oil crisis. In previous years, oil consumption had grown at an astonishing rate in both Denmark and Finland. Due to rapid economic growth and modernization, both countries had about doubled their oil consumption in less than a decade. The 1973 oil shock hit the two small Nordic countries hard, creating an energy crisis that forced both governments to act. As we will soon see, however, the policy responses could not have been more different.

Denmark's energy situation was nothing short of desperate in 1973. The country had no fossil fuel resources to speak of, and yet oil consumption has grown rapidly during the past decade or so. According to Vasi (2011), 88% of the country's primary energy consumption was in the form of oil. Lund (2000) notes that even in the electricity sector, 85% of total power was generated from oil. Because no other country in Northern or Western Europe was as dependent on oil, it is fair to say that Denmark suffered from the oil shock more than any other industrialized country. As Lund (2000, 250) puts it, "Denmark was totally unprepared for the first oil crisis in 1972–1973."

The government's initial response was, as would be typical of the time, to announce ambitious plans for investment in nuclear power generation. At the end of 1973, the Danish government announced a plan to build nuclear plants—the country had none at the time (Vasi 2011). In 1976, the National Energy Plan announced the construction of five nuclear power plants (Hadjilambrinos 2000). However, these plans went nowhere because of public opposition (Lipp 2007). Vasi (2011, 70) writes that the grassroots Danish campaign against nuclear power, organized under the Organization for Information about Nuclear Power, was able to repeatedly postpone the government's "plans for nuclear power for fear of a referendum." Consequently, the nuclear solution to the severe 1973 oil shock was out of the question, and by 1980, the Social Democratic government of Denmark buried the idea for good (Heymann 1998).

Besides some efforts to improve energy efficiency, Denmark's primary response was to be based on wind energy. Two reports—*Wind Energy in Denmark* and *Wind Power 2: Proposal for an Active Programme*—prepared by the Danish Academy of Technical Sciences played a key role in shaping elite ideology about wind energy (Sawin 2001). In them, researchers came to the conclusion that Denmark has excellent conditions for wind energy generation and proposed a five-year program of action. These reports were important because they provided a positive assessment of wind energy already in the short run and proposed a series of concrete steps.

Beginning with the 1976 plan, the national energy plans of Denmark contained a variety of measures to support wind energy. The initial policies to support the construction of windmills were tax credits and generous investment subsidies (Lipp 2007), and Denmark's agricultural cooperatives encouraged, in collaboration with environmental activists, local ownership of wind power (Hadjilambrinos 2000). In turn, local ownership swayed public opinion in favor of wind energy, allowing the government to continue the expansion of wind energy over time. According to Hadjilambrinos (2000, 1122), "the high level of involvement of the public in energy policy-making in Denmark has resulted in policy and technology choices that resemble the 'soft energy path"' that Lovins (1976) was advocating in the United States at this time. By 1980, annual public R&D investment in wind energy had increased to $6.5 million, whereas nuclear R&D was reduced to zero in the following year. In per capita terms, Denmark's investments in wind energy were much larger than those of either Germany or the United States.

Denmark's global significance, of course, is as a pioneer in wind technology manufacturer. Whereas the Danish market for wind power was small, technological development allowed the Danish company Vestas to seize leadership in the wind technology sector (Heymann 1998; Lewis and Wiser 2007). In fact, California's wind frenzy was a key reason behind Denmark's early success as a wind technology manufacturer. As Gipe (1991) notes, half of the wind-generation capacity in California in the 1980s was supplied by the Danish wind industry, Vestas in particular. In 1981, Vestas sold to California 55-KW turbines "having 120,000 hours of operation experience and 95 percent turbine availability," which is more than what competitors were able to offer (Heymann 1998, 664).

As Heymann (1998, 662) notes in his comparison of early wind energy innovation in Denmark, Germany, and the United States, a key reason behind Denmark's early leadership wind turbine manufacturing was the success of small, perhaps even artisanal, production:

> Most Danish commercial wind turbines—and the most successful ones—were developed independently of government R&D programs. Beginning in the mid-1970s, enthusiastic amateurs and skilled artisans committed themselves to the development of wind turbines. One of the first and most influential of these pioneers was the carpenter Christian Riisager. Concerned with environmental problems, he began to build simple wind turbines. . . . . Riisager assembled his first wind turbines from

inexpensive off-the-shelf parts, such as standard asynchronous generators and truck gears, axles, and brakes.

In spite of his limited theoretical background and experience, by 1976 Riisager had produced a surprisingly reliable 22-kilowatt turbine. By January 1978 he had sold six copies; within the next two years he sold fifty more.

The early success of the Danish wind industry thus reflected the combination of small-scale innovation in favorable domestic circumstances and the demand from California.

The Danish government's decisive response to the oil shock is consistent with our first expectation, and the domestic conditions enabling it also support the theory. The historical reliance on decentralized agricultural cooperatives generated public enthusiasm for wind power; elite ideology played a secondary, if ultimately important, role. The Danish political system allowed grassroots interests and local communities to play a much larger role in energy policy than was possible in the United States or Germany. In combination with Denmark's dependence on foreign oil for primary energy, the stage was set for an ambitious and effective policy response. Although Denmark did not even have an energy ministry in 1973 (Lund 2000), the broader societal context was conducive to an effective renewable energy response that had major ripple effects outside Denmark, thanks to the rapid development of wind energy technologies.

To be specific, the early history of the Danish wind energy program suggests a nuanced view of the role of elite ideology. In the United States, elite ideology resulted in a federal R&D program for solar power—one that we consider relatively successful—and generous support for wind energy deployment in California. In Denmark, wind energy R&D did not play such an important role, but the various tax credits, investment subsidies, and regulations that enabled agricultural cooperatives to own wind power created an environment in which the growing Danish wind industry was able to thrive. Elite ideology thus created the favorable conditions for market creation through policies. We also note that without a favorable elite ideology, partially supported by careful academic assessment, the Danish government might have favored nuclear power, although here we must also emphasize the antinuclear sentiment among the public.

The Danish case also shows that there was not really much political opposition to renewables. Although the government's grandiose plans for nuclear power provoked a vigorous environmental backlash, plans for renewable energy expansion were popular from the beginning. From antinuclear activists to local communities and wind energy enthusiasts, a wide range of Danish political constituencies and interest groups were in favor of wind energy expansion. The country's energy sector, which was largely owned by the agricultural cooperatives and municipalities, also secured direct benefits from wind electricity generation, thanks to generous subsidies and tax incentives from the national government. Therefore, the case also supports our third expectation: in the early stages of renewable energy development, opposition to policy is limited because the stakes are not high enough.

The absence of opposition can be seen in the position of Danish utilities (N. Meyer 2007). In the 1970s, electric utilities did not yet share the enthusiasm of the grassroots alternative energy movement or the government for Danish wind energy, but they also did not try to undermine the wind technology industry—after all, the utilities believed that the industry would flounder. Toward the end of the 1980s—years after Danish wind technology exports exploded—the utilities found themselves in disputes with owners of wind mills about grid access, pricing of wind electricity, and so on: "after a number of disagreements between utilities and wind power producers over conditions for grid connections and tariffs, regulations were introduced by the Danish government in 1992" (N. Meyer 2007, 350–351).

To illustrate the Danish approach to wind turbine development, no case is more suitable than the construction of the Tvind windmill, the world's largest at the time (Jamison 1978). Tvind, a Danish community on the country's west coast, has an experimental school. In 1974, the Tvind school teachers and the broader community decided "to apply their educational methods to the problem of energy production" (Jamison 1978, 26). The 53-meter-high windmill, which still produces electricity, was built by hundreds of volunteers and finished by the end of 1975. This grassroots effort made wind power a smash hit in Denmark; in 1976 alone, more than 100,000 people visited Tvind to see the windmill. As Jamison (1978) notes, many of the visitors came specifically for technical and design expertise. Indeed, already in 1976, the government's energy plan put special emphasis on wind energy, showing how a grassroots experiment in a rural community produced larger changes over time. Because of this favorable environment, the Danish wind technology industry thrived and was able to meet the demand for wind energy in California at a time when the US wind technology industry was not yet ready to produce high-quality turbines.

In the meantime, things looked different on the Northern side of the Baltic Sea. In 1973, Finland's energy problems were serious. Similar to Denmark, the Finnish economy had doubled its oil consumption in less than a decade. In 1970, 57% of Finland's primary energy consumption was oil, and all of it was imported, mostly from the Soviet Union. Also similar to the Danish case, Finland had no domestic fossil fuel resources to exploit. Although the country's pulp and paper industry was able to rely on wood for fuel, the burning of wood for electricity generation was not considered a serious policy alternative for the national power sector until much later (Ericsson et al. 2004). As the president of Finland, Urho Kekkonen, stated in a speech given to the elite Paasikivi Society in the city of Tampere on December 19, 1973, the "energy crisis" was a serious warning about fundamental changes in the world economy and politics.[33]

In Finland, solar and wind did not draw any attention outside academic research institutions, and the 1979 national energy strategy stated that these energy sources would only be developed through international cooperation (Helynen 2004). According to Huttunen (2009), no strategy was in place for nonwood biofuels until news regarding Sweden's successful investments in biofuels became a central subject of the debate. To the extent that Finland implemented policies to promote the use of biomass, the early emphasis was only on wood—a traditional mature energy source in the large forestry sector of Finland—and peat, a nonrenewable biofuel with significant negative consequences for the climate (Helynen 2004). Moreover, in the 1960s, the Finnish paper and pulp industry had, due to technological advances, begun using birch and other wood types that had traditionally been set aside for firewood. As a result, the Finnish government's enthusiasm for promoting firewood as an alternative to fossil fuels was reduced (Kunnas and Myllyntaus 2009). Compared with Denmark's large R&D program in renewable energy, peaking at $30 million (2012 constant prices), and ability to generate wind energy patents already during the 1970s, Finland's achievements are decidedly less impressive.

Finland's lackluster response to the 1973 oil crisis originates from a lack of enabling factors. Already before 1973, Finland had initiated major investments in nuclear power, planning to build multiple units for power generation (Sunell 2001). These investments, combined with a pipeline for Soviet natural gas forcefully promoted by President Kekkonen,[34] constituted a somewhat fortuitous response to the oil shock. The massive amounts of capital that the state and the leading industrial interests poured into these projects preempted the need for renewable energy, and they allowed Finland to reduce its oil dependence over time. At the same time, Finland's unique bilateral trade arrangement with the Soviet Union partly shielded the country from oil shocks, although the terms of trade were based on the international price of oil.

In addition to these investments, many other factors predicted a negligible response. In particular, elite ideology in Finland was strongly in favor of nuclear power because heavy industry favored nuclear power to generate inexpensive and reliable electricity to power their operations, and the political leadership saw major advantages in ordering a nuclear plant from the Soviet Union.[35] Indeed, all key players in the Finnish government were in favor of nuclear power already before the onset of the oil crisis in October 1973. In his thesis on the politics of Finland's nuclear expansion in 1972, Sunell (2001, 89) notes that "the decision to build additional nuclear power was itself not difficult, because experts were in favor of the expansion."[36] In particular, he notes that Finland's advisory council on energy, *Energiapoliittinen neuvottelukunta* in Finnish, was unambiguously in favor of additional nuclear power (Sunell 2001).[37] The board was chaired by the minister for trade and industry and included all major governmental and industrial interests of Finland, such as the Bank of Finland, the Ministry of Foreign Affairs, the oil company Neste, and the industry energy corporation Teollisuuden Voima (Sunell 2001). Sunell (2001) also notes that the Social Democratic Party of Finland, which held the position of the prime minister from September 1972 until June 1975 and then again from November 1975 to May 1979, was strongly in favor of nuclear power, although the party's position was more supportive of public than private ownership of nuclear reactors.

To be sure, elite ideology in Finland also reflected the pressures of the Cold War on a small, neutral country that tried to balance the demands of the East and the West. In his doctoral thesis on Finnish nuclear policy, Särkikoski (2011) offers a comprehensive evaluation of the different forces that drew Finland to the path of nuclear expansion. Besides concerns about the cost and abundance of oil and coal, Särkikoski (2011) argues, Finland's interest in nuclear power was related to the more general European interest in nuclear power as a political strategy that would allow Europeans to increase their status in world politics relative to the United States and Soviet Union.[38] Relatedly, Särkikoski (2011) argues that the Finnish government and industrial elites were optimistic about the economic potential of nuclear power as an engine of Finnish industrial growth. Already in 1969, the Finnish advisory committee on nuclear power, *Atomienergianeuvottelukunta*, paved the way to nuclear construction with a report that emphasized the compelling case for nuclear power as an environmental and cost-effective solution.[39] What is particularly striking about the arguments is that, although they were made in public years before the events of October 1973, they already emphasized nuclear power as the primary alternative to fossil fuels. These views did not leave much room for renewable energy, again with the exception of investments into the use of residual products from the booming Finnish pulp and paper sector.

In an interview given to the Swedish newspaper *Göteborgs-Posten* on May 12, 1980, President Kekkonen emphasized Finland's achievements in forestry and the use of peat, as well as the virtues of nuclear energy cooperation with the Soviet Union.[40] This lack of elite interest in renewable energy, with the partial exception of burning wood for heating, gave little impetus to a decisive policy response. As to institutional capacity, suffice it say that it took Finland until 1979 to formulate a national energy strategy, and therein a decision was made to exclude new forms of renewable energy from consideration.

In summary, the Danish and Finnish cases again underscore the importance of national conditions in explaining and understanding policy responses to external shocks. Although both Finland and Denmark were vulnerable to the 1973 oil shock, their responses were different. Finland continued on the path toward increased reliance on nuclear electricity and Soviet natural gas exports, both for economic and political reasons, whereas Denmark started investing in wind energy. National conditions led the two nations to different directions, and one of them became a frontrunner in renewable energy, whereas the other chose to rely on a combination of fossil fuels and nuclear power.

5 Politicization: When the Stakes Grow Too High

One of the most important features of the comparative case studies presented in the previous chapter was the low level of opposition to renewables in the countries under analysis. In Denmark, Germany, and the United States, the initial stages of their growth were mostly uncontroversial. In places where renewables were not considered, there was never a major debate on their merits.

At least to us, these findings were initially surprising. During the past two decades, the renewable energy industry has been at the center of intense political debates in many countries. On the prorenewables front, thousands of people have gone to the streets in Germany when the government has tried to cut support for renewables. On the opposite side, efforts to discredit the value of renewable energy have been relentless. The Koch brothers, who have funded various conservative causes in the United States, have invested significant sums in antirenewable lobbying.[1] The politicization of renewable energy has reached high levels, with opponents and proponents engaging in fierce debates about the merits of renewable sources of energy. How did we get here? When and how did renewable energy become the target of fierce opposition across the world?

The goal of this chapter is to explain variation in the degree and nature of politicization in different countries. We move from the early stages of renewable energy—roughly speaking, the first, short decade of renewables between 1973 and 1981—to the next stage. As oil prices finally stabilized and even dropped, critical voices began to question the merits of renewable energy policy. In the meantime, renewables had grown enough that their future potential no longer seemed trivial to established energy interests, heavy industry, and ideological opponents of government intervention in favor of a greener energy economy.

We attempt to make sense of the reasons that renewable energy became so controversial in some frontrunner countries while remaining relatively uncontroversial in others. Given the multitude of factors that could predict politicization, we combine simple descriptive statistics with somewhat more detailed comparative case studies. The analysis in this chapter is largely limited to the frontrunners identified in the previous chapter: Germany, the United States, and Denmark. The laggards—France, the United Kingdom, and Finland—are less essential for the analytical purposes of this chapter. Where renewable energy did not grow initially, there is little scope for politicization and opposition. Because renewables do not play much of a role in the energy economy in any case, their growth does not threaten any political or economic interests that would find active opposition worth their while. Therefore, we revisit these three cases only briefly.

The next section provides some descriptive statistics on renewable energy growth and policies across nations. We document the transition of renewables from small-scale ventures into larger projects. This was made possible through stable public support, initially public funding for R&D and later policies that supported deployment, such as feed-in tariffs. In so doing, this section sets the stage for the research design of our comparative analysis. After this methodological introduction, we devote one section each to the United States, Germany, and Denmark, highlighting the extent to which our expectations are met in these cases. The final section briefly describes developments in the other three countries (France, the United Kingdom, and Finland) over the same time period.

5.1 Renewable Energy Growth and Policies Across Nations

International shocks generated a strong prorenewables coalition among frontrunners. When oil prices were high and nuclear power looked dangerous, few voices dissented against the need for an alternative energy paradigm. International shocks silenced opposition because they made the shortcomings of the status quo obvious. However, these shocks were not permanent. People soon forgot the fears created by Three Mile Island and Chernobyl. The memories of long lines at the gas stations started to wane. The prospects for renewables were related with their ability to maintain the prorenewable consensus once the shock had passed. This consensus came under attack everywhere, and renewables became politicized. In some countries, the political conflicts suppressed the nascent renewable industry and started a period of decline. In these countries, the shocks turned out too weak to break the stronghold of traditional energy sources. In other countries, however, renewables survived the initial onslaught and became the major source of power that they are today.

The second phase of our political history of renewable energy lasts approximately from 1981—President Reagan's election—to the first decade of the 2000s. During this period, renewables had to show that they were becoming a mature technology. Renewable energy technology had to leave its protective niche and confront market-tested competitors (Smith et al. 2014). The renewable energy industry's ability to do so, however, still depended on unwavering public support. In some countries, such as Denmark, support remained mostly high, with limited and ephemeral phases of opposition. In others, renewable energy faced strong and effective opposition.

Before the period of politicization, in 1980, the United States was the leader in renewable energy. The country was at the forefront of technological development. One reason for this was that the United States had the highest public investments into R&D worldwide, with about $1.6 billion per year (most of which went toward solar research). This number was large in both absolute and relative terms. The United States was the largest investor in R&D per capita. Denmark was following closely but on a much smaller scale in absolute numbers. Germany, still hesitating between nuclear power and renewables, was lagging behind. Similarly, the US private sector carried its weight. Between 1976 and 1980, the United States issued more than 60 patents (0.29 per million people), ahead of Denmark (1 patent, or 0.2 per million people) and Germany (10 patents, or 0.16 per million people) (OECD 2014a). Early commentators, such as Twidell and Brice (1992, 464), could rightly claim that "[t]he impact of the USA on renewable energy has been dominant, especially under President Carter's administration."

Carter's loss marked the end of a united front in favor of alternatives to oil and nuclear power. In fact, US leadership in renewables already began to suffer under Carter. Laird (2001, 178) shows how the influence of the prosolar lobby in the Carter administration began to decline in 1979. Those favorable to renewable energy and solar power lacked access to the top policymaking circles and were unable to mold energy regulations. Of course, these problems only worsened under the Reagan administration. Reagan's hostility to renewables, documented in the previous chapter, translated into slashed R&D figures. From $1.6 billion, public R&D decreased to about $200 million per year by the end of Reagan's presidency, a staggering drop of more than 87%. The end was not in sight, as funding shrank even further under George Bush. Per capita R&D dropped well below Denmark's; although it seems comparable to Germany's, this approach hides Germany's turn toward solar energy. By 1990, Germany invested on an annual basis almost four times as much into solar R&D as the United States. Over a period of about ten years, the United States lost its leadership to Denmark and Germany.

R&D was not the only way in which policymakers helped the nascent renewable energy industry. Unlike the United States, Germany and Denmark made use of feed-in tariffs (FITs). FITs are particularly useful in promoting a new energy source between its early development stages and its full commercialization because they move the focus of policy from research to actual deployment, allowing renewables to overcome the carbon lock-in outside the laboratory and demonstration projects. Germany enacted an FIT for electricity from wind and solar systems in 1991 already. Denmark followed with an FIT for wind in 1992 and for solar electricity in 1998. The United States, again, remained behind. The federal government never enacted an FIT. Electricity governance was in the hands of authorities who, for most part, did not have the interest, means, or will to support renewable energy. In California, for example, wind capacity actually declined in the 1990s, as we show below.

The lack of public support in the United States crippled the deployment of renewable energy. In 1980, the United States was a leader in research as well as output. The share of renewables in electricity capacity, although still small, was higher than even Denmark's. However, by the end of the decade, it became apparent that the US renewable industry was at best stagnating. Figure 5.1 shows the share of renewables in total electricity capacity and depicts how the paths of Denmark, Germany, and the United States diverged. Denmark overtook the United States by the mid-1980s and became the clear leader in actual renewable deployment. Germany's trajectory mirrors Denmark's, although with a five-year delay. By 2000, the share of renewables was about 22% in Denmark, 7% in Germany, and 2% in the United States.

Denmark, Germany, and the United States had all responded to the oil and nuclear shocks of the 1970s by experimenting with renewable energy in a new, "soft" energy paradigm. However, over the next decade, the effects of the shock dissipated, and early proponents among policymakers lost their position. In some countries, the prorenewable consensus did not survive the disappearance of urgency that had been imposed by the oil and nuclear crises. The United States is a case in point. In other countries, renewables became literally and figuratively part of the landscape. There renewables mostly stayed out of party politics and grew enough in importance to become self-sustaining. Denmark is the ideal type of such a country. Finally, renewables did become an object of contention in some countries, but the support that this industry had was sufficient to fend off these challenges.

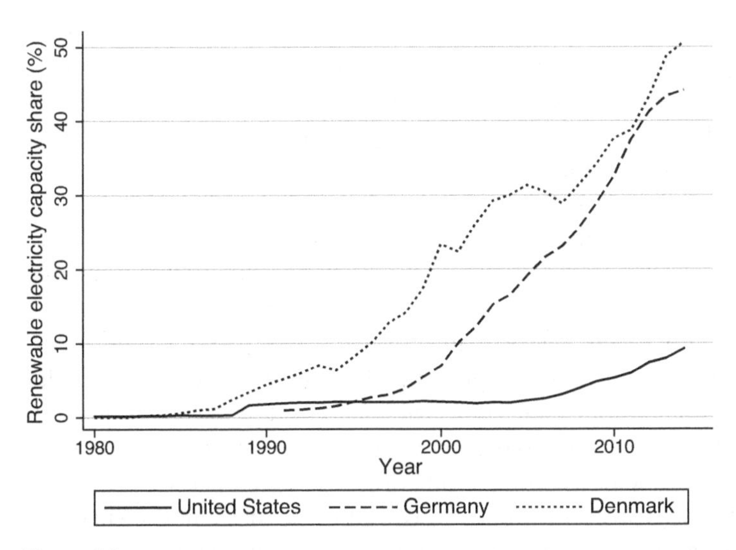

Figure 5.1
Share of renewables in electricity generation capacity in the United States, Germany, and Denmark. Includes all renewable sources except hydropower.
Source: EIA (2017a).

Germany illustrates this scenario. We explain next why these country trajectories diverged and why the second stage of our political history—the politicization phase—is critical for understanding renewable energy today and tomorrow.

5.2 Summary of Comparative Case Studies

To gain a better understanding of the logic of politicization of renewable energy in frontrunner countries, we again rely on the comparative case method. Because we have already shown variation in the trajectory of renewable energy policy across nations, and the degree of political opposition is difficult to quantify for the purposes of more detailed quantitative analysis, comparative case studies are ideal for our purposes. They allow us to account for the context in each individual nation, but the method of structured and focused comparison also enables inferences from comparisons across the cases.

The primary conceptual challenge for the analysis is the measurement of opposition. As we noted in the theory chapter, our analytical approach emphasizes political opposition as the central obstacle to the growth of renewable energy. This notion underscores the idea that renewables depend on government support for their growth, whereas government support depends, in turn, on the general legitimacy of renewables as a rational and socially desirable solution to the problem of carbon lock-in. In conducting the case studies, therefore, we search for traces of increased hostility and criticism of renewable energy.

Keeping this conceptual challenge in mind, we consider multiple expectations on factors that would influence the degree of opposition to renewable energy. We begin with the expectation that public opinion about nuclear power and climate change play an important role. Next, we consider a series of claims on the role of different political factors in determining the ability of the advocacy coalition for renewables to stop the opposition and continue enjoying high levels of support among the policymaking elite. We investigate the role of left-wing, right-wing, and Green parties in the process, and we also consider the role of environmental social movements. Our other expectations focus on economic interests along the lines of Cheon and Urpelainen (2013), examining the evidence for the role of fossil fuel producers, heavy industry, and the clean technology industry.

The expectations and cases are summarized in Table 5.1, one by one. We also provide a concise summary of the strength of the evidence for the different cases. For most expectations, there is evidence across all three cases. Overall, the most striking contrast is between the United States, where early leadership in renewables soon disappeared, and Germany and Denmark, where the outcome of the political struggle was the emergence of renewables as the victors. The one somewhat surprising observation is that Germany, despite a large industrial economy and abundant domestic coal resources, became such a strong advocate of renewable energy. In the German case study, we pay particular attention to this issue.

Across the cases, we try to measure key concepts as consistently as possible. Where possible, we use cross-national public opinion data about climate and nuclear power. When this is not possible, we try to use data from the same years with the most similar question wordings. We also focus on left-right political shifts within each country, instead of trying to evaluate how leftist or rightist the political parties are across countries. To evaluate the economic importance and political clout of the heavy industry, fossil fuel producers, and the clean technology industry, we use data on industrial structure and technology innovation. Whenever possible, we also provide direct quantitative evidence on lobbying expenditures. An additional benefit of using these data is that they help us avoid tautological inferences based on the reasoning that, if an industry managed to secure a positive

Table 5.1
Summary of expectations and cases for politicization.

Expectations	
Expectation 1	The less concerned the public is about nuclear power and climate change, the higher the degree of opposition to renewables.
Expectation 2	Politically powerful clean technology industries and environmental groups reduce opposition to renewables.
Expectation 3	Politically powerful fossil-fuel producers and heavy industries increase opposition to renewables.
Expectation 4	Compared with their left-wing counterparts, right-wing governments are more opposed to renewables. Furthermore, strong Green parties reduce opposition to renewables.
Cases	
United States	Politicization starts in the 1980s, triggered by early successes in California. Less public concern about climate change than elsewhere (**E1**). Weak clean technology industry and environmental groups (**E2**). Strong heavy industry and fossil-fuel producers (**E3**). Reagan's backlash against renewables (**E4**).
GER	Over time, renewables become popular and begin to grow fast thanks to supportive government policy. High level of public concern about climate change and nuclear energy, especially after the Chernobyl disaster (**E1**). Strong clean technology industry in the North, supportive farmers benefiting from renewables in Bavaria, and powerful environmental groups in general (**E2**). Strong heavy industry and fossil-fuel (coal) producers opposed renewable energy policy (**E3**), but their influence waned over time. Helmut Kohl's right-wing government undermines renewables, but a strong Green party promotes them (**E4**).
DEN	Opposition to renewables remains low most of the time largely thanks to success and popularity of wind energy. High level of public concern about climate change, and nuclear energy in particular, throughout (**E1**). Strong environmental groups and clean technology industry, especially in wind technology (**E2**). Weak heavy industry and fossil-fuel producers (**E3**). Politicization increases with right-wing government, but very late (**E4**).

outcome, it must have been influential. By validating our estimation of economic importance and political clout, we can avoid this common pitfall of comparative inference. These comparable measures are then combined with process tracing within each country for a thorough evaluation of the different expectations.

5.3 Politicization of Renewable Energy in the United States

Reagan's inauguration on January 20, 1981, as the president of the United States was a decisive moment in the politicization of renewable energy. The new administration launched an unprecedented assault on renewables and, in doing so, created a wide political gap between the supporters and opponents of renewables. As Narum (1992, 40) puts it, "[t]he Reagan Administration drastically cut funding for the Department of Energy's (DOE) research and development programmes in energy conservation and renewable energy (CORE). . . . As a result, the Administration has delayed the transition to an energy mix more inclusive of CORE technologies."

Reagan's summary dismissal of renewable energy reflects his strong partisan preference against government intervention in the free market—public support for nuclear power notwithstanding, his critics would note. In an election debate with the incumbent Carter, Reagan dismissed the former's energy policy, stating that, "The Department of Energy has a multibillion-dollar budget, in excess of USD 10 billion. It hasn't produced a quart of oil or a lump of coal or anything else in the line of energy."[2] However, Reagan was far from being alone in the increasingly vocal criticism of renewable energy. As the oil shocks of 1973 and 1979 began to fade from the public's memory (the price of oil collapsed in the 1980s), the politicization of renewable energy reached unprecedented levels, and the steady progress of the industry that was characteristic of the Carter administration died away. As Laird and Stefes (2009, 2621) put it, "[b]y the time President Reagan took office in 1981, renewable energy had become a highly politicized and partisan issue." The societal consensus on the need to make modest investments in renewable energy as a safety measure in the event of the continuation of high energy prices disappeared as the Republican Party began to emphasize the cost and modest achievements of Carter's renewable energy program.

The politicization of renewable energy can easily be seen in the voting record of the US Congress on the topic. Lowry (2008) collected data on congressional roll-call energy votes between 1971 and 2007. For each such vote, he recorded the partisan difference, defined as the percentage of Democrats reporting votes that align with the position of the League of Conservation Voters (LCV) less the percentage of Republicans reporting LCV-aligned votes. As the absolute value of this partisan difference increases, so does politicization along partisan lines. This measure is ideal for capturing broad trends of politicization in renewable energy, and we report the Lowry (2008) data on renewable energy in Table 5.2. As the table shows, there is a dramatic difference in the level of politicization between the Carter and other eras. Until 1980, no vote had a larger partisan gap than a value of 31. Since 1981, the gap has never fallen below 37. Our theory predicts just this kind of politicization process, with high levels of opposition to renewables.

The effects of politicizing the renewable energy industry in the United States were destructive. According to detailed data collected by Nemet and Kammen (2007), between 1980 and 1988, federal R&D investment into renewable energy decreased from $1.02 billion to $0.42 billion, a decrease of almost two-thirds.[3] The decline was also swift. Between 1981 and 1982, the renewable energy R&D budget of the Department of Energy decreased from $1.66 billion to $0.94 billion. Although this was a time of decreasing energy R&D overall, no other type of energy research funding was decreased as rapidly and aggressively as renewables. These cuts depressed private research; the number of new renewable energy patents issued every year stagnated and declined after 1981 (OECD 2014a). New patents went from a peak of 25 in 1979 to a mere 6 in 1982; it was only in the 1990s that private research started to pick up again.

Reagan's administration also allowed the Carter tax credits for both solar and wind to expire. In 1985, the tax credit for solar technologies was removed. Together with the decrease in oil prices, the removal of the tax credit resulted in the collapse of the emerging residential solar industry in the United States (Laird and Stefes 2009). The key technology of the time was solar water heating, and Rich and Roessner (1990, 197) report that sales decreased by 70% and more than half of the firms in the business exited the sector. In 1986, the Tax Reform Act "removed special investment and federal tax incentives, and installations slowed, leaving the PURPA system of guaranteed pricing for qualifying renewable energy facilities as the only incentive" (Harborne and Hendry 2009, 3583). As Gipe (1991, 758) shows, California, where the growth of renewables had been fast, saw a decrease of annual installations of wind-generation capacity from a peak of 398 MW in 1985 to 275 MW in 1986 and to 154 MW in 1987. By 1992, wind energy installations had come to a virtual standstill, with 19 MW only.

The Reagan administration's frontal attack against renewables resulted in the stagnation of the renewable energy industry for a long period of time.

Table 5.2
Renewable energy roll-call votes in the US Congress. Data adapted from Table 3 in Lowry (2008). The "yes" or "no" in parenthesis summarizes the proenvironmental position of the LCV. The mean gap is the sum of absolute values divided by the number of votes.

Year	Issue	Vote Gap	
1973	Creation of Energy Research and Development Administration (yes)	−8	**Pre-Reagan Mean Gap: 14.0**
1974	Increase spending for solar demonstration project (yes)	5	
1976	Amendment to remove funding for soslar (no)	31	
1979	Motion to increase funding for wind energy (yes)	9	
1980	Amendment to increase funding for solar energy (yes)	17	
1981	Amendment to increase funding for solar R&D (yes)	67	**Post-Reagan Mean Gap: 46.4**
1983	Disapprove Reagan deletion of money for solar (yes)	55	
1984	Amendment to move money from nuclear to solar (yes)	37	
1992	Motion to kill effort to require alternative fuels (no)	40	
1994	Motion to table shift of money from nuclear to renewables (no)	51	
1996	Amendment to restore money for renewables (yes)	40	
2002	Amendment to require 20% renewables by 2020 (yes)	37	
2007	Renewable Fuels, Consumer Protection Act of 2007 (yes)	44	

During President Clinton's administration, 1992–2000, the politicization of renewables continued, and the federal government did little to promote renewable energy. The federal R&D budget for renewables had reached its low point during President Bush's tenure in 1990, with an investment of $381 million, and only reached $763 million in 1999, again relying on data collected by Nemet and Kammen (2007).[4] Perhaps the only major incentive for renewables created by the federal government during Clinton's tenure was the 1992 production tax credit of 1.5 cents per kWh of wind electricity, and it expired for the first time in 1999, with major negative consequences for wind energy development (Mendonça, Lacey, and Hvelplund 2009).

Between 1992 and 2000, only three major legislative acts were voted on by the House. As Table 5.2 shows, the partisan cleavage between Republicans and Democrats was as salient as ever. At the same time, notes Karapin (2016), legislative interest in climate and energy issues had reached a nadir. Between 1992 and 2000, President Clinton made virtually no references to energy issues in his State of the Union speeches. Although the 1997 Kyoto negotiations made climate change a hot topic in international politics, it was mentioned in fewer than one in ten State of the Union speeches by the president—a striking contrast to President Obama only a decade later.

State politics also turned against renewable energy. As Karapin (2016) writes, California lost its wind energy leadership soon after the Republican Governor George Deukmejian began his tenure. In the 1980s, tax credits expired and wind power capacity stagnated. In the 1990s, California's total wind energy capacity actually stagnated (Sawin 2001). Although California did create some modest renewable energy policies again in the legislative package that produced the fateful 1998 electricity restructuring that soon led to the 2001 Enron debacle (Wiser, Pickle, and Goldman 1998; Kim, Yang, and Urpelainen 2016), these policies were not powerful enough to bring California back to the forefront of renewable energy development. It was only much later, in the 2000s, that California would make major investments in renewable energy as part of a broader climate leadership initiative—a development we shall discuss in chapter 6.[5]

Based on this discussion, we can now evaluate the empirical accuracy of our expectations. Politicization followed the initial success of renewables in the United States, as the increasing cost of the renewable energy investment became clear. Although the decrease in oil prices in the early 1980s was a key factor behind the declining interest in renewable energy, it is equally clear that the growing partisan divide on renewables and President Reagan's firm antirenewables position contributed to the rapid fall of the renewable energy industry in the United States. The initial strength and cost of Carter's policy response were key arguments for the backlash against renewables. The politicization of renewable energy was thorough and swift, and evidence indicates that this was partly because of Carter's initial ability to implement ambitious policies as a response to the oil crises of the previous decade. Of all energy forms, renewables suffered more than others, such as advanced fossil fuels or nuclear energy.

Identifying the effect of public opinion on renewable energy policy in a case is difficult, yet it is clear that public opinion imposed several constraints on policy formulation. Although US public opinion on renewable energy was positive (Farhar 1994a), similar to virtually every industrialized country, and support for nuclear power remained at low levels (Bolsen and Cook 2008), several dimensions of the energy issue shed light on why US public opinion has not encouraged more action on renewables. Despite environmental concern and high levels of support for renewables in abstract, the US public also had a lot of faith in fossil fuels. Between 1977 and 1989, the percentage of people who thought solar could replace foreign oil within five years ranged between 47% and 65%, and the corresponding numbers for wind were between 17% and 43%. However, the numbers for coal were 34% and 71%, and even the support for nuclear power actually ranged from a low of 25% to a high of 50% (Farhar 1994b). These numbers suggest that renewable energy is not particularly distinct from other fuels. If anything, the low levels of enthusiasm for wind energy suggest that the most commercially viable of modern renewables is considered a noncredible alternative to fossil fuels among the public.

More important, US public opinion about climate change has always remained at low levels and been sharply polarized between Republicans and Democrats. According to a 1993 cross-national survey—one of the first on this new issue—only 47% of Americans considered global warming a "very serious threat," whereas the corresponding number in Denmark was 55% and in Germany 73%. Canada (58%) and Great Britain (62%) also reported much more concern, but French respondents were not interviewed (Brechin 2003). Between 1989 and 2003, the number of Americans who worried about global warming "a great deal" decreased from 35% to 28% despite a steady accumulation of new scientific evidence (Brechin 2003). At the same time, the conservative movement took a sharply anticlimate position, denying the validity of the science and the relevance of the threat, thus turning right-wing voters against climate scientists (McCright and Dunlap 2003). Between 2001 and 2010, only 42% of Republicans without and 43% with a college degree believed that the globe was warming (McCright and Dunlap 2011).

The United States also did not have a strong technology industry that would have supported the growth of renewable energy. As we argued in our description of the initial renewable policies in the United States, they were heavily biased in favor of existing energy interests (Reece 1979; Chubb 1983; McFarland 1984; Laird 2003). As Mendonça, Lacey, and Hvelplund (2009, 382) argue, the corporate emphasis on the development of wind energy failed to create and expand a constituency for wind power: "While support for wind in national polls is high, that support does not always translate into easy development on the ground." Unlike in Denmark, where a strong lobby benefiting from wind energy generation and technology innovation emerged, the United States largely remained a consumer of renewable energy technology. The solar industry was so weak that it failed to prevent the expiration of the critical tax credits in 1985 and then collapsed almost immediately on their expiration. The wind technology industry also failed to survive in competition with their Danish competitors, and Gipe (1991, 759) reports that "[a]bout one-half of the California's wind capacity was built in Denmark and imported to the USA by Californian companies." By 1985, the Danish technocommercial dominance was so overwhelming that, in California, "the 52 percent of turbines that operated reliably included 98 percent of all Danish turbines but only 38 percent of the remaining (mainly American) turbines" (Heymann 1998, 646). Another measure of the weak US clean technology industry is patents for renewable energy technology. Whereas the United States was clearly ahead of the competition in 1980, newly issued patents stagnated throughout the 1980s. Research accelerated somewhat in the early 1990s, rising from 29 patents issued in 1990 to more than 90 in 2000 (OECD 2014a). However, the United States was barely on the same path as Germany. The difference between Germany and Denmark, on the one hand, and the United States, on the other hand, becomes clear when renewable energy patents are measured as a fraction of all patents. In Denmark, in any given year, more than 1% of all patents were related to renewable energy. In Germany, this ratio hovered around 0.5. In the United States, it went from 1.5% in 1980 to 0.3 in 1990 and 0.2 in 2000. Renewable energy research lost steam throughout the 1980s and 1990s. The clean energy industry was clearly not playing a similar dominant role in the United States than in Denmark and, to a lesser extent, in Germany.

With regard to the strength and role of environmental groups, we found little evidence of their contributions to the formulation of renewable energy policy. Already in the 1970s, the environmental groups had shown at best limited interest in renewable energy R&D (Chubb 1983). According to McFarland (1984), the role of environmentalists in energy formulation in the 1970s and early 1980s was largely indirect and operated through changes in public opinion over renewable energy. More generally, the environmental groups have not, at least since Carter's administration and early enthusiasm for environmental regulation, faced a favorable political environment for their concerns in the United States. As Vasi (2011, 109) describes: "the social and political context has been less favorable in the United States than in many European countries. . . . The United States has a very powerful fossil fuel industry that has many political allies." Although both Sine and Lee (2009) and Vasi (2011) find that there is considerable variation across US states in renewable energy deployment, this variation pales in comparison to the stark contrast between Germany and Denmark, on the one hand, and the United States, on the other hand.

Although some environmental organizations, such as the Sierra Club, adopted renewable energy as an important issue on their policy agenda (Sine and Lee 2009), they were never able to successfully promote the kind of policy that their counterparts in Germany and Denmark, as documented below, were forging in the political process. In 1982, for example, the *New York Times* reported that Reagan's Interior Department responded to major environmental organizations' complaints by saying, "The Secretary of the Interior will continue to do the job to which he was appointed. We would welcome constructive dialogue with organizations having legitimate interests in the development and protection of resources, but we will not be influenced by a small number of special interest groups and their commercial leadership."[6]

The role of the heavy and fossil-fuel industry in the politicization and opposition to renewable energy was and continues to be central. With the exception of the nuclear power industry, which had been wounded by growing popular opposition and the Three Mile Island incident of 1979, conventional energy interests have been strong, thriving, and adamantly against major investment in renewable energy. In the 1980s, when President Reagan was determined to reduce federal investment in renewable energy and even California gave up on renewables, it is difficult to establish the direct impact of industrial and fossil-fuel interests on renewable energy policy, given that the government never showed any sign of interest in formulating new policies. Although absence of evidence is not evidence of absence, we can use two different data sources to demonstrate that the heavy and fossil-fuel industries remained inactive during the Reagan era.

First, we conducted a LexisNexis search of the *New York Times* with the keyword "renewable energy" and found little reporting on explicit lobbying against renewables by corporations. In 1980, the *Times* reported that Exxon Corporation dismissed renewables as a marginal source of energy at least until the year 2000,[7] and in 1983 the Senior Corporate Vice President of a major engineering company, Burns and Roe, wrote to the editor about the need to focus on nuclear power instead of renewables.[8] In another 1983 article, it was reported that much of the solar R&D had gone to major oil companies, which considered it a competing source of energy and did not use the funding productively.[9] Other than these isolated incidents, we found little discussion of lobbying by the heavy and fossil-fuel industries.

Second, we looked at campaign contributions by industry groups based on data provided by the OpenSecrets.org website. Although the data are only available from 1990, the data from the early 1990s give us a clear sense of campaign contributions by the heavy and fossil-fuel industry at the end of the Reagan period—that is, before climate policy became a central concern and the political economy of renewable energy in US politics changed. Political contributions by the fossil-fuel industry remained extremely small. In 1990, coal interests spent less than $1 million, 60% of which went to Republicans.[10] Even oil and gas, a much larger sector, only provided $12 million on that year, with a similar split between the two parties.[11]

In contrast, in the 1990s, the importance of the industrial opponents of renewable energy became clear. Although there is little evidence for aggressive lobbying against specific policies, such as tax credits, the effort expended by automobile, oil, coal, and gas companies, as well as their allies, to stop energy policies that would constrain carbon dioxide emissions and reduce the competitiveness of fossil fuels was impressive. Measures to impose a price or national cap on carbon dioxide emissions would have enhanced the competitiveness of renewables, but such measures were vigorously opposed by the heavy and fossil-fuel industry. Agrawala and Andresen (1999) document the centrality of the 1989 Global Climate Coalition—a lobbying group formed by businesses and their trade associations—in undermining the Clinton administration's efforts to implement climate and clean energy policies. Prominent members included ExxonMobil, General Motors, Edison Electric Institute, the US Chamber of Commerce, and the National Association of Manufacturers. Levy and Kolk (2002) emphasize the strategic behavior of major oil companies, with a particular emphasis on Exxon's efforts to undermine climate policies. These business interests found a natural ally in the powerful conservative movement, which played a pivotal role in turning Americans against the Kyoto Protocol and climate policy more generally (McCright and Dunlap 2003).

Campaign contributions based on OpenSecrets.org data illustrate this changing logic.[12] By 2000, the oil and gas contributions went from $12 million to $35 million, and now 77% of that amount was targeted at Republicans. Even the smaller coal industry started spending, with contribution of more than $3.5 million (87% for Republicans).[13] Because lobbying disclosure—unlike the disclosure of direct political contributions—was only made mandatory in 1998, we cannot evaluate the relative investments of the relevant interest groups in lobbying before that year. The years 1998–2000, however, are illustrative in this regard. Again, the OpenSecrets.org website provides useful data because it keeps track of lobbying expenditures by sectors.[14] Using their data and converting it into 2010 dollars, we can compare the lobbying expenditures of oil and gas companies, mining companies, and environmental nongovernmental organizations (unfortunately, such sectoral data are not available for heavy industry or clean technology industry; data for electric utilities do not distinguish between fuels). During these three years, oil and gas companies reported lobbying expenditures of $133 million, whereas mining companies spent another $20 million. At the same time, environmental lobby groups only spent $14 million. For every dollar spent by environmentalists, the fossil-fuel industry spent $10.90. This large difference testifies to the lobbying clout of the opponents of renewable energy in the United States.

Not only were large businesses hostile to renewables, but they were also politically powerful. In 1980, the industrial sector as a whole still employed about 31% of the entire labor force and contributed about 34% of the country's GDP (WDI 2013). Throughout the 1980s, this share was between 2% and 8% larger in the United States than in Denmark. What is more, heavy industry remained important to the economy. In 1980, about 15% of the US economy was heavy industry, following the classification in Fredriksson, Vollebergh, and Dijkgraaf (2004). Although this share has decreased over time, it remained at more than 10% until 2000. At the same time, the share of heavy industry in Denmark was only 11% in 1980 and has remained largely unchanged ever since.

Similarly, fossil-fuel production remained—and remains—significant in the United States through the period of investigation. In 1980, the United States was producing about 10,000 barrels of crude oil per day, a number it would not reach again until 2010 after the shale oil boom (EIA 2013b), making the country the world's third largest producer of oil after the Soviet Union and Saudi Arabia. Germany and Denmark, by contrast, produced barely more than 100 barrels per day. More important for the electricity sector, coal has remained the most important source of power since 1980. In that year, the United States relied on coal for more than 51% of its electricity, and in 2000 this share was almost 53% (WDI 2013). At the same time, the United States is a major coal producer. In 1981, the United States produced 747 million tons of coal, and by 2000 production had increased to 974 million short tons. In both years, the United States alone produced about one-fifth of the world's coal (BP 2013). To summarize, the fossil-fuel industry was an economic powerhouse in the United States throughout the politicization period.

The role of the oil industry deserves a special mention here. Although the United States' oil giants—Exxon, Chevron, and others—did not face direct competition from renewable electricity generation at a time when electric vehicles were yet to appear on the market, research shows that the oil industry did play an important role in opposing climate policy, such as in the Global Climate Coalition, which played an important role in stalling federal climate policy in the United States (e.g., Agrawala and Andresen 1999; Levy and Kolk 2002). Given that a robust federal climate policy, such as a carbon trading system, would have significantly enhanced the competitiveness of wind and solar power in electricity generation, we can thus say that the oil industry's efforts, although not focused on renewable electricity generation, played a powerful indirect effect in turning the United States into a laggard.

The role of the partisan shift from left to right in the politicization and opposition to renewable energy, as captured in our second expectation, was even clearer. Reagan's victory in the November 1980 presidential election brought into power the most conservative Republican president in decades. At the time, the Senate elections resulted in a landslide victory for Republicans, as their seats went from 41 to 53. Although Republicans did not yet gain a majority in the House, they increased their seat numbers from 158 to 192. This partisan shift to the right allowed Reagan to launch an attack on renewable energy that set the United States to a path of inactivity for about two decades. Not even President Clinton's inauguration in 1992 turned the tide, as high levels of politicization continued. In 1994, Newt Gingrich led Republicans to a decisive majority in the House, with 230 Republican congressmembers and a divided government that could not implement changes in energy policy. As McCright, Xiao, and Dunlap (2014) note, the end of the Cold War was followed by a sharp polarization of Republican-Democrat opinion about environmental and energy issues.

In evaluating the role of Green parties in the United States, we are forced to speculate. Due to the majoritarian electoral system that favors large parties, the United States does not have a Green party with any realistic prospects of winning elections. In this sense, the expectation is affirmed, as the absence of the Green party coincides with high levels of opposition to renewables. Given the stiff political competition between Democrats and Republicans, a hypothetical scenario with a strong Green party in the House or Senate could have forced the Democrats to put more priority on renewables and defend the renewable energy programs against the Republican attacks more vigorously. However, this scenario also could have failed to materialize. Because the Republican-Democrat divide on renewables was already extreme, as data provided by Lowry (2008) show, adding a vigorous and strongly ideological Green party might not have done much to break the partisan gridlock that suffocated renewable energy for the two decades under investigation in the United States. We cannot use the US case to draw clear conclusion about the significance of Green parties, but under the assumption that Green parties are at least marginally more favorable to renewables than other parties, we can say that the majoritarian political structure of the United States is at least somewhat unfavorable to renewable energy.

To summarize, the United States is an exemplary case of the kind of politicization and elevated levels of opposition that our theory predicts. President Carter's initial policies gave renewable energy a strong start, but the combination of the withering oil shock and President Reagan's antirenewable policy agenda soon cut the wings of the nascent industry. The sociopolitical context of the United States was not favorable to the growth of renewables, and the combination of a strong fossil-fuel industry and a lack of powerful proponents of renewables constrained the growth of renewable energy for decades. US leadership in renewable energy was lost almost as quickly as it was initially gained.

5.4 The German Case

At first sight, Germany was an unlikely candidate for a stalwart of the renewable revolution. Germany is rich in coal and, by the end of the 1970s, had a vibrant nuclear energy industry. Its manufacturing industry was hungry for cheap electricity; if diversification implied higher prices, then this industry was unlikely to play along. Even if these obstacles could be overcome, Germany's political system was highly resistant to change. From the perspective of Germany's Chancellor, veto players include the junior partners of his or her ruling coalition, the Länder authorities, the federal courts, bureaucracies and rival ministries, large utilities (on the electricity supply side), industry lobbies (on the demand side), and European institutions (Tsebelis 2002; Stefes 2010). Potential opponents to renewable energy were numerous and powerful. Therefore, if we could go back to 1980 and ask an observer whether she believed that the initial effort for renewables would be sustained, the answer would likely have been a resounding "no." Instead, one may have expected serious pushback against renewables. In other words, Germany was ripe for strong politicization and high levels of opposition to renewable energy. To be sure, some of this pushback materialized in the first half of the 1980s and on various occasions in the 1990s. However, the renewable energy industry survived this critical phase. The reason, as we argue below, can be attributed to a mixture of shifts in public opinion, relentless political coalition building, and a bit of luck.

By the time Helmut Kohl (CDU, center-right) replaced Helmut Schmidt (SPD, center-left) as Germany's chancellor in 1982, memories of the oil crises had already started to fade, at least among Christian Democrats. One of Kohl's first decisions in office was to slash research expenditures in renewables by half (Stefes 2010). Germany's public R&D in renewables was cut from $224 million in 1982 to $93 million one year later. Most of the cuts were made in solar research (from $170 million to $70 million), although investments into wind research suffered more in relative terms and were cut by two thirds (from $37 million to $13 million). Investments into nuclear energy plummeted, too, but this drop reflected the maturity of this technology. Indeed, Kohl oversaw the construction of all the currently active nuclear power plants.[15] Out of the seventeen nuclear plants ever built in Germany, ten became operational between 1982 and 1989.

Skepticism among the CDU leadership reinvigorated other opponents to renewables. Renewables faced stiff opposition from the Federation of German Industries (Bundesverband der Deutschen Industrie [BDI]), the umbrella organization for industries, the energy utilities, and traditional electricity producers. Incumbent energy sources (mostly nuclear and coal) were solidly anchored. Nuclear power, although not quite as dominant as in France, remained a major source of electricity. In fact, by the mid-1980s, its share was above the OECD average, with about 30% of total electricity produced from nuclear power in Germany compared with 20% in the OECD (Jahn 1992; WDI 2013). Both fossil fuels and nuclear power appeared strongly entrenched. Moreover, investments in the infrastructures needed for both types of energy sources had been significant in the 1970s and enhanced the lock-in problem. The Ministry had long believed that reliable energy and electricity infrastructures were essential for Germany's recovery in the aftermath of the Second World War, and this point of view remained long unchallenged. This development was reinforced by a revolving-door effect between the energy sector and the Ministry of the Economy (Stefes 2010). Unsurprisingly, then, the Ministry of the Economy was a vocal supporter and a reliable ally of traditional energy sources.

In the early 1980s, the renewable energy outlook appeared bleak. The pushback against renewables translated into reduced support for research and regular attacks against prorenewable policies, including lawsuits. Even when antirenewable groups lost in courts, they managed to weaken the renewable industry by introducing uncertainty about its future. In the words of Stefes (2010, 154), renewables "might have been condemned to a niche source of energy for years to come." In fact, if one had asked an observer in 1985 what Germany's energy future would look like, it is probable that her answer would have included the words "nuclear" and "coal." Ruled by a rightist government backed by powerful interests, Germany looked more like Reagan's United States than Jørgensen's Denmark.

Yet Germany's renewable energy industry survived these difficult conditions for two reasons. First, nuclear and coal options lost most of their appeal during the 1980s. To begin with, public support for nuclear power collapsed during Kohl's tenure. Nuclear power was already a controversial issue in the 1970s. The project to build a nuclear plant in Wyhl led to considerable social unrest in 1974 (Wüstenhagen and Bilharz 2006). A few years later, more than 100,000 people demonstrated against nuclear power in the aftermath of the Three Mile Island incident (Jahn 1992). However, it would understate the role of Chernobyl to claim that it only intensified a trend. By the early 1980s, the nuclear question had been "normalized," and other issues, such as the involvement of NATO in West Germany, were more pressing (Joppke 1990). Chernobyl radically changed this situation. The then-chair of the SPD, Hans-Jochen Vogel, declared in the Süddeutsche Zeitung that "nothing is as it was before" (Joppke 1993, 177). Political leaders were taken by surprise, and their uneven responses to public worries undercut their credibility. The Environment Ministry, created in June 1986, was a direct response to the crisis and marked the political urgency of a strong response to the crisis (Aklin and Urpelainen 2014). This forced nuclear plant owners into playing defense.

The public also increasingly opposed coal, the other alternative to oil. Early public opinion polls show that concern about climate change was prevalent in Germany. A survey fielded in 1992 found that 76% of the respondents expressed "very serious" concern about global warming (Brechin 2003). In response, Kohl publicly declared that climate change was the most important environmental problem (Lauber and Mez 2004). Other coal-related issues such as acid rain also received a lot of attention. Experts fueled these fears by making dire predictions about the mass destruction of forests in industrialized countries.[16] In 1985, consumers sued to end coal subsidies; their claim was eventually upheld by the Constitutional Court (Stefes 2010).

The public's opposition to coal and nuclear energy translated into increasing enthusiasm for renewable energy. In a series of public opinion polls, the share of Germans who thought that solar energy would "contribute significantly" to Germany's energy supply went from 42% in 1987 to 58% in 1991, around the time of the EEG. For wind energy, this number went from 17% in 1984 to 38% in 1999 despite the disappointing setbacks, such as the failure of the GROWIAN project (the attempt to build a large-scale wind mill park that collapsed in the late 1980s). Thus, renewables received acceptance as well as respect. These views were further reinforced by increasing fears of environmental destruction.

Politically, opposition to traditional energy sources and support for renewables as an alternative were organized by the newly established Green Party. The Green Party was founded in 1980 and obtained its first seats in the federal parliament in 1983. The Greens became relentless opponents to nuclear power, breaking the pronuclear consensus in Berlin. The previously pronuclear SPD was forced into taking a more nuanced position. Institutionalization of proenvironmental movements also took place outside of parliament. If the 1970s were the years during which environmental groups were founded, the 1980s and 1990s were the era of renewable organizations, with Eurosolar in 1988, the German Renewable Energy Federation (*Bundesverband Erneuerbare Energien* [BEE]) in 1991, and the German Wind Energy Association (*Bundesverband Windenergie* [BWE]) in 1996. These organizations mobilized latent support for renewables and offered a counterweight to the politically powerful proponents of traditional energy sources.

Despite these advantages, the renewable energy industry was still unlikely to challenge the traditional energy sector. However, an unexpected event prevented them from flexing their political muscle: the collapse of the Soviet Union and the communist government of East Germany. This was particularly relevant for utilities, which were too busy with the takeover of the Democratic Republic of Germany's energy sector to pay much attention to renewables. The effort required to absorb a country of about 16 million people and provide energy for its highly industrialized economy was significant. In fact, utilities had little time to pay attention to the prorenewable laws under discussion in Berlin. In the words of a utilities manager, the "StrEG [the 1990 Electricity Law that favored renewables] was an accident. We just did not see it coming and underestimated its importance" (Stefes 2010, 155). Renewables and their defenders were economically insignificant but popular. Thus, the utilities saw little gain from crushing renewables.

The net effect of these two combined elements—the lack of popularity of coal and nuclear, on the one hand, and the unwillingness of utilities to fight for them, on the other hand—prevented the politicization of renewables in Germany. As a result, after 1986, the story of renewables is one of cross-party cooperation and successful implementation of prorenewable laws. Although opponents did not disappear and would occasionally attack renewables again, most notably at the end of Kohl's tenure, renewables have since celebrated significant successes and victories.

Two pieces of legislation stand out. First, the Electricity Feed-In Act (*Stromeinspeisungsgesetz* [StrEG]) proved a watershed (Stefes 2010; Bruns et al. 2011). Under the impulse of two parliamentarians, Matthias Engelsberger from the CSU (affiliated to the ruling CDU) and Wolfgang Daniels from the Green Party, in October 1990, the Bundestag enacted the StrEG. This law required utilities to buy electricity from renewable energy generators at 65% to 90% of the tariff paid by the final customer (Lauber and Mez 2004). Until then, public support took largely the form of public spending in R&D. Now, renewable energy could force its way on the grid. Engelsberger was seeking a way to support small Bavarian hydropower plants. Daniels' support reflected his ideological preferences. Together they built a coalition across the left and the right, notably drawing on new prorenewable networks around people such as Hermann Scheer (SPD), the founder of Eurosolar, and other sympathetic parliamentarians, such as Bernd Schmidbauer and Michael Müller (Bruns et al. 2011). As a result, the law passed with unanimous support in the Bundestag.

By the mid-1990s, the renewable energy industry had grown enough to become a significant political player (Michaelowa 2005; Aklin and Urpelainen 2013b). The wind industry, which pushed for robust public support, proved to be particularly successful. Yet the perspective of the 1998 elections, and the possibility of a coalition of the SPD and the Green Party, encouraged renewed opposition to the StrEG and renewables in general. Interest groups, such as VDEW (now BDEW), asked the Kohl government to drop the FIT. Political uncertainty depressed new investments, and the number of new wind turbines and solar PV systems declined for the first time since the 1980s (Jacobsson and Lauber 2006). Opponents feared the unknown political realism of the Green Party, which affected the confidence of renewable entrepreneurs. Regardless, in 1998, a historical victory by the SPD and the Green Party led to the first red-green coalition in Germany's history.

One of the first energy policies of the new ruling coalition was to introduce the Renewable Energy Source Act, commonly known as EEG, in 2000. The EEG did not depart widely from the StrEG; rather, the former reinforced the latter. One weakness of the StrEG was that renewable producers suffered from a lot of uncertainty about future prices. Their revenue depended on the average tariff paid by final customers and thus were not fixed. Instead, the EEG initially offered a fixed price, guaranteed over twenty years (Bruns et al. 2011).[17] Renewables were now an attractive target for investors. This shift away from traditional energy sources was reinforced by the willingness of the SPD-Green coalition to phase out nuclear plants.

After a brief period of uncertainty in the early 1980s, Germany's policymakers supported with varying degrees of enthusiasm the renewable revolution. Public support for research in renewables, after a brief decrease to $77 million in 1986, remained above $100 million throughout the 1990s. Solar energy was the most favored recipient of this support, with a peak of $125 million in 1991. Research boomed as demonstrated by the increase in the number of patents on renewable energy technology. According to OECD data, the number of patents for renewable energy generation went from less than 10 per year in the 1980s to about 26 in 1995, 89 in 2000, and continued to grow to 407 in 2010 (OECD 2014a). In comparison, the number of renewable energy-generation patents in France over the same period was about 3 (in 1995), 15 (in 2000), and 73 (in 2010), respectively.

Furthermore, the FITs under StrEG and EEG enabled the creation of a market for renewables that was absent in the 1980s. Despite strong opposition from the private sector, some policymakers, and bureaucrats at the Ministry of the Economy, the renewable energy industry survived and flourished. The wind industry benefited most from this benign turn of events, and its output boomed, as Figure 5.2 shows. Solar followed suit and, at current rates, seems bound to overtake wind power in the coming two decades.

We can now evaluate the validity of our four expectations concerning opposition to renewables. The first expectation is that opposition to renewables will be low when the public is concerned about nuclear power and climate change. As we showed, public opinion on energy issues radically changed after Chernobyl (Jahn 1992). We noted that nuclear power was already controversial in the 1970s, but Chernobyl undercut support for nuclear among mainstream voters. In the aftermath of the accident, public support for nuclear power collapsed. One series of polls suggested that opposition to nuclear power in Germany went from 46% to 83% after Chernobyl (Joppke 1993). Other studies broadly confirm these numbers; another opinion poll conducted in 1988 reported that about 70% of all respondents opposed nuclear energy, whereas only about 10% supported it (the rest having no opinion). Chernobyl delegitimized nuclear power at an unprecedented scale.

These concerns reinforced general worries about the environment. For instance, another popular topic of conversation was the so-called *Waldsterben*, the death of the forests caused by air pollution (Markham 2011). For this reason, coal was held in disdain by environmentalists. It is unsurprising that people increasingly ranked the environment among their main

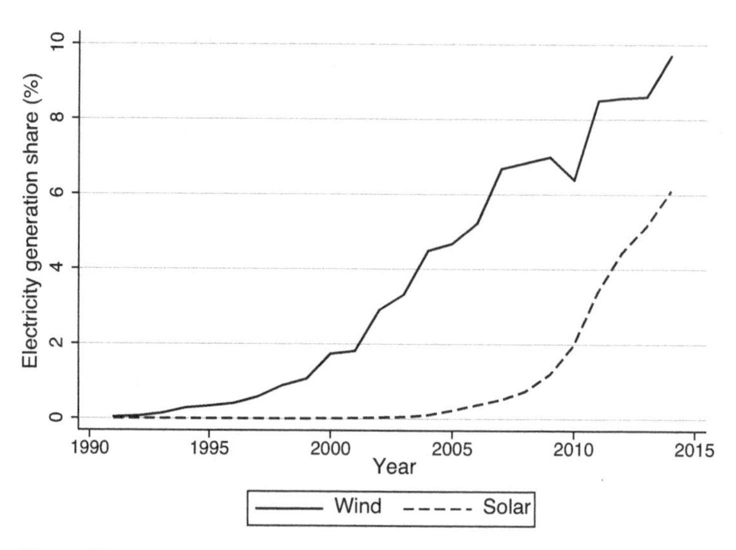

Figure 5.2
Share of electricity generated from solar and wind power in Germany.
Source: EIA (2017a).

concerns over the 1980s (Dalton 1994). The abrupt change in environmental and energy preferences that Chernobyl caused was instrumental in weakening the pronuclear lobby within the CDU and SPD. People's distaste for nuclear energy and, to a lesser extent, coal prevented policymakers from using these resources as the unique response to the oil crises. Importantly, not only did people oppose nuclear power, but they soon started to consider renewable energy sources to be realistic alternatives (Wüstenhagen and Bilharz 2006). According to public opinion surveys, more people believed that solar and wind energy would be long-term sources of energy than nuclear. This shift is remarkable given that often even opponents to nuclear power admit its clout.

Our second and third expectations revolve around the role of the clean technology industry, on the one hand, and the fossil-fuel and heavy industries, on the other hand. Naturally, heavy industry and the coal sector were opposed to renewable energy. Organized around the BDI, heavy industry opposed policies that would increase electricity and energy prices. The StrEG, and later the EEG, did precisely that. Similarly, coal interests—both employers and trade unions—were lobbying the SPD in the 1990s and 2000s against the EEG, although their effort only brought gains in the context of later reforms of the EEG (Bechberger and Reiche 2004).

Opposition by heavy industry could have hurt renewables. The industrial sector remained an important player in German politics, just like it did in the United States, as the share of industry in the German GDP remained above 30% until 2000 (WDI 2013). Industries also continued to employ a large share of the population, ranging from 30% to 40%. Ultimately, these interests secured important exemptions; for instance, they received preferential treatment and paid less on a kWh basis than regular households. In fact, the German FIT regulations strategically ensured that the cost of renewables would be minimal for the state and heavy industries; instead, the burden is mostly born by households.

Although the United States and Germany experienced similar pushback from heavy industry, the influence of fossil-fuel interests was different. Germany, unlike the United States, did not have any large oil companies that could have tilted the scale against renewables. For example, among the 100 largest private employers in Germany in 1973 and 1995, none was in the oil and petroleum business (Fiedler and Gospel 2010). Wintershall, Germany's largest private oil-producing firm, barely ranks among the world's thirty largest oil companies in terms of production and proven reserves (Aiken 2002). The United States stands in stark contrast. Exxon, Texaco, Gulf Oil, Amoco, and Conoco were all ranked in Forbes 500's top-30 US firms by revenue.[18] Because the few oil companies it had remained minor players compared with the large US conglomerates, oil industry opposition was not a major constraint on renewable energy growth in Germany. Whereas US oil companies played an important role in undermining federal climate legislation in the 1990s, their German counterparts had little influence on climate and energy policy in Berlin.

Still, although Germany produced little oil, it had a strong coal industry. In 1980, more than 200,000 short tons were extracted in Germany—about a fourth of the extraction in the United States. Nonetheless, despite the unification with East Germany, the clout of the coal industry gradually declined as production peaked around the end of the Cold War and declined until the year 2010 (EIA 2013b). Although coal producers had traditionally enjoyed considerable influence through their connections with the Ministry of the Economy, their influence declined over time.

Why did the power of heavy industry and the coal lobby not suffice to stop renewables? The reason appears to have been twofold. First, as noted earlier, the attention of Germany's most powerful businesses was turned toward East Germany (Stefes 2010). Heavy industry was undergoing important reforms, including those sectors that are most energy intensive

(Fredriksson, Vollebergh, and Dijkgraaf 2004). In 1991, 317,000 people worked in the mining sector; ten years later, that number had decreased to 130,000 while the sector's value added was cut in half (OECD 2014b). A similar decline occurred in the metal industry, with a decrease of total employment from 522,000 in 1991 to 302,000 in 2000. The rubber industry shed more than 100,000 jobs over the same period. In other words, the momentum was shifting toward other sectors, such as services. Although this development could have led to stronger demands against increasing energy prices, the ultimate consequences of the StrEG and EEG were clearly underestimated.

As important as these factors are, they are not enough to explain Germany's path. The industrial sector's clout and relatively slow decline mirror the American experience, so why did the outcome differ so much? Partly, this is due to the favorable public opinion mentioned previously. Equally important, the antirenewable lobbying was counterbalanced by the increasing strength of the clean energy industry. The SPD and Green Party both received large contributions from renewable energy companies.[19] For instance, Nordex gave about €130,000 to the SPD and Green Party in 2002. Between 2005 and 2007, Solarworld spent more than €100,000 to the SPD, FDP, and Green Party. Overall, between 2000 and 2010, renewable energy companies contributed €870,000 to political parties, whereas heavy industry contributed €1,900,000.[20] The contributions were politically polarized as the renewable energy companies favored the Greens: of total contributions, 48% were made by companies that only gave to the Green Party. At the same time, companies from heavy industry did not make any contributions to the Greens. Although industry trade associations made much larger contributions of approximately €15.3 million, companies from these associations also have a stake in the manufacturing of wind turbines and made no contributions to the Green Party.

These numbers suggest that the ability of the renewable energy companies to exert a political influence is largely through the Green Party. On the opposite side of the political spectrum, small hydro producers located in Bavaria turned out to be influential members of the CSU and CDU. Although Kohl had little to fear from the Green Party directly, he certainly needed the support of the Bavaria's CSU. Bavarian policymakers proved to be particularly savvy in the construction of new alliances. Building on the support among local farmers, they reached out to wind-based industries in the north of the country. Indeed, German wind turbine manufacturers were doing well in the North and increasingly expanding in the former Democratic Republic, drawing on cheap but qualified labor (Michaelowa 2005).

Backed by these two groups, Bavarian policymakers took advantage of this fortunate turn of events to pass key laws through the Bundestag. It is only later, and too late, that industrial interest groups and utilities tried to reverse these unfavorable policies. These efforts drew on local support. Germany had a long tradition (all the way back to the 19th century) of local cooperatives owning and running small-scale energy-generating installations (Yildiz 2014). Although the role of local communities declined in the 20th century, the design of the StrEG ensured the development of similar institutions for modern renewables. It is estimated that about 100,000 Germans either owned or co-owned a wind turbine in the early phase of the StrEG (Bolinger 2001). These so-called *Bürgerwindparks*, or citizens' wind parks, proved politically important. Toke, Breukers, and Wolsink (2008) argue that one of the most important obstacles against wind power was people's worries about the landscape. Community ownership mitigates opposition of this kind because they transform potential opponents into stakeholders. These efforts proved particularly fruitful in bringing together influential actors such as farmers and those, mostly Green supporters, who believed in the value of a lively local economy. Since then, despite the increasing industrialization of the renewable energy sector, private ownership remains an important component of Germany's renewable landscape. This is especially the case for solar power (Schreuer and Weismeier-Sammer 2010). According to some estimates, as of 2012, about 46% of renewable energy capacity belonged in varying degrees to noncorporate actors (Leuphana Universität 2013).

In the late 1990s, industrial interests tried again to politicize the issue, notably by taking legal action against the StrEG (Wüstenhagen and Bilharz 2006).[21] However, by that time, the balance of power had already started to shift. Other organizations, such as the German Engineering Association (VDMA), had decided to support renewable energy (Laird and Stefes 2009). Renewables also became a serious topic with growing mainstream support among important segments of the labor force (Hustedt 1998; Salje 1998). Altogether the "political and legal challenge to renewable energy incited massive campaigning by its supporters" and said supporters had a significant amount of political capital (Laird and Stefes 2009, 2623).

Our last expectation is that politicization is higher under right-wing governments than under their left-wing counterparts but is decreased in the presence of a strong Green party. Incidentally, Germany experienced a rightward shift in the early 1980s as Chancellor Helmut Schmidt was replaced by Helmut Kohl. In turn, this shift was followed by a leftward shift from Kohl to Gerhard Schröder in 1998. The consequences of both shifts are consistent with our expectations. As the initial shock from the two oil crises receded, renewable energy became the target of a pushback. Kohl's accession to power in 1982 planted the seeds for the politicization of renewable energy. He simultaneously supported the construction of nuclear plants and the reduction of support for renewables. Although his predecessor, Helmut Schmidt from the leftist SPD, was not particularly prorenewable, Kohl's CDU was more closely aligned with the industrial sector's interests. The rightward shift that occurred a few years after the second oil crisis could thus have been fatal to the nascent renewable energy industry. The fate of GROWIAN, a large wind turbine project located on the north coast of Germany, is symptomatic. Built at large cost in 1983, this wind mill turned out to be a spectacular failure and was shut down in 1987. GROWIAN's fate was partly due to poor technological planning but also to "half-hearted political support" (Michaelowa 2005, 194).

The growing opposition to renewables under Kohl experienced a spectacular reversal in the mid-1980s. There were two reasons for this. First, as noted earlier, Chernobyl considerably changed the rules of the game. Even traditional conservative voters became hesitant in their opposition to renewables. Second, both the SPD and CDU had to deal with the emergence of a new political force. The emerging Green Party undoubtedly shook the pronuclear consensus in Berlin. Their rapid growth redefined German politics. In 1980, the Greens won about 560,000 votes, or about 1.5%. Three years later, their vote share was 5.6%, enabling them to capture 27 seats in the Bundestag. In 1987, they won more than 3 million votes and obtained 42 seats. Since then, they have never received less than 5% of the vote. In 1998, they participated for the first time in the government under their charismatic leader, Joshka Fischer. The immediate result was the strengthening and renewal of prorenewable policies, such as the EEG, as well as more cabinet support for the environmental ministry. The Greens' influence was also apparent at the local level. In 1985, they joined the SPD to rule the Land of Hesse. Since then (and as of the time of this writing), they have been part of 31 additional Länder cabinets. Overall, the effects of the Greens were threefold. First, they forced the SPD to adopt a more proenvironment position. Second, the Greens were able to construct alliances on both the right and left at the Bundestag to obtain support for policies in favor of renewables. Third, after finally becoming part of the executive, the Greens directly designed policies and laws that would provide resources to the renewable energy industry.

Indeed, the second political shock—the victory of the SPD and the Greens in 1998—similarly shows that leftward shifts are conducive to higher support for renewable energy. About twenty years after it entered the parliament, the Green Party joined the government for the first time. The Greens' Jörgen Trittin replaced a then-junior minister, Angela Merkel, as the Environment Minister. Joshka Fischer, another Green leader, took a central role in Schröder's government. To some degree, the fears of the heavy industry and the traditional producers were thus warranted: the Green Party clearly adopted a prominent role in the policymaking process. The results, such as the adoption of the EEG, showed the government's commitment to renewables. The EEG's success was credited as the inspiration for the widespread use of FITs across other countries (Frondel et al. 2010). Of course, the government also proved to be pragmatic, especially when dealing with industrial interests. Nonetheless, the SPD-Green government clearly signaled its preferences to markets; as a result, investments picked up after a period of uncertainty before the 1998 elections.

The triumphs of the Greens reflect the influence of the civil society (Dryzek et al. 2003). Institutions such as Greenpeace, World Wildlife Fund, the German Nature Protection League (*Naturschutzbund Deutschland* [NABU]), or the German League for Environment and Nature Protection (*Bund für Umwelt und Naturschutz Deutschland* [BUND]) all played a key role in mobilizing different segments of the population (Markham 2011). Many of these organizations have a long history, some of which go back to the early 20th century. Furthermore, these organizations' membership exceeds the Green Party by a vast amount. Thus, the Green movement was not captured by its leftist elements. Instead, it could draw on latent support on the center and right too. Undoubtedly, this explains why Kohl never disparaged environmental concerns. In fact, by the end of the 1980s, Kohl was willing to acknowledge the importance of environmental and climate issues.

Overall, Germany illustrates how dialectic forces can shape renewable energy policy. A rightward shift under Kohl and the presence of an important heavy industry that supported nuclear and coal interests raised the level of opposition to renewables. However, Chernobyl solidified environmental concerns among the broader population and strengthened an emerging Green Party, which later became instrumental in the design of renewable energy policies. Furthermore, the clean energy industry had the political capital to create a niche for its development, thanks to its supporters in rural Bavaria and northern Länders. Local particularities, such as the end of the Cold War, further helped to reduce pushback from utilities. Nonetheless, industrial interests needed to be appeased through preferential treatment. Opposition in Germany was thus lower than in the United States despite similar industrial interests.

5.5 Limits to Politicization and Opposition in Denmark

Was renewable energy ever politicized in Denmark? Based on the literature that praises Denmark for exceptionally rational and effective energy policy, one might imagine there was never a lot of political debate about renewables in Denmark. This rosy picture turns out to be somewhat misleading. Although Denmark achieved success in renewable energy growth, the issue was not uncontroversial in national politics. Over time, renewable energy enthusiasts had to parry and thwart several attacks on renewables. Although it is true that Denmark survived without major political challenges to renewables for long periods of time, there were periods of intense politicization. It is not the absence of politicization but the decisive victory of the renewables advocacy coalition over the opponents that is remarkable about the Danish case.

In the Danish case, it is important to extend the time period of analysis somewhat beyond the year 2000. Although we initially expected key instances of politicization to occur already during the 20th century, the research revealed that, in Denmark, the most important episode of high levels of politicization came only later. For this reason, we extend the time period of analysis by several years. The key period of politicization was from late 2002 to early 2008 or so, and the effects of politicization on renewable energy growth were immediately observable.

In the 1980s, Denmark used several policies to support wind energy (Agnolucci 2007). Since 1979, Danish citizens had enjoyed a 30% investment subsidy for wind turbines. Wind energy was also prioritized in the national Energy Research Programme, and in 1997, a separate fund was established for offshore wind energy deployment. Since 1984, Danish utilities also voluntarily agreed to offer a feed-in premium to wind electricity, and in 1992, a formal FIT was passed after utilities and generators failed to resolve their disagreements about the appropriate premium. Even during the difficult years of 1987 to 1991, when technology exports to California dropped due to the expiration of the wind energy tax credits in the United States, the Danish policy managed to keep the wind turbine industry alive and vibrant (Hvelplund 2013). It was only in 1999 that the major political parties agreed to replace the FIT with a green certificate system as part of a more general effort to modernize and liberalize the Danish power sector. Until the end of the 1990s, the Danish government's support for wind energy was consistent, generous, and effective.

Denmark's achievements in promoting wind energy were unsurpassed by any other country in the world. Agnolucci (2007) estimates that the cost

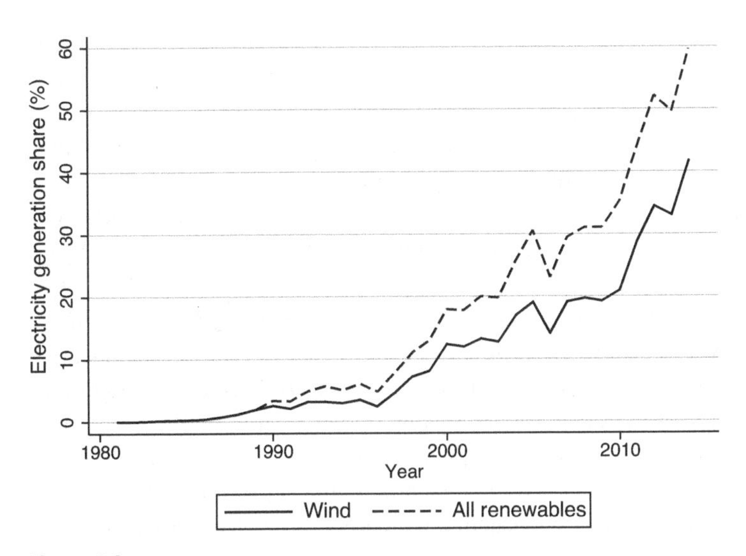

Figure 5.3
Share of wind and all renewables in the generation of electricity in Denmark.
Source: EIA (2017a).

of wind electricity generation decreased from about $0.17 per kWh in 1981 to $0.11 by 1987 and less than $0.07 in 1999.[22] At the same time, the cost of wind power in the United States remained higher. There, the cost of a kWh of wind was about $0.60 in 1980; in 1989, it was still about $0.20 per kWh (US Department of Energy 2013). It was only by the mid-1990s that the costs converged to less than $0.10. In Denmark, the effects of these improvements on the deployment of renewables in general, and wind in particular, is clear, as shown in Figure 5.3. The 1980s marked a period of progressive growth. In fact, wind remained a minor source of electricity until the mid-1990s. It then grew rapidly, passing the 10% mark in 2000.

Before Danish wind energy deployment really took off around 1995, a short period of uncertainty and controversy occurred that highlights the potential for politicization of wind energy in Germany. According to Agnolucci (2007), the reason that an FIT was enacted in late 1992 was the disagreement between utilities and generators on appropriate premia for wind energy, with utilities complaining about high expenses. At the time, wind energy deployment came to a standstill. The Danish utilities delayed the implementation of the law and "successfully created an atmosphere of insecurity among potential investors" (Lauber 2002, 301). In the coming years, however, the wind energy advocacy coalition achieved substantial victories. In 1993, the ministerial portfolios for energy and environment were merged. Svend Auken, a Social Democrat widely known for his enthusiasm for wind energy (Auken 2002), became the minister for energy and environment. In 1994, the new Planning Law required that municipalities set aside land for wind energy deployment (Toke 2002).

Once the FIT became fully operational and investors began to believe it was there to stay, the number of wind energy installations exploded. Agnolucci (2007) estimates the average price for renewable electricity to have been as high as 0.60 Danish kronors (i.e., $0.072), in excess of the 1999 generation cost, even ignoring inflation and further cost reductions.[23] Such a price was more than adequate to cover the cost of renewable electricity generation, enabling rapid growth in wind energy deployment. Thus, in 1980, Denmark generated no electricity from wind, whereas it generated about half a billion kWh in 1990, 1.2 billion kWh in 1995, and 4 billion kWh in 2000. Over the same period, the United States generated barely more electricity from wind than Denmark.

The next challenge to renewables in Denmark pertained to the 1999 agreement to replace the FIT with Green certificates. The change of policy was intended to make Danish renewable energy policy compatible with the liberalization of the power sector. However, the system provoked opposition among the Danish wind energy industry and was abandoned by the fall of 2001 (Toke 2002). In November 2001, the liberal party Venstre won the elections, and Prime Minister Anders Fogh Rasmussen began to dismantle policies that had allowed wind energy to grow so fast in Denmark. According to Ryland (2010), Rasmussen's tenure between 2002 and 2009 was a time of slow growth and uncertainty for the Danish wind energy sector. In 2002, the energy portfolio was given to Bendt Bendtsen, the leader of the Danish Conservative Party and a staunch opponent of wind energy. According to Backwell (2014, 16), in February 2002, Bendtsen told Reuters in an interview that, due to concerns about cost, he was "of the opinion that Denmark shouldn't continue to subsidise installation of new wind turbines after 2003." This anecdote illustrates the roots of the left-right dispute about renewable energy. It is not that Danish right-wing politicians disagree about the value of clean power sources; rather they are more worried about the costs to consumers and taxpayers.

In 2009, wind energy began to grow again. The Danish right-wing government, which was to host the Copenhagen conference in December 2009, brought back an environmental premium for wind energy and saw rapid increases in the deployment of wind power (IRENA/GWEC 2012). Although virtually no growth occurred in wind-generation capacity between 2003 and 2008, capacity increased from 3,163 to 3,871 MW between 2009 and 2011. In 2012, the prime minister and leader of the Social Democrats, Helle Thorning-Schmidt, secured a major victory for renewables when the Danish Parliament passed a bill committing Denmark to 50% electricity generation from wind by 2020.[24]

In light of our argument, it is surprising that the move back to wind energy began already during the right-wing government's rule. Ryland (2010) offers some insights into why, after seven years of opposition to wind power, the governments of Fogh Rasmussen and Løkke Rasmussen moved toward the left's position on wind power. First, in January 2008, the European Parliament had put forward a proposal for a European climate policy that would require Denmark to produce 30% of its energy from renewables by 2020. Second, the upcoming Copenhagen conference in December 2009 drew a lot of attention to Denmark's role as a frontrunner in renewable energy, and the minister for climate and energy, Connie Hedegaard, was an "enthusiastic environmentalist" who put a lot of emphasis on making Copenhagen a success for Denmark. Besides these two international factors, one should note that, by that time, the cost of renewable energy had rapidly decreased and the era of politicization was beginning to end in Denmark. Indeed, Ryland (2010, 85) finishes her analysis in 2010 by concluding, "recent developments appear to be more favourable for the long-term goals of the country's wind power industry, environmentalists and opposition parties"—a prediction that held true during the next five years.

In summary, Denmark's renewable energy industry did face political opposition, but the advocacy coalition was strong enough to survive the different challenges during the past four decades. Compared with the United States and Germany, renewable energy was late to be politicized in Denmark, and there was never opposition to the same extent as in these two larger countries. As we show below, however, there is a natural explanation for the relatively low level of opposition in Denmark. Because virtually every other factor we have considered was aligned against opposition to renewables in Denmark, it is ultimately understandable that the level of politicization remained low and was delayed by decades, compared with the German and US cases.

To begin with, in Denmark, public opinion has always been in favor of the development of renewables. Already in the 1970s, a key reason that Denmark chose not to follow France down the path of nuclear energy was the large size of the grassroots movement against atomic power (Vasi 2011). In surveys conducted in 1978 and 1982, only 37% and 25% of the

Danish respondents considered nuclear power a "worthwhile" pursuit, whereas 34% and 49% found the risks "unacceptable." Compared with West Germany, a hotbed of antinuclear activity, the public opinion had turned much more strongly against nuclear by 1982—among the Germans, 37% found nuclear power worthwhile in 1982 and only 30% found the risk unacceptable (Van der Pligt 1985). In 1985, a year before the Chernobyl accident cast a dark shadow of doubt on the nuclear industry worldwide, the Danish government had scrapped yet another plan for nuclear power (Backwell 2014). Chernobyl turned the public against the splitting of the atom even more strongly, and by the 1993 Eurobarometer poll, Denmark was among the four most antinuclear countries, with opponents being in the majority (Renn 1994).

Although public opinion did not express as much concern on climate change as in Germany, the 1993 cross-national Gallup poll showed that 55% of Danes considered global warming a "very serious" threat, as compared to 47% in the United States—but much less than Germany's 73% (Brechin 2003, 110). The significance of public opinion about climate change in Denmark was not important, however, because Denmark's key choice had been between nuclear and wind power ever since the 1973 crisis; as we see below, domestic deposits of fossil fuels were too limited to be a realistic prospect. Although our expectations focused primarily on climate change and nuclear power, Denmark's public has also shown high levels of enthusiasm for wind energy in particular. According to a 1993 poll, four out of five Danes said that wind energy should be a higher priority on the national policy agenda despite the government's generous subsidies, which had been in force for more than a decade (Krohn and Damborg 1999). In large part, we can interpret this strongly favorable opinion as evidence for successful and popular policies implemented in the past years.

As Vasi (2011) shows, the environmental movement also played a significant role in the evolution of Danish wind policy. In the previous chapter, we showed that already in the 1970s, environmentalists had turned the public opinion against nuclear power and forced the government to abandon plans for the rapid construction of nuclear power plants. In 1975, the Organization for Renewable Energy (OVE) was formed. It published a bimonthly newsletter, built a large wind turbine in 1978, lobbied the government, and raised public awareness about wind power. The organization had "a decisive impact on the growth of the Danish wind energy industry by using a practical approach. . . . Their determination became a source of inspiration for other people" (Vasi 2011, 71–72). The organization grew steadily over a period of time, spanning more than three decades, along with other environmental organizations. According to Jamison, Eyerman, and Cramer (1991), Denmark has had an exceptionally vibrant grassroots environmental movement with a large membership and a proven track record of policy influence.

The most important reason behind the low level of renewable energy politicization in Denmark is probably the strength of the clean technology lobby. For one, the Danish tradition of agricultural cooperatives offered a natural form of local ownership for windmills. According to Agnolucci (2007), already in 1979, when the government began to provide an investment subsidy, agricultural cooperatives were in the forefront of wind energy deployment, whereas traditional electric utilities were reluctant. As Lipp (2007, 5486) writes, "[t]he local ownership helped create widespread support for [renewables], especially wind, because benefits were distributed across a wide group of people." Vasi (2011, 167) estimates that, of all wind energy capacity in Denmark, "about 25 percent . . . has been developed by wind turbine guilds or cooperatives." According to Kim and Urpelainen (2013b), synergistic strategies among environmentalists, local communities, and political parties contributed to the success of renewable energy in Denmark. Because the ownership of wind energy was distributed across the population, especially in rural areas, the number of small-scale producers with a direct material interest in wind energy was large and contributed to the enduring political support of the wind energy industry.

The other leg of the Danish clean technology lobby was wind turbine manufacturing. In the 1980s, while the Danish market was still small, manufacturers from Denmark dominated the large California market due to their practical and robust designs (Gipe 1991; Heymann 1998; van Est 1999; Hvelplund 2013). By 2002, wind technology had become so important for Denmark that even the Confederation of Danish Industry publicly opposed the removal of the wind energy FIT (Agnolucci 2007), and in 2008, energy technology exports almost overtook agricultural products as Denmark's largest export sector, with wind energy being responsible for 70% of said energy technology exports (Ryland 2010). As Lewis and Wiser (2007) note, many of the early Danish policies, such as quality certification requirements and customs duties, were specifically designed to create and protect a local wind technology industry. Indeed, Heymann (1998) attributes the success of the Danish wind technology export industry to the graft—and grit—of Danish small-scale entrepreneurs as opposed to large-scale R&D programs. Echoing the logic in Aklin and Urpelainen (2013b), Danish policies increased the political clout of the Danish wind technology industry while the industry then supported and protected those policies. Indeed, wind turbine manufacturing also became a highly popular issue among Danes, as the continuous flow of visitors to the Tvind experimental wind turbine (Jamison 1978), which continues to this date, shows. Denmark's success in wind energy technology directly contributed to the positive image of the industry among the public and policymakers. Indeed, innovation activity also remained at high levels. In Denmark, in any given year, more than 1% of all international patents registered with the PCT were related to renewable energy, a percentage much higher than in Germany, let alone the United States.

Unfortunately, we do not have quantitative data on lobbying and other political efforts by the Danish clean technology industry. However, there are other clear indicators of the political muscle of the industry. Most important, according to N. Meyer (2007, 351), by 2001, "[a]bout 150,000 Danish households were registered as owners of shares in wind turbines in 2001." There were 2.45 million households in Denmark in 2001, meaning that 6% of all households had a direct stake in wind electricity generation.[25] The general importance of the wind industry for the Danish economy is also significant. Between 2006 and 2014, wind turbines were on average 5.03% of all Danish exports, whereas the annual average of the number of full-time employees serving the wind turbine industry was 29,421 (DAMVAD Danmark 2015). Indeed, Agnolucci (2007) notes that the Danish Confederation of Industry supported the wind turbine industry when the latter convinced the Danish government to abandon a Green certificate scheme, which would have significantly reduced the profitability of the wind industry, in June 2002. The timing here is particular significant, as Denmark was governed by an antiwind, right-wing government at the time. Earlier, the preparations for the Planning Law of 1994, which made Danish municipalities "ensure favourable planning environments for wind power," mobilized three large interest groups: the Danish Wind Turbine Owners Association, the Organisation for Renewable Energy, and the Danish wind turbine industry (Toke 2002). Taken together, these observations show that, especially at the last stage of the politicization period, the Danish clean technology industry had become both economically central and supported by the broader industry association in its efforts to promote renewable energy.

Another reason that the Danish government and society put so much emphasis on renewables was the virtual absence of fossil fuel interests. As documented earlier, at the onset of the 1973 oil crisis, Denmark had no fossil fuel production at all and was chronically dependent on foreign oil for virtually everything, including electricity generation. According to EIA (2013b), Denmark started to produce natural gas only in 1983 after new deposits were discovered in the North Sea. In 1995, when the share of wind electricity in generation began to grow fast, natural gas was only responsible for 10% of electricity consumption, whereas the share of oil—another North Sea product—in the power sector was only 4%. Coal was responsible for 74% of electricity generation, but it was imported, and Denmark did not have any large coal mining companies that would have opposed wind energy deployment. Although electric utilities have expressed many concerns about wind power over the past four decades (Hvelplund 2013), there was no powerful fossil-fuel industry to support them.

Despite the growth of the Danish oil and gas business, it never achieved the large size of the fossil-fuel (i.e., coal) industry in Germany, let alone the massive sector in the United States. The largest Danish oil and gas company is Maersk Oil, which by 2011 had produced 85% of all oil and gas in Denmark (Maersk Oil 2012).[26] Although Denmark produced more than twenty million cubic meters of North Sea oil and more than five million cubic meters of North Sea gas in the early 2000s (DEA 2013), the country only became a net exporter of oil and gas in the mid-1990s (IEA 2014b), well after wind energy began to prosper. What is more, foreign oil companies play a major role in the Danish oil/gas sector. The Danish oil and gas production is managed by the Dansk Undergrunds Consortium, half of which in 2012 was owned by Shell and Chevron.[27] Overall, the export-oriented Danish oil and gas producers have too little at stake in Denmark and are too small to influence international climate policies.

The Danish economy also did not rely on heavy industrial sectors that would have lobbied aggressively for low electricity prices. The industrial sector was smaller in Denmark than in Germany or the United States (WDI 2013). In 1980, the value added by Danish industries was about 27% of GDP, compared with 41% in Germany and 33% in the United States. In the 2000s, the US industrial sector would become smaller than Denmark's. By 2000, industry's value added was 27% of GDP in Denmark, 31% in Germany, and 23% in the United States. Similarly, the number of people employed was also relatively smaller: by 1990, relatively fewer people were employed by the industrial sector in Denmark than in the two other countries. Due to this economic structure, the kind of lobbying against renewables that was characteristic of Germany and, in the 1990s, the United States was of limited importance in Denmark. It is also notable that the Danish energy price difference between households and industry is today among the largest in industrialized countries. In 2010, Danish industries paid only one-fourth of the price that households paid (IEA 2011). This sharp divergence in pricing is yet another reason that heavy industry has not opposed renewables in a more vigorous fashion. Because electricity prices have mostly increased for households over time, the industry has not been under a major threat.

The expectation on left-right partisan politics is also supported by the events. Although the conservative Poul Schlüter headed four successive governments between September 1982 and January 1993, the real boom of wind energy was achieved during Social Democrat rule under the government of Poul Nyrup Rasmussen from January 1993 to November 2001. When the liberal Venstre government gained power in the November 2001 elections, the tables were turned, and the stagnation of wind energy began under Anders Fogh Rasmussen's coalition. Although wind energy began to grow again in 2009, Ryland (2010) ascribes this resurgence to mounting international pressures related to climate change and clean energy.

Similar to the United States, the role of Green parties was limited in Denmark. In the Danish political system, De Grønne, which was founded in 1983, never achieved electoral success in the Folketing elections. By 1990, they gained no seats at all, and in 1994, they formed a Red-Green alliance with the socialists, gaining only six seats. Since then, the Danish Greens have been a small party with a socialist agenda. In Danish politics, the impetus for environmental policies has largely come from larger, more traditional parties. In the case of wind energy, the leadership of the Social Democrat Svend Auken was decisive.

To summarize, the Danish case provides evidence for two important arguments. First, even the most favorable conditions cannot fully avoid dealing with the issue of politicization. Denmark is widely regarded as the pioneer of wind energy, yet the issue was politicized in a major fashion almost three decades after the 1973 oil crisis first created demand for renewable energy policies. Second, a combination of favorable interests and institutions can allow renewable energy to prosper. Denmark's agricultural cooperatives, strong movements and public opinion against nuclear energy, dependence on foreign fossil fuels, and remarkable successes in wind technology development, along with progressive leadership by the Social Democrats, underpinned the remarkably successful growth of wind energy. Even compared with Germany, another renewable energy pioneer, Denmark's success has been exceptional.

5.6 The Dogs That Did Not Bark

In France, the United Kingdom, and Finland, the period between 1980 and 2000 was characterized by quietness. In these countries, the equilibrium was not perturbed. Renewables never became a serious political option, except among a few marginalized groups led by idiosyncratic figures. France is a case in point. Ruled by François Mitterrand (Socialist, center-left) from 1981 to 1995 and by Jacques Chirac (RPR, center-right) afterward, France did not show any signs of interest in supporting renewable energy. In 1985, France was spending a paltry $45 million on renewables R&D, compared with Germany's $121 million (IEA 2014c). In the same year, French investments in nuclear research amounted to $1.16 billion, or 25 times more than renewables research. Despite how little was invested in renewables research in France compared with other countries (or to nuclear power), the early 1980s were actually a golden time for renewables R&D: the amount would actually regularly decrease from its 1985 peak and reach $6 million in 1998. Unsurprisingly, then, the contribution of renewables remained negligible. By 2000, the share of renewables (excluding hydropower) in total electricity generation amounted to about 0.8% for a total of 4.2 billion kWh. In contrast, Germany generated about 3.6% of its electricity from renewables, or 19.5 billion kWh, at the turn of the century.

One of the key reasons for the lackluster performances of renewables in France was obviously the technological and political triumph of nuclear power. Throughout the 1980s and 1990s, France continued its ambitious nuclear program. Supported through generous public funding for research—France spent $18 billion on nuclear R&D over the 1980–2000 period—the nuclear energy industry developed into one of the largest in the world. In 1980, nuclear power provided about a quarter of total electricity generated in France. By 2000, that number had reached 77%. France was now firmly established as the second largest producer of nuclear electricity, behind the United States. In fact, from 1990 onward, nuclear energy contributed almost continuously about 80% of total electricity generated in France, as Figure 5.4 shows.

The success of nuclear energy and the failure of renewables are, of course, related. From the onset, French elites had decided to favor the deployment of nuclear plants as a response to the threats to France's energy security. As we documented earlier, the elites were unanimously supportive of the nuclear option. Two institutions were particularly important. First, EDF, the main French utility, controlled most of the electricity market and was an enthusiastic proponent of nuclear energy. It was staffed with engineers who had the skills and incentives to build large-scale plants. In the 1980s alone, France built 43 nuclear reactors.[28] Eight more were added in the 1990s. Second, the government was key in nurturing EDF's development and in many ways encouraged it. Here, as in the 1970s, geopolitical preferences dictated Paris' choice (Jasper 1992). Nuclear power was seen as the

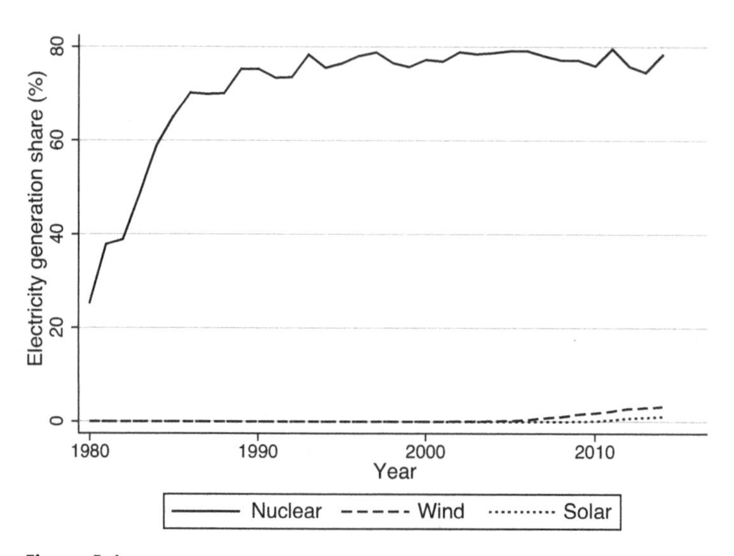

Figure 5.4
Share of nuclear, wind, and solar electricity generated in France.
Source: EIA (2017a).

key to avoid having to rely on oil from foreign countries and in enabling the buildup of a national nuclear weapon capability. Both EDF and the government were instrumental in the nuclear project. Using the myth of Prometheus as a metaphor, Jasper (1992) likens the French government to the gods and EDF to the Titan and notes that "the god and the Titan worked very closely together" (655). The mortals—the citizens—remained largely excluded from energy policy.

Indeed, neither the Socialists around Mitterrand nor the Gaullists around Chirac saw any reason to abandon nuclear power. Soon after getting into power in 1982, Mitterrand faced an economic crisis and was thus wary about adding to the industrial sector's woes. Providing resources to renewables would have been costly. Furthermore, it might have created unnecessary frictions between the different wings of the left. Finally, EDF technicians had little interest in spending resources on research on new technology that would not be as profitable.

Despite the timid emergence of an antinuclear (and presumably prorenewables) movement in the late 1970s, the pronuclear consensus prevented any departure from the atom. The Green movement remained largely impotent. There are two reasons for this. First, the French Green movement, unlike its German counterpart, failed to gather enough electoral power to force the Socialists to depart from their pronuclear position. Second, even when it was successful, like when the Green Party obtained more than 10% of the votes at the 1989 European parliamentary elections, the French policymaking process was largely based on the sole executive branch's preferences (Kitschelt 1986). The French decision making remained largely isolated from these new pressure groups and the will of the National Assembly. Again, we can draw a parallel to Germany: while the industrial sector enjoyed privileged access in both countries, German leaders were more easily influenced by new movements and the preferences of parliamentarians. In France, the government was EDF and EDF was the government. As a result, renewables failed to take off.

Similar conclusions apply to the United Kingdom. In 1980, renewables did not make any contribution to the generation of electricity, and the United Kingdom did not contribute much to research either. In 1979, when Margaret Thatcher (Conservative) became prime minister, the United Kingdom was spending $60 million on renewables R&D. In comparison, investments in nuclear research were about $800 million per year. Renewables R&D expenditure would continue to hover around $40 million per year until John Major's stint. Shortly after Major's reelection in 1992, economic conditions rapidly deteriorated, and unemployment reached double digits in 1993. Spending on research for renewables was slashed and reached $7 million in 1998. Tony Blair (New Labour, center-left), who returned to power the previous year, would only later reallocate funding to R&D. In the meantime, the results were abysmal. Twenty years after Thatcher's accession to power, renewables represented a meager 1.5% of total electricity generation or 1.5 billion kWh—even less than France in absolute values.

Although both the United Kingdom and France failed to develop their renewable energy industries during the 1980–2000 period, the context differ in some respects. First of all, the United Kingdom never deployed nuclear energy to the same extent as France. Admittedly, the share of nuclear power increased markedly from about 12% in 1980 to a peak of about 28% in the late 1990s. However, the crucial resources remained fossil fuels: first coal and later gas. Domestic coal covered most needs for electricity generation in the 1980s. In 1980, coal provided more than 80% of total electricity generated in the United Kingdom.[29] This number gradually decreased to 30% in 2000, although by then most of the coal was imported. Over the same period, natural gas became the predominant source of electricity, increasing its share from zero in 1980 to about 37% at the turn of the millennium. Furthermore, utilities were dealing with different issues. In France, giants such as EDF and GDF had a quasi-monopolistic position and had little interest in engaging with a sector like renewables that was not lucrative. In the United Kingdom, utilities were mostly kept busy by the impending liberalization and its effects on their business model. Again, this left them with few incentives to invest in a new model but for different reasons than France.

As we are writing this chapter, renewable energy has not been politicized in Finnish energy policy. The modern renewable sources of wind and solar for electricity play virtually no role at all. By 2010, Finland's wind energy capacity was only 200 MW (Stenberg and Holttinen 2011) and solar energy capacity negligible. Although environmentalists have every now and then called for policies to promote renewable electricity, various governments have done little to promote these sources of energy. Their role in national politics has been marginal.

Researchers agree that the reason that modern renewables have been marginalized in Finnish energy policy is the continued emphasis on nuclear power (Litmanen and Kojo 2011; Teräväinen, Lehtonen, and Martiskainen 2011). As Finnish electricity demand grew, the influential heavy industry continued to demand more nuclear power as a stable and reliable source of power. In 1993, a nuclear power application was rejected by the parliament largely due to concerns about the 1986 Chernobyl accident. In 2002, however, a new reactor in the Olkiluoto area, where two reactors were already operating, was approved by the parliament—the first one in Europe in a decade (Teräväinen, Lehtonen, and Martiskainen 2011). When Finland was formulating a national climate strategy in 2005, when the Kyoto Protocol entered into force after Russia's ratification and critical energy policy decisions were made, environmentalists and clean technology advocates lobbied for measures to promote renewable energy and conservation as an alternative to more nuclear power and fossil fuels. When the official strategy was circulated among environmental organizations in June 2005, Greenpeace Finland, the Finnish Association for Nature Conservation, and Friends of the Earth Finland wrote a public letter to the government.[30] The letter sharply criticized the government for failing to support renewable energy despite earlier promises not to rely on nuclear power only. However, this criticism and other campaigns by activists did not have much effect on Finland's energy policy. The 2005 climate strategy relied almost exclusively on nuclear power and the European Emissions Trading Scheme (KTM 2005). In 2010, the Parliament of Finland approved the sixth and seventh nuclear reactors.[31] If these nuclear power plants are ultimately built, there is little scope left for renewables to meet Finnish electricity demand.

Today wood biomass plays a major role in the forestry industry, with a share of 21% of primary energy consumption in 2008, up from 15% in

1990.[32] However, little growth took place in the use of biomass outside the forest industry. The use of wood biomass in Finland began well before any concerns about energy scarcity, and in 1960, the share of biomass in the Finnish energy mix was about 50% (P. Lund 2007). Decisions to increase the use of biomass have been consensual and supported by the politically influential forestry industry. Incremental improvements in the policies that support biomass and punish the use of fossil fuels, especially in cogeneration for centralized district heating, have allowed the use of biomass to maintain its share and even grow slightly during the past two decades in Finland (Ericsson et al. 2004; Helynen 2004). As Kivimaa and Mickwitz (2011, 1812–1821) put it, "[i]ncreased focus on fuel resources and the supply system will not change energy systems if the bottlenecks are related to use and distribution. Similarly, many significant possibilities may be forgone if biofuel use is seen as just traditional centralised heat and power though profitable options may exist in other areas, such as air fuels."

The Finnish case also highlights the importance of Green parties. Although Finland is a proportional system, the Finnish Greens have not done as well in elections as their German peers. Although the German Greens already won 5.6% of votes in 1984, the Finnish Green League broke the 5% threshold in 1992, with a vote share of 6.82%. Although the Finnish Greens already made it to government in 1995, they had to resign in May 2002 because the left-right coalition government overrode them and decided to accept the construction of a new nuclear reactor. Although the Greens were in the government, the coalition held a clear majority (129/200) of the seats even after losing the Greens. The inability of the Greens to stop nuclear power expansion and direct Finnish policy toward modern renewables shows that a proportional system does not guarantee a prorenewables policy unless Green parties can play an important role in policy formulation, perhaps because they are pivotal to maintaining the majority of the government coalition.

In summary, renewable energy never became a political priority in Finland. Biomass notwithstanding, modern sources of renewable energy never had a chance thanks to the strong pronuclear consensus in Finland. The development of biomass continued as a matter of course due to the abundance of the resource and the large size of the forestry industry, but newer renewable energy technologies were not seriously explored or considered.

6 Lock-In: An Era of Rapid Growth

Once modern renewables, led by wind and followed by solar, survived the political clashes described earlier, the path was clear for rapid growth. During the past decade, renewables have maintained a sustained path of development in frontrunner countries, especially Germany and Denmark. The bulk of this growth is found in Europe, where nonhydroelectric renewables have increased their share of electricity generation from less than 5% to more than 15%. Just as important, many laggard countries have made significant progress. Renewables are winning market share on all continents.

Throughout this period of growth, renewables have been successful in three areas. First, they have grown quickly in the original frontrunner countries, Denmark and Germany. Denmark almost doubled its number of wind turbines between 2000 and 2014. By 2015, about 60.7% of electricity generated in Denmark came from renewables (WDI 2017). Germany does not lag far behind; 31.6% of its electricity in 2015 came from renewables (AGEE 2016). Hydropower plays only a small role in this growth, as its share among renewables is less than one-sixth and decreasing. In these countries, renewables have reached a critical mass that makes them "mature" technologies (Victor 2011). Wind and solar power have survived the challenges from their opponents. Their contribution to the power sector is such that an energy paradigm without them is now unthinkable in Copenhagen and Berlin.

Second, renewables are now expanding in countries in which the political struggle was the toughest. The United States, for instance, is making a comeback. As of 2015, just under 30 states have a renewable portfolio standard, and a handful of others have instituted an FIT.[1] In the first eight months of 2014, nonhydro renewables already surpassed hydropower in electricity generation (EIA 2014a). The most rapid growth comes from solar and geothermal sources, although wind power remains the largest contributor of renewables. In 2015, renewables represented a share of 14% to 16%

of total electricity generation, the bulk of which came from hydropower (about 6%–7%) and wind (about 5%).[2] Total nonhydroelectric renewable generation reached more than 260,000 GWh, mostly from the West Coast and Texas, of which only 12% came from biomass.

Even countries such as France and the United Kingdom are making rapid progress. For instance, in France, nonhydro renewables generated 1,700 GWh of electricity in 1990, most of which came from biomass (about 0.5% of total production; WDI 2017). Ten years later, that number had increased to a bit less than 3,000 GWh (less than 1%). In 2010, renewables excluding hydro produced 25,000 GWh (more than 3%). The latest figures available indicate that nonhydro renewables have generated more than 28,000 GWh in 2014. Solar and wind contribute to about 16,000 GWh and biomass about 7,000 GWh (EIA 2017a).[3] As renewables have become increasingly competitive, French energy heavyweights such as EDF have also become increasingly attracted to this lucrative market.[4]

To be sure, opposition to renewables remains fierce in some quarters. In the United States, special interests around the fossil-fuel industry have continued their struggle against political support for renewables. Furthermore, the investment environment for renewables is not quite settled. Rapid changes in technology and production can hurt domestic producers. For instance, China's mass production of solar panels have sunk costs and undercut European producers. To some extent, this is endemic of any new technology: innovations and competition increase the volatility of profits. New competitors such as shale gas may replace renewables, especially in the long run, and renewables still experience pushback even in countries such as Germany, where the public has recently complained about the high cost of the FIT.

The third and most important advance is that renewables are expanding in countries that had, so far, shown little interest in these energy sources. For example, China has become a major player in the production of solar panels and as a consumer of solar power. In 2013, China led all countries worldwide by installing 12.9 GW of new solar capacity (REN21 2014). By that time, China was second behind Germany worldwide in total installed solar capacity with about 20 GW. Although this figure is relatively small compared with total capacity, the share of solar in new capacity is more than 10%.[5] Other countries are making rapid progress. India's Jawaharlal Nehru National Solar Mission seeks to install 20 GW of solar power. Although this goal is possibly overly ambitious, it does signal the strength of the government's support for new technologies (Harish and Raghavan 2011).[6] More generally, renewables in the developing world have caught the attention of investors. Ernst & Young, a consultancy firm, estimated in a report published in 2014 that five out of the leading twenty investment destinations are developing or emerging countries (Ernst & Young 2014). China ranked second and India ninth. Brazil, Chile, and South Africa were all ahead of industrialized countries such as Finland. This change is significant because it implies that renewable energy is not just a luxury good that only the wealthiest and greenest countries can afford.

These developments constitute what we call a global lock-in of renewable energy. The renewables lock-in, which we discussed in chapter 3, does not mean that competitors have been wiped out. Rather, a renewables lock-in describes a new equilibrium in which renewables are no longer under threat. In Denmark and Germany, with occasional bumps on the road, renewable energy has become a new status quo. Even more significantly, these successes have enabled laggards to resurrect—or, for some, kickstart—their own renewable energy industries. This is true for the United States, France, the United Kingdom, and emerging economies such as China and India. Even in the least developed countries, such as Kenya, renewable energy is increasingly important to the overall energy sector. Still, renewables face a number of challenges, and it is not possible to say that they are crowding out conventional fossil fuels. The goal of this chapter is to summarize the progress and status of renewables in the context of general energy development across the world.

6.1 A Global Boom in Renewable Energy Investment

The 1980s and 1990s marked a period of experimentation and limited deployment of renewable energy, whereas the 2000s was a decade of massive investments. Renewables evolved from a venture of subsidized startups into a major industry. In 2004, worldwide new investments in nonhydro renewables amounted to a little less than $40 billion (BNEF 2014b, Figure 3).[7] This number includes investments from sources such as the financial sector, venture capital, and governments. Ten years later, annual investments reached $227 billion, with a peak of $279 billion in 2011. According to the IEA (2014d), the share of wind and solar alone accounted for about 45% of total power plant investments in 2012. Neither the global financial crisis nor China's mass production of renewable energy devices has dented this enthusiasm, as global investments have remained above $200 billion every year after the financial crisis. Even accounting for the decrease of investments due to these two events, the annual growth rate of renewable energy investments over this decade has, on average, been greater than 20%.

Perhaps the most remarkable transformation of the renewable energy industry since 2010 relates to the sources of capital. Early renewable firms were generally funded by government subsidies and the limited cash flow coming from their operations. Private funding remained rare because profits were small and dependent on government goodwill. Since 2010, however, markets have grown increasingly enthusiastic about providing capital to renewables. "Green bonds," which include bonds for renewable energy investments, have grown from virtually nothing in 2009 to about $40 billion in 2014, which would represent about 3% of the US corporate bond market (Bloomberg 2014a). These bonds are not charity. Their returns are generally on par with or higher than those of other corporate bonds. Some of these investments are tied to ambitious projects. For instance, Berkshire Hathaway Energy invested $1 billion in the Solar Star, a 579 MW solar park in California.

Renewables are thus undergoing a further transition. In countries such as Germany, renewables have survived the "valley of death" (Wüstenhagen and Menichetti 2012), that is, the period between technological innovation and commercial competitiveness. Although figures vary, prices of solar modules have continuously declined. According to some estimates, the cost of photovoltaic modules went from about $80 per watt in 1976 to less than $1 per watt in 2013 (Bloomberg 2014b). Indeed, many renewable energy sources are becoming competitive. To show this, we can compare them using the levelized cost of electricity (LCOE), which measures the price at which an electricity source becomes profitable.[8] In the United States, the estimated ranges of LCOE for biomass and wind, for instance, overlap with the LCOE of coal, nuclear, and natural gas (Bloomberg 2014b), with solar PV closing in. Off-shore wind and tidal energy, in contrast, remain costly.

Thanks to their increased competitiveness, renewables have become a global industry. In 2004, investments on renewable energy in developing countries amounted to about 8 billion, or one-fifth of the worldwide investments (BNEF 2014b). Ten years later, developing countries received about $93 billion, a share now representing more than 40%. Forecasts confirm the trend. A study conducted by Bloomberg New Energy Finance predicts that cumulative investments in renewables between 2014 and 2030 will be higher in Asia than in Europe. At $3.6 trillion, Asia will capture more than half of worldwide investments ($5.1 trillion).[9]

At the same time, production has also shifted to emerging economies. China is the undisputed leader in production of solar modules. The two largest producers of these modules in 2013 were Chinese firms, Yingli and Trina Solar (Earth Policy Institute 2014). JA Solar and Jinko Solar, two other Chinese firms, were also among the leading ten firms. The situation is not much different in the manufacturing of wind turbines and related equipment.

Developing countries have also become consumers of renewable energy. India, for instance, has placed much hope in the deployment of both solar and wind power. Even China, whose electrification levels have long been high, has used small-scale wind turbines to provide electricity in remote areas such as Inner Mongolia (REN21 2014). Africa is following suit, with most countries now providing public support to renewable energy. South Africa has invested vast amounts into solar and wind energy (almost $5 billion in 2013) and ranked ninth among G20 countries in 2013 (Pew 2014). As of 2014, South Africa had 295 MW of installed renewable capacity but 2.4 GW under construction (total capacity was about 44 GW; WDI 2013). Even the poorest countries have shown an appetite for renewables. Kenya, for instance, is a leader on geothermal power and has a vibrant private solar off-grid market.

As a result of this rising demand, the renewable energy industry has become a large employer. According to estimates for 2013, about 2.6 million people in China are employed directly (e.g., engineers) or indirectly (e.g., installers) by the renewable energy sector (REN21 2014). In Brazil, more than 900,000 individuals work in this industry, mostly in the biofuel sector (IRENA 2015). Whereas Danish wind turbines flooded the Californian market in the 1980s, the largest wind turbine producers are now located all around the world, including the United States, Germany, India, and China (Navigant 2013). Internationalization leads to lower production costs and a better flow of technologies and intellectual property (Løvdal and Neumann 2011). Overall, we thus observe tremendous changes since 2000. Investments and production are now global and have reached unprecedented levels. Consumption is picking up in most countries, even outside of the usual suspects such as Denmark.

This is not to say that temporary or even long-lasting reversals are impossible. For instance, in Brazil, the biofuel industry came under intense pressure when the Bush administration increased tariffs on ethanol.[10] Indeed, governments remain key players: the correlation between public support and renewable energy generation, as well as new capacity, remains high, and investments are still sensitive to changes in the regulatory regime (Schmalensee 2012). The intense debates around the reforms to Germany's EEG testify to the importance of adequate regulations for the renewable energy industry to prosper. These debates are over large sums. Estimates vary, but some believe that Germany subsidizes renewables by about €20 to 24 billion per year.[11] Prorenewable US policymakers have attempted to

Figure 6.1
Feed-in tariffs (FITs). Shaded countries have implemented an FIT at the national level as of 2012.
Source: REN21 (2013).

push their country on the path, followed by Germany and other leading countries. In France, the government has encouraged its industrial champions to develop their interests in renewables.

This evolution is partly a result of an increasing number of favorable public policies. Whereas virtually no incentives existed in developing countries in the early 2000s, policies have become common on all continents (REN21 2014). In Figure 6.1, we map countries that have adopted a feed-in tariff (FIT) as of 2012. They are found in all continents; most of Southeast Asia has adopted this policy tool, as have many countries in Latin America. Africa remains a bit behind, but even there prorenewable policies have been expanding.

6.2 Summary of Comparative Case Studies

Our first expectation for the cases is a massive increase in the deployment of renewables, driven by decreasing costs and higher investments. We anticipate that frontrunners—mostly Denmark and Germany—continue to experience sustained growth. In these countries, renewables have survived the pushback from the politicization phase. Renewables become increasingly competitive while still benefiting from the general support of political elites. Opposition, if it arises, does not present a fundamental threat to renewables because decision makers have political and economic incentives to maintain high growth rates for renewables. The predictability of the policymaking process encourages investors to finance the addition of new capacity. Investments and innovation, in turn, reduce the cost of renewables, making them more competitive. The rest of the world benefits from the progress made by frontrunners. Denmark, Germany, and, to a lesser extent, the United States have operated since the 1970s as political entrepreneurs. Years of testing various policies and investing directly or indirectly in the various renewable energy industries have paid off. Newcomers, shielded from the costs of these trials, can now reap the renewable energy premium. We show how the diffusion of renewables occurred across both industrialized and developing countries. The major change during the past ten years is that developing countries have embraced renewables as a solution to their own issues.

Our second expectation is that as renewable energy industries grow, they acquire political clout. Entrenched energy interests in industrialized countries continue to oppose renewables, although with less success than in the past, and we expect less opposition in developing countries. This is our third expectation. The politics of renewable energy are fundamentally different in developing countries because of the pressing energy needs they face and the virtues of renewable energy. We should, therefore, see less opposition to renewable energy across the board, with particularly low levels in developing countries.

In earlier years, we expected to find systematic differences between frontrunners and laggards in renewable energy policy and deployment, but we now expect a significant degree of convergence in renewable energy output. Given this expectation, we cannot use France as a counterfactual to Germany anymore. Instead, we rely on within-country changes. Our main prediction is that countries should become more friendly to renewables over time. With regard to the politics of renewable energy, we do expect differences between industrialized and developing countries. As we conduct our case studies, we explicitly contrast the politics of renewable energy in the industrialized and developing world. Here, we can conduct comparative analysis similar to that in the previous chapters. We document our claims by relying on a wide range of data sources. Besides energy data, we have compiled information on investments and policies at both the national and subnational levels.

The three expectations and cases are summarized in Table 6.1. We first examine the cases of the pioneering countries, Denmark and Germany. We next explore the laggards, beginning with the comeback of the United States. We then turn our attention to France and the United Kingdom, noting the lack of meaningful renewable energy policies in their past. France

Table 6.1

Summary of expectations and cases for lock-in. Industrialized country cases are presented by country category due to space constraints. Because the developing country cases are of particular importance, they are presented individually.

Expectations	
Expectation 1	As the cost of renewable energy decreases, supportive policies spread among countries, promoting global convergence in both policy and outcomes.
Expectation 2	As the cost of renewable energy decreases and the number of people producing or using it increases, the political controversies surrounding it are mitigated.
Expectation 3	Opposition to renewable energy decreases more rapidly in countries that have earlier made large capital investments into an energy infrastructure based on fossil fuels.
Cases	
DEN, GER, USA	Increasingly cost-effective power generation drives rapid growth of renewable energy (E1). Investment and public support remain high overall (E2), but politicization continues (E3).
FRA, UK, Spain	Investment in renewable energy increases (E1); support for renewables increases among elites (E2) despite some politicization (E3). In Spain, economic recession presents a major challenge to renewables.
China	Elites seek energy security and good infrastructures to generate high growth rates, triggering large investments in renewables (E1); government and public support remain high (E2), with much less opposition than in industrialized countries (E3).
India	Investment in renewable energy increases (E1); high support for renewables among BJP and Congress, the country's two major parties (E2), with little evidence of opposition (E3).
South Africa	Renewables are part of an effort to improve electrification and reduce the negative effects of fossil fuels (E1). Strong support from the ANC (E2), with virtually no opposition from the public utility or coal producers (E3).
Brazil	Investments in renewables (ethanol) since the 1973 oil crisis designed to increase energy independence despite setbacks; since the 2000s, public support for other renewables such as wind (E1). Prorenewable position is shared across all major political parties (E2), and there are no major opponents (E3).
Kenya	Long-standing investments in off-grid solar; increasing capital investments and policy support (FIT, net metering) in grid-connected wind and geothermal projects (E1). Policies unaffected by political turnover (E2).

was under the political domination of its nuclear power industry, and the United Kingdom generated little electricity from renewable sources, but today both are investing in renewables. We then expand our horizon by considering the case of Spain, a key contributor in the 2000s. In all these cases, we see some political opposition, and yet renewables have continued to grow and survive even in difficult economic conditions.

Finally, we discuss a range of emerging and developing countries. The sampling strategy here is slightly different from what it has been so far. We first deliberately select the largest emerging countries: Brazil, China, and India. For renewables to have an effect on global energy markets and environmental outcomes, they must become meaningful in these countries. Therefore, it is important to examine whether the mechanisms we articulated apply to these countries. In addition, to improve the geographical representativeness of our sample, we also select two less familiar cases from Africa: South Africa and Kenya. We begin with China and India, given their importance as consumers and producers, and then explore the situation in Brazil, a major producer of biofuels. We conclude with South Africa and Kenya, both of which have made tremendous progress since 2000. In these case studies, we emphasize strong national interests in renewable energy and the limited degree of political opposition. We also demonstrate that renewables are growing alongside other sources of energy, such as coal and natural gas.

6.3 Rapid Growth in Pioneer Countries: Denmark and Germany

Denmark's wind energy boom has continued unabated since the threat posed by the right-wing Rasmussen government subsided by the year 2009. Denmark is continuously adding new turbines, with their number going from about 1,000 in 1995 to 5,000 by 2010. These turbines are increasingly powerful, as the mean capacity of a single turbine increased from less than 1 MW in 2000 to about 3MW ten years later. This underscores the significant technological progress that has been made in the lock-in era. The most powerful and largest turbines are now located off-shore. At the time of this writing, there are more turbines with a capacity of 2 MW in off-shore than in on-shore installations (398–307) (DEA 2014). In 2012, the government announced a target of 50% of electricity generated from wind power by 2020 (Global Wind Energy Council 2013a). In other words, the government wants the share of wind to increase by 17 percentage points in eight years. By 2050, Denmark expects to be able to cover all of its energy need from renewables. These goals are not a policy platform of the Social Democrats but a consensual policy objective. For instance, Lykke Friis, a leader of the Liberal Party, was quoted as saying, "No matter what we do, we will have an increase in the price of energy, simply because people in India and China want to have a car, want to travel. . . . That is why we came out with a clear ambition to be independent of fossil fuels: so we are not vulnerable to great fluctuations in energy price."[12]

Public support has triggered the aggressive expansion of Danish renewable firms. DONG Energy, Denmark's largest energy company, has invested vast sums in wind energy and biomass. For instance, DONG invested between $1 and $2 billion in the Anholt Offshore Wind Farm, one of the world's largest wind farms (400 MW).[13] DONG has also invested considerable sums to transform its coal plants into biomass plants. In 2012, it announced plans to spend $795 million to do so for three plants.[14] In 2012, exportation of Green goods and technology—including everything from turbines to biomass technologies—yielded about DKK 60 billion ($11 billion), or about 10% out of total exports (DKK 606, or $111 billion) (DEA 2012).

All of this has led to the improved economic competitiveness of renewable energy. In general, Denmark has the lowest capital cost worldwide for new onshore wind projects (IRENA 2012). Only China is similarly placed. In turn, low capital costs make wind energy lucrative. According to a 2008 study by the IEA, wind energy costs in Denmark were the lowest among the seven industrialized countries studied (Schwabe, Lensink, and Hand 2011).[15] The LCOE for wind electricity was about $85 per MWh. Furthermore, wind was profitable: revenues from electricity sale, FIT, and subsidies exceeded costs by about $8/MWh. Only in Sweden was wind electricity more profitable. These changes took place in a mostly favorable context. Danish public opinion remained strongly convinced about the danger of climate change. In 2008, a Eurobarometer survey found that 71% of Danes considered climate change to be one of the most important problems (Eurobarometer 2008). In comparison, the EU-wide average was 62%. In 2015, 73% of Danes still considered climate change a major problem, placing Denmark second in Europe behind Sweden (Eurobarometer 2015), whereas average support for this claim in the European Union had decreased to 47%. Few data directly support renewable energy, but available numbers—often commissioned by industrial interests and therefore to be taken with a grain of salt—tend to suggest that it is high. For instance, a 2015 survey by the Danish Wind Industry Association suggests that about 92% of the respondents wanted to see an expansion of renewable energy generation.[16]

Although public support remained high, the structure of the wind power environment started to change over the same period. In 2000, 84% of total wind capacity was owned by communities, which represents about 175,000 households (Bolinger 2005). In the following years, local involvement slowed down considerably. This seemed to be triggered by saturation and reforms that relaxed rules requiring local involvement (Sperling, Hvelplund, and Mathiesen 2010). The consequence was a form of "professionalization" of wind power, with larger actors and utilities playing a more important role (Bauwens, Gotchev, and Holstenkamp 2016).

Overall, Denmark shows that frontrunners have reaped the benefits of two decades of heavy investment in renewable energy technology. They have also had positive side effects on greenhouse gases. Carbon dioxide, for instance, has decreased from a peak of 3.56 million metric tons of carbon per capita in 1996 to 2.28 in 2010. Moreover, renewables have become a political force on their own. The major Danish parties now see the renewable industry as an important component of the country's economy. Even when political leaders, such as Anders Fogh Rasmussen, attempted to curtail public support for renewables, the clout of the renewable energy industry and its importance for Denmark's economy precluded him from doing so.

This is not to say that renewables do not face any political or economic obstacles. By 2010, Denmark's electricity bill was among the highest worldwide. Furthermore, Denmark will have to figure out how to efficiently use its electricity because at times its wind farm production exceeds demand. This has led Denmark on a few occasions to pay other countries for taking its electricity.[17] Issues related to storage will have to be solved. One possible solution may be increased cooperation with Norway, which could provide pumped-storage hydropower (Gullberg, Ohlhorst, and Schreurs 2014).

Despite these modest setbacks, renewables have achieved a lock-in and are now the status quo in Denmark. Consistent with our expectations, renewables have continued to grow in Denmark and, despite some political opposition, are no longer under a serious threat of reversal. The reduced cost and improved performance of renewable energy have changed the political landscape, making wind energy in particular a popular issue in Danish politics.

If Denmark shows how successful renewables can be under favorable conditions, Germany exemplifies the triumph of unwavering political support. Since 2000, the progression of renewables has been smooth. Between 2000 and 2014, installed renewable electricity capacity (excluding hydropower) jumped from 8 to 87 GW, a tenfold increase (EIA 2017a). Electricity generation followed a similar path, from 20,000 to 90,000 GWh. The growth in capacity was sustained by wind and, in more recent years, solar electricity. In terms of actual generation, wind and biomass are the main renewable contributors. Solar, a more recent technology and one that is less reliable in Germany, lags slightly behind.

To put these numbers in context, in 2000, renewables contributed 6.2% of the electricity consumed in Germany and 4.4% of its heating needs (AGEE 2016). Fast forward to 2015, when renewables generated about 31.6% of Germany's electricity demand (+25.4 percentage points) and 13.2% of its heating (+8.8 percentage points). The main contributors were wind, with a share of 13.3%, and solar PV (with 6.5%). Hydropower, the traditional renewable energy source, came fourth, with a share of 3.2%. Overall, the share of renewables in gross energy consumption increased from 3.7% in 2000 to 14.9% in 2015.

Just like in Denmark, these changes took place in a favorable context. Public concern about climate change remained strong despite the crisis. In 2008, Germany was tied with Denmark, with 71% of the population indicating that climate change was a major problem facing the world (Eurobarometer 2008). In 2015, despite the attention given to issues such as the economic and financial downturn in Europe, 65% of the German population continued to hold this view, good for third in the European Union behind Sweden and Denmark (Eurobarometer 2015). This support expanded to renewable energy in general. A 2013 study by a consumer protection nonprofit found that 82% of the German population supported the *Energiewende*, the energy transition (Verbraucherzentrale 2013). Although many were critical on how the transition was being implemented, many of the critical voices considered that it was too slow.

It seems likely that a high level of public support can be traced back to the role of local communities. By 2010, about half of the renewable energy capacity in Germany was owned by private citizens (Nolden 2013; Yildiz 2014). Another 40% or so was in the hands of institutional actors. Utilities and energy producers only owned about 10% of total capacity. Given the role played by local actors, it is interesting to note that even those who are critical of the energy transition do not believe that it is being driven by lobbies. According to the consumer survey mentioned earlier, only 1% of the respondents explained their criticism by noting the role of the renewable energy industry (Verbraucherzentrale 2013). Instead, most of the worries were about energy prices.

Indeed, the cost of the energy transition became the biggest challenge. Mirroring the Danish case, Germany's CDU-SPD coalition government under Angela Merkel expressed doubts about the sustainability of the EEG. She believed that public support for renewables was too costly; instead, nuclear power was seen as cheap and safe.[18] As late as 2010, Merkel expressed the desire to expand Germany's nuclear plant lifespan, a move that would have been a direct threat to renewables.

The 2011 Fukushima catastrophe upended Merkel's hopes. Just as in 1986, the nuclear question unraveled German politics. After Chernobyl, people remained unconvinced by the argument that nuclear only failed there because of Soviet inefficiency. Convincing them after a similar catastrophe in Japan, a much wealthier country, would have been a Herculean task. As Merkel reportedly told an advisor, "Fukushima has forever changed the way we define risk in Germany."[19] Her environment minister, Norbert Röttgen, said, "[the Fukushima disaster] has swapped a mathematical definition of nuclear energy's residual risk with a terrible real-life experience," further underscoring that nuclear power was out of favor.

The political reaction was swift and cut across partisan lines. Merkel declared that she wanted to shut down Germany's nuclear plants (Gullberg, Ohlhorst, and Schreurs 2014). Eight nuclear plants, run by powerful utilities such as EOn and RWE, were closed in 2011.[20] The nine remaining ones were scheduled to close by 2022. Merkel's plans were met with approval across most parties. The Bundestag decisively supported her decision, with a vote of 513–79 (Gullberg et al. 2014). In fact, the opposition mainly came from the Linke (radical left), which argued that the plan was too slow. Thus, "there is now basically a cross-party consensus on abandoning nuclear energy making a future reversal of the policy highly unlikely. Germany is embarking on an energy revolution of enormous proportions and significance" (Gullberg et al. 2014, 218). This plan is an acceleration of the *Energiewende* and a reform of the EEG.

The German renewable energy program received a welcome boost after Fukushima. In 2014, the coalition government under Merkel announced that it wanted renewables to contribute 40% to 45% of total electricity generation by 2025 and 55% to 60% by 2035.[21] Despite the challenges that Germany will face, even utilities seem resigned. Hildegard Müller, the head of the energy sector's main lobby, has said, "the question is not whether the Energiewende will be implemented, but how."[22] Furthermore, the industrial sector remained fairly subdued in its criticism. The fact that large industrial firms were exempt from the higher energy costs from the FITs has played a role in this respect, ensuring that both Merkel and SPD leaders would avoid opposition from corporations and unions. Industrial interests also lobbied intensively to prevent taxes on coal (the largest source of electricity), keeping prices in check.[23] The way was open for renewables.

External shocks also helped renewables. Without Fukushima, renewables would have faced considerable competition from nuclear power. However, debates about the EEG reform in the aftermath of Fukushima underlined the deep-rooted opposition to renewables among some segments of the business sector and trade unions. In fact, the EEG reforms implemented in 2014 cap the deployment of solar and on-shore wind energy. These reforms also include a reduction of subsidies to renewables. Thus, the renewable energy industry will undergo changes: some firms will disappear, whereas others will emerge. However, nobody expects renewable energy to become less important in the future. Since 2000, Germany's renewable energy industry has experienced an era of tremendous growth, in terms of both output and political clout. Renewable energy represents a significant contributor to the energy mix. In 2015, about 31% of all electricity came from renewable sources (including hydro).[24] As we documented earlier, public support has remained strong despite initial doubts under the conservative leadership of Merkel. Admittedly, both the CDU and SPD contain factions that are hostile to policies that could threaten the industrial sector. Furthermore, the level of support, mostly through the FIT, has been the object of a contentious debate. However, the end of nuclear power ensures that demand for electricity will remain high. By closing half of their nuclear power plants, policymakers have effectively tied their own hands.

Of course, the closure of nuclear plants also means that dirtier alternatives, such as coal, may fill the void. Yet even utilities that have decided to invest in coal in the aftermath of Fukushima understand the risks involved. After RWE, a major utility, announced its decision to invest in coal, one of its main unions responded, "RWE can't rely on conditions for lignite [brown coal] not to deteriorate politically. . . . Almost all activities [undertaken] come with political risks attached."[25] The struggles of the main utilities are symptomatic of the change in the political environment.

Once all-powerful, German utilities now face an uphill battle. Their collective value suffered under the combined loss from their nuclear revenues and the uncertainty from their other sources of income such as coal.[26] RWE's leader candidly said, "[c]onventional power generation, quite frankly, as a business unit, is fighting for its economic survival." German utilities have slowly attempted to join the renewables game. In 2015, nonhydro, nonbiomass renewables represented about 5.5% (5.25 GW) of RWE's global electricity capacity.[27] EOn, another major utility, claimed that its renewables operations have a capacity of 9 GW.[28] Since this announcement, EOn indicated that it wished to sell off its nonrenewable assets and focus on renewable energy generation.[29]

This favorable political environment has triggered an investment boom in renewables. In Germany alone, investments in nonhydro renewables reached about €11.7 billion ($13.3 billion) in 2005. The latest data for 2014 suggests that these investments climbed to €18.8 billion (more than $21 billion) (BMWi 2015). (The bulk of these resources goes toward solar and wind research.) This total does not even include other investments in the electricity sector that were triggered by the deployment of renewables. If those additional investments are included, then the total reaches almost €50 billion in 2010 ($68 billion) (AGEE 2013). Although utilities invested in coal plants over the same period, oversupply of electricity meant that they soon began to lose money from coal-based electricity generation. The head of RWE admitted that coal plants "make no return on capital" (Heinrich Böll Stiftung 2014, 16). Similarly, an energy researcher at DIW, a think tank, judged that "investments in new coal plants are bad investments."[30]

In summary, the pioneering countries have experienced a period of steady growth since 2000. Both Denmark and Germany experienced a political pushback, but in both cases it revealed a cross-partisan consensus in favor of renewables. In Denmark, the wind sector has become a major export industry. In Germany, the strong grip of the largest utilities is weakening, and renewables are making inroads. Thanks to public support, private investment has gone up, and renewables have become increasingly competitive.

6.4 The Comeback of an Early Leader: The United States

The United States was a frontrunner before Reagan's presidency began in 1981. California, in particular, was key to the development of the global wind energy industry. However, due to political opposition to renewables, the United States entered a period of stagnation. Federal support was lacking and the competition from fossil fuels too fierce. Despite these setbacks, renewable energy is now making a comeback. Consider the general trends in electricity generation in the United States since 2000 (EIA 2012, 2013a, 2014b). The most important sources remain fossil fuels, namely, coal (39.1%) and gas (27.5%). The contribution of nuclear power has remained stable, just below 20%. Although renewables remained mostly stable between 2000 and 2004, the period since 2005 marks an era of steady growth. Renewable electricity generation expanded threefold between 2005 and 2013. In terms of relative contribution, renewables went from 2% in 2000 to more than 6%. Wind has driven this growth. In 2013, wind accounted for two-thirds of all renewables, compared with less than 10% in 2000. In contrast, solar remained a fairly minor contributor, although photovoltaics have expanded rapidly since 2010.

The return of renewable energy in the United States was driven by several changes in the political and economic landscape. First, public policies played an important, if unreliable, role. At the federal level, a few measures were introduced in the 2000s to spur the growth of renewables. These policies included tax credits, both for individuals and corporations, and funding support for local projects (mostly through green bonds).[31] Although these policies may be credited with some of the growth of renewables, they are also highly volatile. Tax credits, for instance, needed to be renewed by Congress on a regular basis. Given the gridlock experienced in Washington, DC, these tax credits have occasionally not been renewed. Tax credits expired in 2000, 2002, and 2004. In each of these years, new capacity collapsed (by 93, 72, and 77%, respectively) (Congressional Research Service 2011). In contrast, FITs have remained unused at the federal level. Growth of renewables has been possible despite, not because of, federal energy policy.

Although federal authorities have been unable and unwilling to intervene in renewable energy policy, state authorities have been more active. Delmas and Montes-Sancho (2011) claim, "U.S. states have taken a leading role in establishing renewable energy policies since the late 1990s." There is considerable variation in state policies. As of 2014, 36 states had implemented some kind of a renewable portfolio standard. Some states, such as California, have mandated that 33% of retail electricity sales should come from renewables by the end of 2020.[32] California's ambitious renewable energy policies are part of an ambitious climate policy agenda that has, again, made the state a forerunner in renewable energy (Karapin 2016). Other states, such as South Dakota, had more modest ambitions, targeting 10% by 2015.[33] Finally, states such as New York have put forward tentative plans to decentralize the generation of electricity through the expanded use of solar and wind energy.[34] Overall, all states had implemented at least one prorenewable policy, often motivated by a combination of economic and environmental concerns.

The output of these policies varied greatly. As of April 2014, the leading renewable electricity producers came from the so-called West North Central states.[35] Iowa, for instance, generated 33.6% of its electricity from renewable sources, excluding hydropower.[36] Similarly, renewables contributed more than a quarter of the electricity generated in Kansas, Maine, and South Dakota. Texas, a leader in wind electricity, ranked 18th with 11%. On the opposite side, Kentucky, Ohio, and Tennessee barely had any electricity from these sources. On average, renewables represented about 7.4% of total power generation.

These investments have spurred the development of renewable technology firms. GE Wind has become the largest seller in the US market (US Department of Energy 2012). Investors such as Warren Buffett have committed capital to various renewable ventures. Berkshire Hathaway invested $15 billion through MidAmerican Energy, its energy branch.[37] Another giant, Google, invested $1 billion in various solar and wind projects.[38] In turn, competition and investments have pushed prices down. Drawing on data from EPA (2014), we computed the average retail price for electricity generated from wind turbines. This indicator is useful for the end-user cost of electricity from the most mature renewable energy source. The price went from about $60 per MWh in 2000 to $40 per MWh by 2008, a one-third decrease. In comparison, the cost of electricity from gas turbines ranged from $61 per MWh in 2005 to $36 in 2012.[39]

How does the US case fit with our expectations? Although it is clear that renewables are doing better than in the past, our second expectation about reduced political opposition requires further investigation. The United States certainly remains a battlefield of sorts for renewable energy. In spring 2014, for example, the *Los Angeles Times* reported on the Koch brothers' campaign to stop the spread of residential, rooftop solar policies.[40] These schemes are a threat to the profitability of traditional electric utilities, as homeowners who generate solar power consume less electricity from the grid. Although residential solar is still a small business and, even under the most conducive conditions, years away from truly eroding electric utility profits, the Kochs' political machinery is attacking rooftop solar energy policies across the country. In April 2014, for example, the Republican Oklahoma Governor Mary Fallin signed a legislative act that adds a solar fee to the electric bills of homeowners who have installed either solar panels or wind turbines on their property. This kind of active attack on renewable energy is different from the benign neglect of solar and wind power by the electricity utilities during, say, President Carter's tenure.

The effectiveness of today's political opposition is quite different from the past, however, as renewable energy is starting to benefit solar and wind technology companies, people working in the installation business, and homeowners looking to reduce their electricity bills. According to the 2013 Solar Jobs Census (Solar Foundation 2014), for example, there were more than 142,000 US jobs in the solar energy business in 2013. This figure indicates an increase of 50% over 2010. Although wind energy is less labor-intensive than solar, the number of jobs supported by wind electricity generation was also notable. According to the American Wind Energy

Association, there were 50,500 jobs in the wind energy field at the end of 2013.[41] People working in these businesses have a strong interest in the continuation of renewable energy policies, such as renewable portfolio standards, in their states. Indeed, the state renewable portfolio standards are an excellent example of how the renewables advocacy coalition can survive and thrive even under pressure. In 2014, the American Legislative Exchange Council (ALEC), a conservative organization, launched a major attack on renewable energy across the states. So far, however, ALEC has had limited success in repealing existing portfolio standards.

The case of Texas, which leads America's wind energy production, illustrates this notion (Rabe 2004; Hurlbut 2008). A staunchly conservative state with influential fossil-fuel interests, Texas became an early leader in renewables in the United States in 1999 when the state government enacted a renewable portfolio standard for a 3% renewable electricity share by 2009. The standard was part of a legislative package for power sector deregulation initiated a few years after the 1992 Energy Policy Act, which allowed states to set their own power sector policies. As Rabe (2004) notes, the enactment of the standard was essentially an issue linkage. In 1995, the Texas Public Utilities Commission had initiated deliberative polling—large but structured public meetings to discuss energy policy—to include the public in energy planning. In the polling sessions, it became clear that many citizens, along with consumer and environmental interest groups, expressed a strong interest in cost-effective renewable energy policies. True to the conservative spirit, the Texan government enacted a policy that allowed trading of renewable electricity credits. By 2013, Texas had installed 12,000 MW of wind electricity-generation capacity,[42] with no end in sight to renewable energy expansion despite a simultaneous shale gas and oil boom. Clearly, the United States lags behind Denmark and Germany. The political obstacles to renewable energy remain many. Yet the tendency is toward convergence with these two countries. First, output is growing. Second, investments are large and backed by influential players. The power balance between renewables and traditional utilities is tilting toward the former. In 2014, Barclays announced that it had "downgraded the entire electricity-generating sector of the US high-grade corporate bond market because of the challenges posed by renewables and the fact that the market isn't pricing in those challenges."[43] Furthermore, the bank predicted that "[o]ver the next few years . . . we believe that a confluence of declining cost trends in distributed solar photovoltaic (PV) power generation and residential-scale power storage is likely to disrupt the status quo." Such claims would have been unimaginable in the 1990s.

6.5 Renewables Arrive in France and the United Kingdom

By the end of the 1990s, the situation for renewables in France and the United Kingdom was desolate. In France, the dominance of the nuclear power industry precluded investments into alternative energy sources. Political elites were united in keeping nuclear research going for military and civilian reasons. Large utilities, such as EDF, ensured that competitors would not be allowed to enter into its lucrative market. French politicians, knowing that their electricity was among the cheapest in Europe, saw little reason to do anything that would hurt their industrial sector's competitiveness. In the United Kingdom, energy policy was still reeling from years of privatization under Thatcher in the previous decade.

Renewables contributed little to the energy output of these two countries. In 2000, the share of renewables in total electricity generation was less than 1% (4,200 GWh) in France and 1.5% (5,400 GWh) in the United Kingdom (Figure 6.2). Capacity was also lacking, with about 0.8 GW available in each country. Despite this hostile environment, the first decade of the 21st century marked a period of strong growth in both countries. In ten years, generation increased to 17,000 GWh in France and 24,000 GWh in

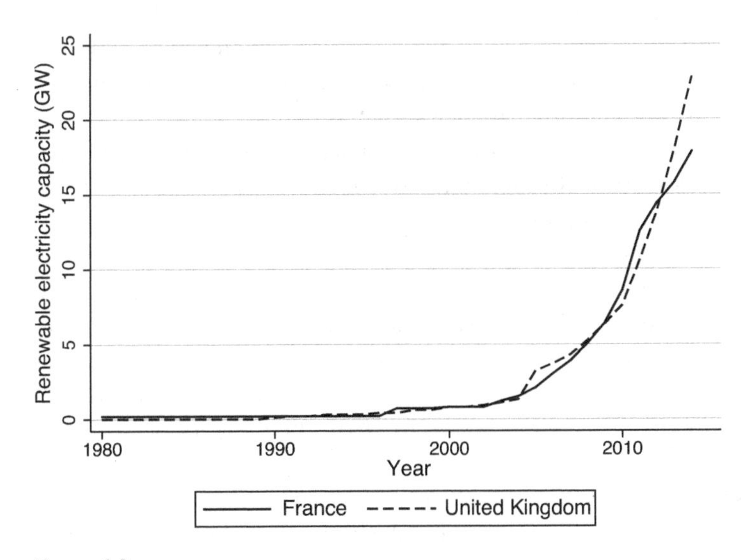

Figure 6.2
Renewable electricity capacity (in GW) in France and the United Kingdom, excluding hydro, 1980–2014.
Source: EIA (2017a).

the United Kingdom. At the same time, capacity increased to about 8 GW in both countries. This represented 7% (France) and 8% (UK) of total electricity capacity in 2010 (EIA 2017a). Wind electricity was the main driver in both cases because it represented on its own a share of about 5.7% in the United Kingdom and 4.8% in France. In France, electricity from wind, which was nil in 1990, represented 92 GWh in 2000.[44] At the end of the decade, wind had boomed to 10,000 GWh, and the number is still increasing. The latest figures for 2014 indicate that wind generates 17,000 GWh (EIA 2017a).

Of course, these figures remain modest compared with heavy hitters such as Germany, which has more than six times as much renewable capacity. By 2010, France and the United Kingdom had only slightly more capacity than tiny Denmark. Nonetheless, this reversal remains remarkable and has, at least in part, been driven by favorable policies. In the United Kingdom, a number of approaches were used to improve the position of renewables. First, research funds were boosted from a paltry $9 million in 2000 to $276 million per year in 2010. Biofuels received particular attention under this system. Since then, R&D expenditures have decreased but remain significantly higher than they were before 2000. Second, the British government more directly supported the deployment of renewables by implementing a portfolio standard ("renewable obligation") in 2002 and an FIT in 2010 (Haas et al. 2011; Cherrington et al. 2013). The FIT has proved to be particularly generous because it includes paying those who generate renewables even for the energy that they produce and consume.[45]

Prorenewable policies have also been enacted in France. Early attempts to trigger investments through tenders proved disappointing because many projects were abandoned (Cochet 2000). A more serious attempt to promote renewables was the 2000 Electricity Act, which saw the creation of an FIT. Even then, the effects of the tariff were initially muted partly because the largest utilities, such as EDF and GDF, were already generating enough electricity to cover French demand (Szarka 2007). Policymakers hoped at the time that the FIT and tenders would suffice to raise the capacity of wind to 10 GW (Cochet 2000). In 2010, the capacity of wind was about 6 GW.

Besides designing direct policies, France has provided indirect support to renewables by focusing on innovation. Like in the United Kingdom, public expenditure on research has significantly increased in France. In 2000, public R&D in renewables amounted to about $21 million per year (two-thirds going toward solar research, the rest being split between wind and biofuels). This is one-fifth of the amount spent in Germany in the same year. By 2011,

R&D had jumped to $202 million per year. A gap between France and Germany still remains, but the investments became meaningful. Most of these resources went into solar and biofuel research, with wind receiving only a fraction of the total funds.

Overall, renewables have made significant progress in France and the United Kingdom. From marginal contributors, they have become a non-negligible part of the energy mix. The change was rapid, with most of the progress made after 2005. Why this sudden transformation in these two countries? Our expectations focus on the role of affordability of renewables and the development of political constituencies. In both countries, economic problems increased the demand for renewables. British voters worried about energy prices, and the French sought new jobs. In both countries, a significant share of the population demanded action on climate change. Because cutting carbon emissions through taxes or regulations was not necessarily popular, policymakers responded by pushing renewable energy.

In the United Kingdom, decreasing prices made renewables clearly more affordable. Cherrington et al. (2013) show that the price of a solar module in the United Kingdom declined by 26% in 2010 alone, and this change occurred after years and years of decreasing costs. This decline turned out to be much larger than the government anticipated, leading it to consider reducing the level of FITs. Even the plans of the Conservative government to scrap on-shore wind subsidies in 2014 reveal the progress made by this technology: by that time, wind had become cheaper than nuclear power and only slightly more expensive than fossil fuels.[46]

In addition to renewables becoming more affordable, alternatives have actually become more expensive. The cost of energy has become one of the main policy concerns among the British public (Toke 2011). A main driver for these worries is the rapidly increasing cost of the electricity bill. With gas reserves in the North Sea dwindling and coal becoming a minor contributor, Britain's electricity cost is now a major source of concern. In 2009, the British Energy Minister of State was quoted as saying, "We need to bring about a revolution in the way energy is produced. . . . Imagine you are pin-striped revolutionaries in the spirit of Che Guevara on the Sierra Madre" (Toke 2011, 528).

Similarly, prices have declined in France. The cost of solar photovoltaic, for instance, decreased by a factor of 2 to 3 between the mid-2000s and 2014 (Commission de Régulation de l'Énergie 2014). The sunk costs in nuclear power will make it difficult for renewables to reach parity, but renewable energy may become competitive enough to convince firms such as EDF to jump in. The market for renewable electricity has already begun to attract the French titans, pitting EDF and Alstom against GDF in the wind sector and Total dominating the solar market.[47]

By extension, this also makes it easier for prorenewable politicians to support these ventures. The renewable industries in these two countries are becoming valuable sectors. The implementation of the FIT in the United Kingdom is estimated to have created 25,000 jobs (Cherrington et al. 2013). RenewableUK, a lobby for the wind and marine energy industries, claims that more than 34,000 individuals were directly or indirectly employed by these sectors (RenewableUK 2013). Furthermore, their total turnover amounted to GBP 2.8 billion in 2010 ($4.79 billion) and GBP 8.1 billion ($13.85 billion) in 2013.

The promise of renewables appears to encourage their politicization in the United Kingdom. The Conservative government under David Cameron has pushed back against new and existing support policies. His constituency being mostly rural and particularly hostile to wind turbines, Cameron has little incentive to cater to a constituency that is closer to Labour and the Liberal Democrats. Unsurprisingly, opposition to the Conservatives' plans was clearly articulated in terms of the economic importance of the clean tech industry. For instance, a Liberal Democrat MP responded, "Putting the brakes on onshore wind would be disastrous for business and jobs in our growing green economy."[48]

We expect that political controversies surrounding renewables will decrease if the number of people depending on it increases. Contrasting the British and French cases, it is striking that the British renewable industry has remained smaller. Thus, political controversies may be the result of its lack of power. We can then conduct the following thought experiment: had Labour under Blair and Brown been more proactive in pushing renewables, or had Brown remained in power a bit longer, then the renewable energy industry might have been strong enough to avoid Cameron's pushback. As it is, Labour did not support renewables forcefully enough, leaving it vulnerable to the change of power to the Conservatives. The lack of influence of the renewable community is apparent when we explore campaign contributions: virtually no renewable energy company gave money to either party in the run-up to the 2010 and 2015 elections.[49] We made a similar argument in Aklin and Urpelainen (2013b), where we underscored how government turnovers could have adverse effects on energy policy.

In France, the lack of jobs has been used to explain the limited initial deployment of renewables. Although some estimates expected more than 16,000 positions to be created due to the policy reforms of the early 2000s, the reality was that most French renewable firms (especially in the wind sector) were small and specialized (Szarka 2007). Nonetheless, policy reforms enacted at the end of the 2000s, referred to as the *Grenelle de l'environnement,* were partly motivated by the creation of jobs. The Grenelle refers to a series of meetings of policymakers, industrialists, and nongovernmental organizations, with the aim of solving pressing environmental issues such as climate change. These meetings led to various proposals to deploy more renewable energy. Overall, the project was expected to create 600,000 jobs between 2009 and 2020. Studies by the *Agence de l'Environnement et de la Maîtrise de l'Énergie* (French Environment and Energy Management Agency [ADEME]) similarly forecasted that prorenewable policies may lead to the creation of 169,000 new jobs. Even if these figures appear optimistic, the key insight is that French policymakers were tying renewable energy to labor markets. A report published in 2012 suggested that about 100,000 people were directly employed in the renewable energy industry (ADEME 2012).

In summary, we find evidence for the first and second expectations. Both France and the United Kingdom increased the generation of renewable energy. Governments in both countries implemented policies designed to do so, and output from these sources has generally increased since 2005. We also find evidence that the competitiveness of renewables played a role in enabling these countries to join the movement started by Denmark, Germany, and the United States. Furthermore, we find that renewables have been more strongly supported in France, and this coincides with a stronger development of the renewable industry. Whereas the French industrial sector has grown increasingly interested in entering this market, British renewable firms remain smaller and politically more isolated. In line with our argument, political controversies have been fiercer in the United Kingdom, although the political environment has become decidedly more favorable to renewables in both countries.

6.6 The Case of Spain

In the 1970s, Spain was still under General Franco's rule and, compared with the rest of Western Europe, a relatively poor country. Since then, Spain has come a long way. By 2012, Spain had become a top-5 country in solar thermal power and wind power capacity (REN21 2014). Why did renewables pick up in Spain? Admittedly, the country is ideally situated for wind and solar energy. Nonetheless, it is only thanks to strong public support that renewables began to grow fast. Socioeconomic considerations played a key role in the political calculus. The two main considerations that led to prorenewable policies were economic development and environmental concerns. In the words of del Río and Unruh (2007, 1504), public support for renewables "has its roots in its perceived environmental and socio-economic benefits, especially job creation." Spain was in a similar position to other southern European countries. The country wanted to keep high growth rates, support its construction boom, and position itself on the production of modern technologies. Furthermore, Spain had ratified the Kyoto Protocol and thus needed to position itself on climate change mitigation. National environmental movements were enthusiastic about wind energy (Vasi 2011).

In the 1990s, Spain began enacting a series of prorenewable policies. Various regulations have sought to expand renewables, with wind being the main beneficiary of public support. The government pushed renewables through both prices (FIT) and quantity (capacity targets) (del Río and Unruh 2007). These policies enabled the rapid deployment of renewables. Wind capacity increased from 0.1 GW in 1995 to about 8 GW in 2004 (del Río and Unruh 2007, Table 2). According to Lewis and Wiser (2007), the government has used policies such as local content requirements to support the domestic wind industry, with impressive results. According to their data, in 2004, the Spanish wind turbine manufacturer Gamesa was second only to Vestas, showing that the Spanish strategy worked (Lewis and Wiser 2007). By 2012, Spain had the fourth largest capacity worldwide, just behind China, the United States, and Germany but ahead of giants such as India, Japan, or the United Kingdom (Pew 2014). Spain had thus become a global leader.

The Spanish case, however, also underscores the fragility of these recent developments. The 2007 financial crisis plunged many countries, including Spain, into a painful recession. Amid general cutbacks in public spending, many countries decided to scale back their level of support for renewable energy. Spain is such a case. It initially froze its FIT and eventually reduced the renewable electricity premium in a retroactive fashion (REN21 2014). Thus, new projects were not eligible for the FIT, whereas existing ones had to operate under worse conditions. These decisions have had dramatic consequences on new investments, which collapsed to $2.9 billion in 2012 (70% year-to-year decrease) and $400 million in 2013 (80% year-to-year decrease) (Pew 2014). These unfavorable policies coincided with the return to power of the center-right Conservatives. Like in the United Kingdom, Spain experienced a late case of politicization of renewables largely due to extreme economic distress.

Yet Spain remains a prime location for renewables. By 2013, total electricity capacity from renewable sources amounted to about 34 GW out of slightly more than 100 GW (Pew 2014). Most of this capacity was provided by wind (23 GW) and solar (6.8 GW) energy. The latter can be expected to grow because solar power is now the largest recipient of investments, with 80% of all capital going into this energy source. Gamesa remained a top-10 global firm in the wind turbine industry, with a sizable market share of about 6.1% in 2012 (REN21 2014). Overall, renewables will remain a hotly debated policy issue in Spain. Production has reached a high enough level to be economically important. According to IRENA (2014c), about 114,000 people work in Spain's renewable energy sector. However, the growth of renewables has not been decisive enough to prevent the politicization of energy policies. The backlash following the financial crisis increased political uncertainty and showed that renewable investments remain vulnerable to adverse conditions. Redundancies have reduced the number of people employed in the sector, and the government remained keen to cut subsidies further.

In summary, the Spanish case provides evidence for the progressive global lock-in of renewables, although the local lock-in remains partial and fragile in the face of an economic crisis. Consistent with our first expectation, Spain's investments in renewable energy clearly reflect decreasing prices and increased attractiveness over time. We also see reduced political opposition and strong public support for renewables, as our second expectation would have it, but the fact remains that Spain's economic recession has undermined renewables and brought back political opposition. Given the severity of Spain's economic difficulties and the generous renewable energy policy of the past, this is not entirely surprising.

6.7 The Global Diffusion of Renewable Energy

Thus far our political history has focused on a small set of countries. This list of countries is by no means exhaustive and ignores the rapid growth of new markets in developing and emerging countries. Here, we first provide a general overview of the state of renewables in the rest of the world. We then examine a number of cases in greater detail. Specifically, we explore the expansion of renewables in China, India, Brazil, South Africa, and Kenya.

Before conducting the country case studies, we contrast the broad patterns of modern renewables in industrialized and developing countries. Because the cost of renewables has gone down across the board, one way to see whether renewables gained in importance is to use simple econometric methods to assess the significance of these trends. We explored this question in Figure 6.3 by regressing the share of (nonhydro) renewables on year

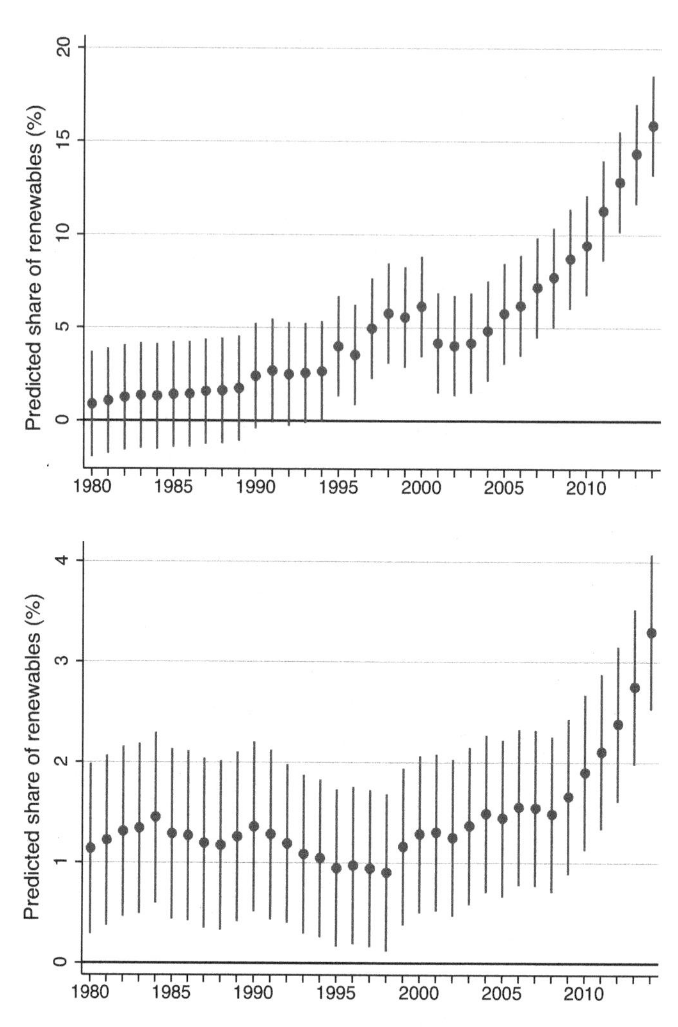

Figure 6.3
Renewables fixed-effects analysis, 1980–2014. These graphs plot the estimates of the regression of the share of renewables (nonhydro) on year fixed effects for OECD (Above; N=1,048) and non-OECD countries (Below; N=5,063). The vertical bars indicate the 95% confidence interval.
Source: WDI (2017).

fixed effects. Consider first the OECD countries at the top. As we can see, the share of renewables was fairly stable until the second half of the 1990s. Their expected share changed little from year to year, remaining close to nil. Then toward the turn of the millennium, their share rapidly increased. It soon cracked the 5% bar and continued to grow. At the end of our sample, it had almost reached 15%.[50] More remarkable, a similar pattern holds for developing countries. Once the same analysis is performed for non-OECD countries only, we observe that renewables entered a period of rapid increase in the early 2000s after two decades of stability. The progression starts later and is less pronounced, yet clearly visible. Are these changes significant? To ensure that there has indeed been a change somewhere between 1995 and 2000, we conduct a Chow test for structural breaks. We examine whether the slope of the linear time trend changes for different periods, specifically 1995 to 2000 and then after 2000. The result is clear: there is a structural break between the periods.

6.8 Renewables in Emerging Giants: China and India

China and India offer an ideal setting for evaluating renewable energy. Their large size notwithstanding, the two countries are very different. China is an industrialized and an increasingly wealthy autocracy, whereas India is a poorer, less industrialized country governed by democratic rules. Therefore, we can evaluate our expectation that, despite these differences, renewables become increasingly popular in both, and little political opposition to them exists.

China's energy trajectory provides a remarkable example of how rapidly a country can transform its energy sector. In 1990, virtually none of the electricity generated in the country came from renewables (WDI 2013). Fifteen years later, the share of renewables remained small at about 0.3%. Since 2005, however, the transformation has been astonishingly rapid. Within five years, renewables increased their contribution to about 1.7%. In terms of actual generation, China drew 3.2 GWh in 2000 and was still at a meager 7.4 GWh in 2005. However, in 2012, renewables contributed a stunning 136 GWh.[51] What is more, modern renewables lead the way in China: a 2014 report by Pew (2014) notes that more than 90% of the nonhydroelectric capacity is wind or solar power.

What were the drivers of this transformation? It appears that energy policy is strongly determined by the general socioeconomic policy followed by Chinese officials. Chinese leaders have long sought to use economic growth to tame dissidence in the population. In turn, generating high growth rates necessitates immense amounts of energy as inputs. This is particularly true for the manufacturing sector. Furthermore, it also requires providing electricity to rural areas, which are often neglected. Yet the energy market often comes up short. In fact, Cherni and Kentish (2007) claim that China faces "important socio-economic irregularities" with respect to its energy situation. Despite adding massive amounts of capacity, China regularly suffers from blackouts and production regularly has to shut down (Cherni and Kentish 2007). Although China's electrification rate is high, millions of people are still without electricity. To prevent social unrest, Chinese leaders are thus heavily reliant on obtaining cheap and vast amounts of energy. Renewables can help China meet the society's growing demand for energy.

Although China still relies heavily on coal-fueled thermal plants to power its economy, there are two reasons that renewables have played a key role in the late development of China. First, coal plants are polluting, and the Chinese population has started to take notice of the poor air quality around the country's major cities (Cherni and Kentish 2007). Writing more than a decade ago, Wu (2003, 1420) claims, "air pollution has recently become the main concern of urban residents. As a result, in many Chinese cities, the quality of air is monitored and reported every day. Environmental damages also affect the growth of agriculture and forestry. It is estimated that the direct cost of environmental damage in the 1990s amounted to 7 per cent of China's GDP." Along with other sources, such as nuclear energy and shale gas, renewables are well placed to alleviate these concerns.

Another compelling reason is that the production of technology for renewable energy generation has become a lucrative business. China's highly competitive manufacturing sector is ideally placed to mass produce components such as turbines or solar modules. Unsurprisingly, Chinese firms, led by Yingli Green Energy, have become global players in the market for renewables. In 2012, nine out of the top-15 solar panel producers were from China, representing about 30% of world market share (REN21 2014). In the wind industry, which is older and thus less open to newcomers, China still managed to place four of its firms in the top 10, for a world market share of almost 17% (REN21 2014). Together these firms are similar in size to the two largest producers, GE Wind (United States) and Vestas (Denmark).

As China's development has continued unabated, policymakers have increasingly perceived renewable energy as an important component of their long-term political plans. The Eighth Five Year Plan first mentioned renewables in 1991 (Peidong et al. 2009). By the mid-2000s, time was ripe for policymakers to support a series of decisive prorenewable policies (Cherni and Kentish 2007). Two main policy reforms were made. First, Chinese lawmakers enacted the Renewable Energy Promotion Law (Cherni and Kentish 2007), which instituted FITs and renewable portfolio standards. The government now possessed the legal tools to push renewables. Second, Chinese authorities instituted the "dual system," which consisted of a limited privatization effort. Vast amounts of capital could thus be directed toward utilities and renewable technology firms. The results were remarkable, and investments spiked. In 2012, they reached $54.2 billion, more than any other country. This number represented about 29% of all investments in the G-20 countries (Pew 2014). Investments in wind alone represented almost as much as the combined investments in wind and solar in the United States. Wind energy was the prime recipient of this money, with a share of about 60%, whereas solar took the second spot with 30% (Pew 2014).

This flow of capital produced concrete results. In 2004, China had announced targets of about 10 GW of capacity from nonhydroelectric renewables by 2010 (Martinot and Li 2007). In reality, renewable capacity had reached 36 GW by 2010. In 2013, capacity from wind, solar, and biomass had reached about 120 GW, putting China slightly ahead of the United States and Germany (Pew 2014). In fact, the main challenges now pertain to financing and access to capital to expand production.[52] Thus, within a decade, China became the biggest renewable player worldwide, further confirming how quickly renewables were deployed in the country (REN21 2014). This revolution was largely driven by wind energy, which had a capacity of about 90 GW. Solar was trailing at about 20 GW. Needless to say, these numbers are bound to evolve rapidly given how quickly China raised to become a giant in the field of renewable energy. Overall, China has now become a major player in the renewable energy business, as both a consumer and a producer. It tops most important lists: investments, capacity, and generation. China appears to be in a position to determine world prices for various renewables.

China's story clearly shows that, as renewable energy has become more competitive, it has become increasingly useful. China uses renewables to meet a rapidly growing demand, improve energy security, and improve air quality. These observations support our first expectation. What about the degree of political opposition? Although China's political dynamics are difficult to evaluate because of the closed nature of the main political bodies, the evidence overall appears clear. From one energy plan to another, China has consistently emphasized the importance of renewable energy.

In the latest energy plan, titled China's Energy Policy 2012, Beijing states,

> Vigorously developing new and renewable energy is a key strategic measure for promoting the multiple and clean development of energy, and fostering emerging industries of strategic importance. It is also an urgent need in the protection of the environment, response to climate change and achievement of sustainable development. Through unswerving efforts in developing new and renewable energy sources, China endeavors to increase the shares of non-fossil fuels in primary energy consumption and installed generating capacity to 11.4 percent and 30 percent, respectively, by the end of the 12th Five-Year Plan. (China 2012)

Besides hydro and nuclear power, the document summarizes goals for wind, solar, and biomass energy—along with specialized plans for distributed renewable energy generation.[53] A review of the government's official website for documents related to climate policy, China Climate Change Info-Net, quickly reveals that the Chinese government repeatedly emphasizes the importance of renewable energy as a strategy for sustainable economic development.[54] Because the website contains documents dating back to 2002, it is easy to see that the political support for renewables at the highest level has remained strong. As the world's largest emitter of carbon dioxide, China is under tremendous international pressure to adopt low-carbon growth strategies and contribute to the implementation of the Paris Agreement on climate change—and renewable energy is a key component of this strategy.[55]

Furthermore, the politically influential urban middle class strongly supports renewable energy. Chen, Cheng, and Urpelainen (2014) conducted a June 2014 online survey of more than 2,000 Chinese inhabitants across the major urban areas of the country and evaluated their support for renewable energy on a Likert-type scale (1–5), with higher values indicating more support. Varying the framing of the problem, they find that the mean level of support is 4.62 despite the fact that one-half of the sample is subjected to a negative frame emphasizing the economic cost of renewable energy. The most important reason for supporting renewables appears to be concern with energy security, highlighting the variety of rationales for increased renewable energy deployment. Moreover, when requested to indicate their preference for nuclear versus renewables on a 1 to 3 scale (1 meaning preference for nuclear and 3 meaning preferences for renewables), the average response is 2.66, showing a strong preference for renewable energy instead of nuclear power. What makes these results particularly notable is the high number of Communist Party members, as 34% of the sample report being members. Not only does the Chinese government emphasize renewables, but this strategy has the firm backing of the politically critical urban middle class.

Overall, China's increased use of renewable energy has contributed to the country's gradual decarbonization, but much remains to be done. Although China's progress in renewables is impressive, in 2013, the country generated 74% of its electricity from coal, whereas wind generated 2.6% (Cornot-Gandolphe 2014). These numbers are becoming more favorable to renewables every year, but wind, solar, and other renewables are far from replacing coal as China's primary fuel. Renewables cannot be dismissed because they play a central role in China's energy future, but China is going to remain dependent on coal for electricity generation for decades. China's political and economic environment is favorable to renewables, but the path to decarbonization is a long one.

India is another gigantic market for renewables but for reasons different from China's. India, to a much higher degree than China, suffers from massive electricity problems that impede the country's development (Chakrabarti and Chakrabarti 2002). The situation is dire: about 400 million Indians do not have reliable access to electricity (Government of India 2011). Furthermore, when it is available, electricity is of poor quality: it is unreliable, and the grid regularly suffers from blackout. Overall, electricity consumption is extremely low. We mentioned earlier that China generated about 3,300 kWh per capita in 2011. Indians, in comparison, generated a mere 684 kWh per capita. Each Chinese inhabitant consumed almost four times as much electricity as each Indian. The problems are not limited to electricity. Three quarters of rural Indians still use firewood as their principal cooking fuel (NSS 2010).

India's energy issues go back to its independence. Historically, part of the issue can be traced to poor regulation. In particular, subsidized electricity prices meant low profits and insufficient investments. Because the cash-strapped government was generally unwilling to put up the difference, grid extension turned out to be extremely slow (A. Singh 2006). In recent years, the government attempted to reform the country's energy market. Some of these reforms were designed to attract private capital (Dubash and Rajan 2001). Others were project-oriented and targeted specific populations (Bhattacharya and Patel 2007). For instance, the Rajiv Gandhi Grameen Vidyutikaran Yojana has sought to expand electrification in rural areas and has generally been a success.[56]

Despite these reforms, the scale of the problem remains staggering.[57] Even when villages are officially connected to the grid, actual household connections (the so-called "last mile") are often lacking. Although village electrification rates are now high, 45% of Indian households living in rural areas lack access to electricity (Gambhir, Toro, and Ganapathy 2012). Even when this obstacle is set aside, the quality of electricity fluctuates widely; the numbers of hours per day during which electricity is available highly depends on a household's location (Aklin et al. 2015). Rural areas and slums are at a clear disadvantage. Furthermore, the supply side is plagued with problems of its own. The power sector as a whole is regularly accused of being mismanaged (Dubash and Rajan 2001). The coal industry, one of the main sources of electrical power in India, remains highly inefficient despite having vast reserves and is more often than not embroiled in corruption scandals.[57] The results of this dysfunctional market are regular blackouts and underprovision of electricity. The 2012 crash of the grid, which left an estimated 700 million people without power, is one of the most recent examples.[58]

In this context, observers have suggested that renewable energy may play two roles. First, renewables can fuel the grid. This is the way most industrialized countries use renewables. India offers good conditions for both wind and solar power (Carolin Mabel and Fernandez 2008; Pillai and Banerjee 2009). Renewables could offer reliable electricity to the extent that the supply chain bypasses the collection and distribution of actual fuel. This is not the case for coal, which needs to be extracted and transported to power stations. As we indicated earlier, this chain is fraught with inefficiencies and regularly suffers from failures. Despite abundant coal reserves, Indian coal is left idle. The coal industry is riddled with accusations of corruption and has been unable to reform.[59]

Second, renewable energy can be used to fuel off-grid electricity generators. For instance, solar microgrids can power rural households that are too remote to be connected to the grid. The difference with grid solutions is essentially one of scale. This application of renewables is particularly relevant for the most rural areas of the country, given their geographical distance to the country's main grid (Kanase-Patil, Saini, and Sharma 2010). These decentralized systems have a number of advantages over the grid, such as being more flexible and technologically easy to implement. Furthermore, simple off-grid systems are fairly cheap, although usage costs can vary (Urpelainen 2014). In practice, however, distributed renewable electricity generation remains on a small scale in India: in August 2014, the total capacity of off-grid solar photovoltaic power in India was only 200 MW (Goyal and Wiemann 2015).

Policymakers have noted the potential for renewables to play a role in solving India's energy woes. India benefits from ideal conditions for various renewables, in particular for solar energy. Thus, the Indian government gave the green light to the Jawaharlal Nehru National Solar Mission

(JNNSM) in 2009 (Ummadisingu and Soni 2011). This ambitious program seeks to add 20 GW of solar capacity to the grid as well as 2 GW of off-grid solar power by 2022 (Deshmukh, Gambhir, and Sant 2010). Since then, despite a change in the ruling government from Congress to the Bharatiya Janata Party (BJP), solar power expansion has continued. Prime Minister Modi has hinted at the possible acceleration of the project.[60] Whether the targets will be met on time remains to be seen, but by making the targets public, Modi has ensured that he would face a reputational cost if his government failed to invest into solar energy. Furthermore, investments can bring the cost of solar power down in the coming years. As of 2013, wind's LCOE ranged between $70 and 200 per MWh, whereas gas ranged between slightly more than $50 and 150 per MWh. Yet according to Bloomberg New Energy Finance, both solar and wind could be competitive with coal by 2020.[61]

Political support for renewables freed up much needed capital. Investments reached about $7 billion in 2012 and $6 billion in 2013, good for eighth among the G20 countries (Pew 2014). About half of these resources went into wind energy, and a bit more than 20% funded solar deployment. Output is following suit. Estimates vary, but it is believed that renewables capacity reached about 30 GW in 2013, most coming from wind (about 18 GW) and only a small share (5 GW) from hydro (Pew 2014). Wind represents about two-thirds of this amount, whereas solar lags behind, with estimates ranging from 1 to 2 GW. In parallel, Indian firms became serious industrial competitors. Suzlon Group, for instance, controls about 5.3% of the world solar market (REN21 2014). Firms such as Tata Solar Power and SunEdison India are well-known brands in India. Of course these firms benefit from deep ties with the political system. As in the Chinese case, they would not have been successful without some degree of public support. As a result, India, like China, has been accused of breaking international law. The subsidy program under JNNSM, for instance, requires part of the production of solar module to occur in India. These protectionist policies have drawn opposition from the United States, which has brought the case to the World Trade Organization's attention.[62]

An important feature of the Indian case is the degree of consensus among political elites about the rationality of pursuing renewables as part of the solution to chronic electricity shortages. In 2009, the then-ruling Indian National Congress party promised in its manifesto that "the country [will add] at least 12,000–15,000 mw of capacity every year through a mix of sources—coal, hydro, nuclear and renewables" (Indian National Congress 2009, 19). Five years later, Congress continued on the same line and announced, "[w]e will give a new thrust to new and renewable energy, including hydro, solar and nuclear energy" (Indian National Congress 2014, 39). It also committed to the "accelerated [implementation] of the Jawaharlal Nehru National Solar Mission" and the creation of "a major new "National Mission on Wind Energy" to harness the power of this clean and renewable source of energy, which has tremendous untapped potential in India." What is remarkable is that the main opposition party, the BJP, adopted a similar position. In 2009, BJP promised "to invest heavily in developing non-fossil fuel-based clean energy sources, especially for electricity production. [BJP's] goal will be to add at least 120,000 MW of power over the next five years, with 20 per cent of it coming from renewable sources."[63] In 2014, BJP wanted to "[g]ive a thrust to renewable sources of energy as an important component of India's energy mix" and "[e]xpand and strengthen the national solar mission" (Bharatiya Janata Party 2014, 34). BJP also wished to encourage "research and innovation in areas of relevance for the economy and society. . . . Sectors such as . . . energy including renewable energy . . . would be done with special importance." The BJP, which replaced Congress after the 2014 elections, has since then not shown any sign indicating that it would renege on these promises. Thus, although they disagree on a range of issues, India's two main political parties find common ground on renewable energy policy. This is in sharp contrast to the situation in many industrialized countries, such as the United States, in which renewable energy remains a divisive issue.

India's progress toward increased use of renewable energy is impressive against the low baseline, but, similar to China, the road to a clean energy system is a long one. According to WDI (2013), India's dependence on coal for electricity generation was 71.1% in 2012. In 2015, the Indian government set an ambitious goal to increase coal production (mining) from approximately 600 million to 1,500 million metric tons—an increase of about 150%.[64] Although renewable electricity generation is growing faster than this rate and some of the production increase will replace imports, coal use may continue to grow in India. Again, renewables are playing an important role in India's energy future, but they are only partially crowding out conventional fossil fuels. As the international focus in climate change negotiations begins to shift from China to India, which holds the largest potential for future emissions growth thanks to its massive population and bright economic prospects (IEA 2015), renewable power offers the Delhi government a way to supply energy for economic prosperity without explosive growth in carbon dioxide emissions.

In summary, the cases of China and India are consistent with our expectations. In both countries, renewables have become popular because of their increased competitiveness. Due to the dramatic infrastructural needs of the two countries, renewables play an important role in energy policy. There has been much less political opposition to renewables in these two countries than in industrialized countries because the urgent need for abundant energy in an emerging but still poor economy makes renewables attractive to policymakers and powerful interest groups. In contrast, the fact remains that these countries remain dependent on coal for most of their electricity generation, a dependence that is unlikely to change any time soon.

6.9 Beyond China and India

The deployment of renewable energy is not limited to Asia. In fact, renewables are making inroads into Africa and Latin America. South Africa, for instance, has become a hub for renewable energy. In 2012, it became the ninth largest renewables market, with a turnover of about $5 billion (about two-thirds of which went to solar), and has been one of the fastest growing markets worldwide (Pew 2014). Policies such as auctioning of projects have produced impressive results. It is also one of the few countries that focuses mostly on solar energy. More than 60% of its investments are made in this sector. Although capacity remains low (less than 0.2 GW of solar), many projects are being developed at the time of this writing.

Why does a country such as South Africa focus on renewables? First, renewables may enable it to reduce its dependence on fossil fuels such as coal (Winkler 2005). Thus, renewable energy could alleviate the environmental impact of current electricity generation. Furthermore, South African officials realized in the late 1990s and early 2000s that investing in renewable energy would be helpful in climate negotiations (Department of Minerals and Energy 2002). South Africa has embraced cooperative multilateralism in climate negotiations (Urpelainen and Vihma 2015), and renewable energy allows the country to reduce its reliance on coal.

Second, South Africa still suffers from electrification issues. Although the country has made decisive progress (two-thirds of the population did not have access to electricity in 1994; Winkler 2005), eight million people are still without electricity as of 2013 (IEA 2013). The problem is particularly acute in rural areas, where the electrification rate is about 67%. More generally, the root of the problem is the high level of unemployment and inequality that plagues South Africa (Pegels 2010). In the context of the end of apartheid, the question of equality of access became particularly salient and a key social issue (Spalding-Fecher, Williams, and van Horen 2000).

These issues led the South African government to prepare a new energy policy in the late 1990s. In 1998, the first White Paper defined the main themes of South Africa's future energy policy (Department of Minerals and Energy 1998). The key axes were availability and affordability. Four years later, the second White Paper focused more narrowly on renewable energy (Department of Minerals and Energy 2002). The focus now was on the development of renewable energy and the need for adequate public policies. Based on encouraging potential, the South African government has embarked on ambitious plans to develop local renewable energy capacity (Bugaje 2006). This has fueled investments and encouraged the growth of a local renewable energy industry, with firms such as Abengoa growing rapidly.[65]

Of course, South Africa's future is not free of problems. Corruption remains a major problem in the energy market (Krupa and Burch 2011). For years, weak institutional capacity and conflicts among various government agencies have undermined effective renewable energy policy (L. Baker 2011). Eskom, the country's largest utility, dominates the market and tends to be close to powerful politicians. Furthermore, people are concerned about the cost of renewable energy policy. Coal is cheap and abundant, making the redistributive effects of renewable energy policies particularly salient. Similarly, tensions exist between the country's need to grow to improve its unemployment rate and its willingness to see wholesale prices of electricity go up. Still, in December 2014, the government announced a doubling of the target for renewable electricity generation capacity through auctions, saying that bids for another 3.9 GW would be announced in 2015 (existing capacity is about 44 GW).[66] If the government's ambitious plans are realized, then South Africa's dependence on coal will begin to decrease. With 93.8% dependence on coal in electricity generation, however, the change will be slow and gradual (WDI 2013).

Overall, South Africa offers clear support for our expectations. South Africa is facing major issues of quality and equality of access to electricity. Furthermore, its historical use of coal has led to environmental and health issues that accentuate the differences between the wealthy and the poor. Increasingly competitive renewable energy provides a solution to many issues faced by the country. Given that the process is in its early stages, it is still early to assess the degree of pushback against renewables that South Africa will experience in the long run. So far, however, renewables have been broadly popular, which is initially surprising in a country that both relies almost entirely on coal and is Africa's largest coal exporter. The political power of the African National Congress has been stable since the end of apartheid, and the government has announced increasingly ambitious renewable energy targets over time. South Africa's success with auctions of renewable energy projects has been widely recognized in the international community. The government's solid support for renewables has also ensured that the public monopoly utility, Eskom, has not voiced opposition to renewables. Major coal producers, such as Exxaro, have begun investing in renewables as a strategy of diversification and in an effort to exploit the opportunities of the "new energy economy" (L. Baker 2011, 17). In stark contrast to the fierce political struggles of industrialized countries, coal and renewables seem to coexist peacefully in a context of rapidly growing demand for power. Despite South Africa's economic struggles, renewable energy continued to grow in the early 2010s (REN21 2015).

Latin America has also become a contributor to the global diffusion of renewables. A case in point is Brazil, which has invested large sums in renewable energy. Its commitment to biofuels (mostly ethanol) is well known. However, Brazil's contribution goes beyond fuels; for instance, the country is also a leader in the generation of electricity from biomass. In fact, Brazil is the second largest generator of electricity from renewables in the Americas (behind the United States) and is larger than countries such as France (EIA 2013b). In 2012, Brazil invested $3.1 billion in renewables, 40% of which was targeted to biofuels (Pew 2014). Only the United States surpassed Brazil in this respect.

As in other developing countries, the rationale for renewables in Brazil is tightly linked socioeconomic issues. Coincidentally, scholars have characterized Brazilian renewable energy policy in terms that are similar to South Africa's. Geller et al. (2004) state, "energy policy objectives reflect broader national goals . . . including maintaining strong economic growth, fostering social development in impoverished communities and regions, and reducing environmental degradation." The political objectives are to tackle social issues through the development of a domestic renewable energy industry.

In many respects, Brazil's situation differs from other developing countries. Its energy policy is rooted in the issues the country faced in the aftermath of the 1973 oil crisis. Brazil's military dictatorship was dealing at the same time with three issues (Hira and de Oliveira 2009). First, the country was highly dependent on foreign oil, importing about 80% of its needs. In line with the reactions of industrialized countries, Brazilian officials sought a way to reduce the country's dependency on foreign fossil fuels. Second, the oil crisis triggered a financial crisis that threatened the stability of the local currency. A currency crisis would further hurt economic stability and imperil the dictatorship. Third, the price of sugar collapsed, leaving vast number of farmers on the verge of bankruptcy. Biofuels offered an ideal solution to these three issues. They would reduce demand for foreign oil and thus reduce the pressure on the exchange rate. At the same time, they would increase national security, and, finally, they would bail out struggling farmers. Biofuels were a silver bullet.

Since then, the biofuel industry has gone through ups and downs. Public support decreased in the 1980s and 1990s, leading to a collapse of demand for biofuels (Pousa, Santos, and Suarez 2007). Similarly, the broader renewable energy industry was mostly nonexistent before 2000. In that year, nonhydro renewables generated about 7.8 GWh compared with 86 GWh in the United States (EIA 2013b). The situation changed in 2002. New tax incentives revitalized the demand for cars using biofuels (Hira and de Oliveira 2009). The support for sugarcane in turn led to increasing amounts of waste ("bagasse"), which in turn could be used for the generation of electricity from biomass (Martinot et al. 2002). More generally, Brazil has implemented a number of policies that are favorable to renewables. Notably, the government designed Proinfa, an FIT, which was applied between 2002 and 2006, followed since then by an auction system (Barroso 2012). The results were remarkable: the price of renewable electricity from wind and bio electricity rapidly decreased. The cost per MWh of wind went from about $150 under Proinfa to $53 in 2011 while capacity kept increasing. The generation of nonhydro renewables grew rapidly to reach about 35 GWh in 2011. According to REN21 (2014), Brazil had 3.5 GW of wind capacity alone in 2013, with another 10 GW already under contract. In turn, the renewable energy sector has become a large employer, with about 890,000 people on its payroll, including those directly and indirectly employed by these industries, most of whose income is tied to biofuels (REN21 2014). Brazil has become the clear leader of renewables in Latin America.

The Brazilian case buttresses our argument concerning the attractiveness of cost-effective renewables for developing countries. Investment has increased, prices have decreased, and production has gone up. Brazilian officials saw in renewable energy, especially after 2002, a tool to tackle issues pertaining to poverty, rural electrification in remote communities, and rapidly growing power demand. At the same time, support for renewables was a convenient way to offer support to important constituencies, including farmers. These findings are in line with our first expectation.

What about the expectation that renewables should have met much less political opposition in Brazil than in the industrialized countries? Again, a survey of Brazilian energy politics supports this expectation. The high-profile controversies in Brazil have, instead, recently revolved around the massive Belo Monte Dam project in the state of Pará (Carvalho 2006). Brazil's electricity consumption is expected to grow by more than 50% in the next decade, meaning that there is a clear need for new sources of power.[67] In such a context, modern renewables are understandably not generating heated political controversies because they present a solution to an important problem.

The popularity of renewable energy in Brazil can also be seen in public events. The 2014 World Cup, for instance, can be considered the "greenest . . . in football's history" because of the extensive use of solar power in the stadiums.[68] Large amounts of solar generation capacity were commissioned by the federal government of Brazil to bolster the Cup's green credentials and present Brazil as a modern economy with a strong commitment to sustainability.

Brazilian political parties agree on the importance of renewable energy. President Dilma Rousseff, who was Brazil's head of state from January 2011 to August 2016, has used state banks to offer generous credit to wind energy developers and can be considered a wind energy advocate.[69] Of her primary challengers in the October 2014 elections, the runner-up, Aécio Neves of the Brazilian Social Democracy Party, also campaigned on prorenewable energy policies, although with an emphasis on market-based solutions that focused on business development (Edelman Significa 2014). Marina Silva, who ran for presidency in 2014 under the Socialist Party, is an environmentalist who is even more strongly committed to renewable energy than the other candidates. Brazil's Intended Nationally Determined Contribution for the Paris Agreement codifies these positive sentiments and declares the goal of increasing nonhydrorenewable electricity generation to 23% by 2030—from a baseline of less than 10% in 2015.[70]

Compared with the emerging economies analyzed earlier, Brazil's challenges are different. Because of the country's large hydroelectric potential, Brazil generated in 2012 only 2.6% of its electricity from coal and 8.5% from the less polluting natural gas (WDI 2013). Looking to the future, however, Brazil's electricity demand is growing rapidly, and the potential for large hydroelectric facilities is being exhausted. For Brazil, then, modern renewables offer an opportunity to prevent an increased dependence on fossil fuels in a previously hydro-based power sector. At the same time, distributed renewable electricity generation offers a possible solution to energy access problems in the vast Amazonian rainforest, although progress so far has been quite slow (e.g., Zerriffi 2011).

Thus far we have focused on emerging countries. What about the least developed countries? There, too, renewable energy has become an increasingly popular way to deal with socioeconomic issues. Mostly, renewable energy is often perceived as a solution to the endemic electrification problem (Karekezi 1994). This is particularly true for Sub-Saharan Africa and its rural regions (Deichmann et al. 2011). Not only is electrification an important determinant of quality of life, but it also has many indirect effects. Studies show that reducing the energy access gap between rural and urban regions would contribute to the reduction of poverty (Barnes 2007). Notably, this is because electrification enables small businesses to flourish, gives more time for children to study, and improves people's health.

Given the obstacles for grid development, off-grid solutions have attracted much attention in recent years. The deployment of off-grid devices based on renewable energy is particularly salient, given that they are cheaper to operate than off-grid devices that run on fuels such as kerosene. A microgrid powered by solar energy, for instance, generates low marginal costs. If somehow the problem of high capital costs could be solved, then such a device could in theory be used by both poor and isolated households. Unsurprisingly, many hope that small-scale solar micro-grids or small wind turbines could alleviate the issue of rural electrification.

Similar to many other countries, Kenya suffered tremendously in the aftermath of the two oil crises, with virtually no economic growth. This weak performance created a mass of unemployed youth and other social problems related to a stagnating economy. At the same time, Kenya, like many other Sub-Saharan African countries, suffered (and still suffers) from serious electrification issues. According to the IEA (2013), about 34 million Kenyans do not have access to electricity. The problem is particularly acute in rural areas, where the electrification rate is about 7%. As such, it is significantly below the already low average Sub-Saharan electrification rate (18%).

To address these issues, government officials considered the idea of promoting the use of solar photovoltaic grids already in the 1970s. These plans were soon abandoned, however, because the prices were simply too high for a poor country. The situation gradually changed in the 1980s, with the development of a private market for solar systems (Acker and Kammen 1996). The boom was possible thanks to two favorable developments. First, local entrepreneurs enthusiastically pushed for the development of new solar devices. Second, world prices decreased, making solar panels affordable to the relatively poor population (Acker and Kammen 1996). Since then, the sale of solar systems has outpaced the number of new connections to the grid (A. Jacobson 2007). An interesting feature of the Kenyan case is that the government's role has remained limited throughout. In fact, even in the 2000s, the vast majority of off-grid solar photovoltaic systems in circulation were unsubsidized (A. Jacobson 2007). The results are impressive. In 2014, for example, Kenyans bought 940,000 branded solar off-grid products (BNEF 2016).

In parallel, Kenya is increasingly investing in the use of renewable energy to power its grid. The government introduced an FIT in 2008 and net metering in 2012.[71] Kenya's grand strategic development plan for the upcoming decades, *Kenya Vision 2030*, lists a number of large-scale wind and solar projects as priority targets. These include solar PV but also and mostly wind farms, such as the 100 MW Isiolo project (Government of Kenya 2012). Most of these are placed under the tutelage of KenGen, a state-owned utility. These projects aim to provide power through the grid and reduce the bill from fossil fuels. They are not small. For instance, the Lake Turkana Wind Power project is expected to generate about 15% of the country's electricity and reduce fuel imports by about $186 million per year.[72] These numbers are not negligible to the extent that the plans materialize. Even discounting these projects, Kenya achieved the remarkable feat of investing more than $1 billion in renewables, joining a small club that includes Chile, Indonesia, Mexico, South Africa, and Turkey (REN21 2015). To put these numbers in context, this is about one-half of the investment Kenya Power plans to make between 2013 and 2018 to upgrade its entire electric grid.[73]

The government's influence also shows in the development of geothermal power, which has grown faster in Kenya than in most countries around the world (REN21 2015). Early successes could not mask the slow pace of development of these projects. Hence, in the early 2000s, the government decided to accelerate the process by creating the Geothermal Development Company. The aim was to encourage further investments in geothermal power. The result was a period of rapid growth and one of the world's largest geothermal parks, Olkaria (Bertani 2005). To ensure the influx of private capital, the government enacted a series of policies designed to reduce uncertainty about profits. These included FITs as well as guarantees against losses. This was enough to make Kenya the second largest African investor in renewables after South Africa (REN21 2014).

The renewable energy transition has thus affected relatively poor countries such as Kenya. The general decline in global prices since 2000 coincides with the flourishing development of the local solar industry. This is particularly true for the solar off-grid market but also increasingly for the grid business (mostly through wind and geothermal power). Although electrification rates remain abysmal, the transition has provided Kenya with the opportunity to reinvent its energy infrastructure. The political elites clearly believe that renewables are part of the solution. The ruling Jubilee coalition, in its manifesto for the 2013–2017 legislature, says, "[w]e believe that every Kenyan should have reliable electricity supply in their homes and that all businesses and industries should be able to operate safely with the knowledge that they will have access to energy they need. Kenya must promote diversification and generation of alternative forms of renewable energy such as solar, biogas and wind" (Jubilee 2013, 46). The business angle is reiterated when Jubilee discusses its plan to create one million new jobs and create "a reliable energy infrastructure by extending the national grid network and promoting renewable energy. This will enable the private sector to plan and invest for growth" (Jubilee 2013, 36). On the side of the opposition, former Prime Minister Raila Odinga exhorted his fellow citizens to invest in renewable energy: "[l]et us create a conducive environment for partnerships, let us find investment for power generation projects based on renewable projects."[74] Odinga further argued, "there is no better way to ensure growth than to invest in clean, reliable and affordable energy." Kalonzo Musyoka, another opposition leader, was part of the government when the FIT and net metering regulations were enacted. A former trade minister and yet another leading opposition figure, Moses Wetangula pushed for increasing cooperation with industrialized countries on environmental and energy issues, with particular attention to Kenya's ties with Denmark.[75] Incidentally, Vestas, a Danish firm, would in later years be the biggest contributor to the Lake Turkana project mentioned earlier. Thus, across parties and coalitions, political leaders who otherwise disagree on many important policies shared similar views on the benefits of renewable energy. Obstacles to the deployment of renewables do not stem from political opposition but from other issues that are endemic to developing countries, such as capital constraints and corruption.

Although Kenya is a poorer country than Brazil, their energy sectors are similar in that Kenya generated 75% of its electricity from hydroelectric and other renewable sources in 2012, whereas the remaining 25% came from expensive oil (WDI 2013). As Kenya's economy grows, the country needs more energy. Renewable energy can reduce the pressure to invest in polluting coal. Although the Kenyan government awarded in 2014 a contract for a 1,000-megawatt coal power plant in Lamuo on the north coast of the country,[76] renewable energy can reduce the need to make additional investments into coal-fired generation capacity.

Overall, the picture is clear. Renewable energy has become a global market that attracts the attention of a large number of governments. Prices have declined and investments increased. The competitiveness of renewables makes it easier for local governments to join this technological transition. Their incentives to participate in this race vary. Spain wants to develop its high-technology industry, China wants to avoid social upheaval, and India needs to grow. Others, such as Brazil, possess a comparative advantage in traditional oil fuels and seek to boost their economy on the biofuel wave. South Africa is coping with a range of inequality concerns, while Kenya seeks a solution to its electrification problems. Regardless of their idiosyncratic issues, many governments seem to view renewable energy as a useful solution to their problems. Renewables remain far from crowding out conventional alternatives, but they offer a promising approach to reduce national dependence on coal in countries such as China and India, while allowing Brazil to supplement hydroelectric generation and Kenya to eradicate energy poverty without massive increases in the use of coal.

III The Road Ahead

7 Improving National Policy

Although renewables have become increasingly competitive in global energy markets, their future remains subject to great uncertainty. Enthusiasts envision a future of nothing but renewable energy by the year 2050, a revolution in dealing with climate change and the problem of energy security (Vad Mathiesen, Lund, and Karlsson 2011). More moderate commentators, such as the 2013 *World Energy Outlook* (IEA 2013), expect renewables to contribute an increasing share of the world's total energy consumption, but in their telling, fossil fuels remain dominant for the next 50 to –100 years or so. As the energy analyst Daniel Yergin put it in an interview with *Boston Globe*, "It's a contest, or even a battleground. . . . Renewables will grow a lot, but they will still be, 20 years from now, a relatively small part of the overall mix."[1]

Although these predictions are interesting, their relevance for practice is limited. The outlook for renewables is not a question of prediction but of action. The role and contribution of renewable energy depends on the wisdom of the national and international policies that policymakers are so frantically designing around the world. The potential for the growth of renewable energy depends on policy design. Renewable energy policy must strike a balance between economic cost-effectiveness and the stark political realities that we have emphasized in the previous chapters. A careful consideration of renewable energy policy and outcomes can help policymakers avoid pitfalls and exploit opportunities through analogical reasoning. External shocks, political institutions, interest groups, and public opinion will continue to influence the success and failure of renewable energy policy in the future.

This chapter translates the theory and empirical evidence from previous chapters into guidelines for better renewable energy policy. The goal of our policy analysis is to identify and summarize lessons from our empirical analysis for formulating policies that survive under political pressure.

We begin by emphasizing the need for governments to identify and seize opportunities for new renewable energy policies, reflecting on the previous case studies and offering new examples. We then discuss the challenge of using policy to create advocacy coalitions that help more or less interested governments sustain renewable energy policies for extended periods of time. The third section focuses on the challenge of sustaining this enthusiasm and momentum over time, given changing circumstances and new societal challenges. The fourth section addresses the question of how governments should deal with maturing renewable energy technologies that are undergoing their lock-in, and the last two sections discuss the role of international cooperation in the future development of renewable energy.

Our argument can be condensed into a series of policy prescriptions. To begin with, external shocks create a window of opportunity for new policies that, although seemingly trivial, can contribute to new innovations in the renewable energy field. For example, there is clearly a need for policies that promote renewable energy outside the power sector. Over time, exogenous shocks may allow governments to build new constituencies in favor of renewables and ensure the growth of policy. As recent political backlashes against renewables in countries such as Germany show, however, it is essential to move from rapid growth to the management of lock-in. For electricity generation with renewables, this shift is now essential for long-term success. With careful institutional design, international cooperation on renewables can help governments overcome current challenges.

7.1 Seizing the Opportunity: Tailored Policy Responses

External shocks open a window of opportunity for governments interested in experimenting with renewable energy even under carbon lock-in. While governments cannot manufacture an external shock, the nature and strength of the policy response is in their control. An interested policymaker can respond to an external shock by implementing policies that are politically feasible due to the external shock and allow renewable energy to grow. If anything, the implementation of such policies is easier in today's environment than in the past, given that the cost of renewable energy generation has decreased and the availability of technological solutions greatly improved. Advances in the quality and cost-effectiveness of renewable energy technology allow governments today to break their national carbon lock-in much more easily than in the cases we analyzed. According to the US Energy Information Administration (EIA 2014c), a conventional coal power plant constructed in 2019 would cost $95.6 dollars (2012 prices) per MWh, whereas a conventional natural gas plant would cost $66.3 and an onshore wind facility $80.3. According to this method, wind is already less expensive than coal and only 21% more expensive than natural gas. This is a dramatic improvement over the situation in, say, 1979.

The key challenge for the government is to select policies that are simultaneously an effective response to the external shock and a good fit to the domestic political and economic conditions. In today's policy landscape of the United States, renewable portfolio standards, which were adopted by 29 states by March 2013,[2] are a politically viable way forward (Rabe 2006). The initial policy responses in our case studies were rarely surrounded by the political controversies that renewable energy provokes in the later stages of its development. A common feature of the cases analyzed above is that the initial renewable energy policies were not perceived as threats to the established energy interests. The potential losers from the growth of renewable energy had more important concerns, and the external shock made it politically costly to mount an aggressive attack against renewables at that stage. The former condition of there being larger problems than renewables certainly applies in today's world, as the power sectors of different countries face difficult problems related to their fiscal viability (Bhattacharyya 2007; Victor and Heller 2007), and external shocks may generate even more support for renewables now that these energy technologies are no longer that expensive. At the same time, opponents of renewable energy are now clearly aware of the potentially disruptive consequences of the lock-in of renewable energy. Therefore, in contrast to the conditions that prevailed in 1973, the "element of surprise" is no longer a factor that favors renewables.

For renewable energy advocates, this brief discussion has two implications. First, the reduced price of renewable energy makes today's policies much more cost-effective. Renewable energy leaders can now expect much better returns from their policies than those available only a few years ago. Second, the cost-effectiveness of today's renewable energy policies means that political attacks against them will be more vigorous. The large potential of renewable energy implies that new renewable energy policies present a salient threat to electric utilities, and today's policymakers cannot expect the kind of benign neglect that characterized the early years of renewable energy.

The case studies also emphasize the importance of elite ideology for exploiting the opportunities afforded by external shocks. With the exception of Governor Brown's California, great enthusiasm among the policy elites does not appear to be necessary for successful renewable energy policy. Federal governments in the United States and Germany were not enthusiastic about renewable energy, and even the Danish government only turned to wind after efforts to promote nuclear power had already failed, following years and years of political pressure by environmentalists, wind energy enthusiasts, and antinuclear campaigners. Renewable energy is a practical solution to a host of energy policy problems, and building an advocacy coalition for investment in renewable energy is easier than one might initially imagine.

More important than high levels of enthusiasm at the highest levels of government were the lack of vocal opponents and the perception of an opportunity. The relatively innocuous and inexpensive nature of the initial renewable energy policies mitigated political problems in the critical early stages of industry development, and the positive public image of renewables generated political support. Institutional capacity was critical because it determined the nature of appropriate political responses and policy instruments. The powerful, influential fossil fuel lobbies in the United States and Germany did not try to undermine renewable energy policies. If anything, corporations either ignored or tried to benefit from available funding to develop their technologies and ensure their standing in renewable energy markets.

In addition to the historical examples we have already provided, the more recent cases of Japan and India shed a lot of light on the role of external shocks in promoting the growth of renewable energy. Consider Japan's experience with the Fukushima disaster. On March 11, 2011, a tsunami caused by a powerful earthquake killed 20,000 and resulted in a nuclear meltdown in a nuclear facility in Fukushima.[3] In February 2014, when the cabinet unveiled a new energy policy strategy, all of Japan's 48 commercial nuclear reactors remained offline.[4] Although the new energy policy, which is intended to reverse the moratorium on nuclear power imposed immediately after the meltdown, does not aim to remove nuclear power from the country's energy mix, the plan emphasizes the importance of investing in renewables to drive the use of nuclear power "as low as possible." At the same time, the Fukushima incident was decisive for Germany's decision to phase out nuclear energy. Following the largest antinuclear protest in Switzerland's history, the cabinet also banned the construction of new nuclear power. In Italy, a referendum also banned the construction of any new nuclear power. These policies create openings for renewable energy, as any nuclear power generation phased out or planned but not constructed in Germany, Switzerland, and Italy will have to be replaced with other electricity capacity.

Compared with the international response, Fukushima's own response was much more aggressive. In February 2014, the Fukushima prefecture announced a goal of 100% renewable energy by 2030.[5] The prefecture had already started investing in solar and offshore wind power in 2013.

Because the Fukushima population was directly affected by the meltdown of the nuclear reactors, this external shock created enough political support for the announcement of an aggressive target over a relatively short time horizon, along with concrete investments to meet the target. Fukushima's own reaction shows how the severity and salience of the external shock in different parts of the world trigger different policy responses, although it remains to be seen how far Fukushima can go with respect to renewable energy.

In the United States, renewable sources of energy are now facing an altogether different external shock. Technological advances in the extraction of tight natural gas through hydraulic fracturing, or "fracking," have prompted rapid and large decreases in the domestic price of natural gas in electricity generation, although the price remains much higher in other regions of the world. Between 2003 and 2013, the share of natural gas in electricity generation increased from 17% to 27%.[6] Because natural gas competes with wind power, the direct effect of the low natural gas prices may be a slowing of the growth in renewable electricity generation. It bears remembering, however, that hybrid systems based on renewables and natural gas hold considerable promise (Lee, Zinaman, and Logan 2012).

Our political analysis suggests that the shale gas boom may produce some other benefits in the long run. Because natural gas has already displaced a lot of coal generation capacity, the most vocal opponents of climate policies and carbon abatement in the power sector are losing their political clout. As the importance of coal in the US power sector decreases, the political impediments to carbon abatement may be reduced. This is good news for renewable energy because, whereas natural gas produces one-half of the carbon dioxide of coal per unit of electricity, carbon emissions from wind and solar power are negligible. Therefore, even if shale gas is undermining the growth of wind and solar power in the short run, political logic suggests that the declining significance of coal may allow a surge of renewable energy generation in the future. In 2014, President Obama put limits on mercury emissions from power plants and began implementing policies that cap carbon dioxide emissions at the coal plant level under the Clean Air Act.[7]

There are several plausible candidates for future shocks that policymakers should consider. One scenario is a rapid increase in the prices of fossil fuels. These prices have fluctuated over time, with an unexpected collapse in late 2014 and early 2015. Even under these relatively low and stable fossil fuel prices, forecasts predict that renewables will become more and more important in the world economy (BP 2014). However, these predictions assume there are no major shocks in global energy markets, such as a major conflict in the Middle East. If the prices of these fossil fuels prove higher than expected, then renewables will become even more attractive. High oil prices could encourage efforts to use biofuels or electrification in the transportation sector, and high natural gas or coal prices could promote renewable electricity generation.

Positive shocks are also possible through technological innovation. One important reason that wind energy paved the way for renewables was the rapid cost reduction achieved in Denmark through commercial innovation. From the perspective of other countries, the Danish achievements can be considered an external shock that created an opportunity for renewable energy. As the cost of wind electricity generation decreased over time, governments around the world could expect more returns to renewable energy policies. Today, offshore wind electricity is at the cusp of commercialization, and some countries, such as Germany, the United Kingdom, and China, are investing in the commercialization of offshore wind resources.

Although the future of renewable electricity seems rather bright even without additional external shocks, the situation is quite different for nonelectric renewables. Of particular import is the transportation sector. Efforts to replace oil with more sustainable alternatives have largely been unsuccessful, with few exceptions such as Brazil, where a favorable natural environment allows efficient production of ethanol from sugar cane. According to the IPCC (2007), transportation is responsible for 13% of global greenhouse gas emissions. What is more, emissions from the sector have been growing fast. As long as the demand for motorized transportation continues to grow, technological breakthroughs are necessary for sustained emissions reductions. Advanced biofuels and electric vehicles hold promise, but they have yet to achieve the kind of technological momentum that has allowed wind and solar to grow exponentially in the power sector. With enough technological progress, advanced biofuels could replace oil without demanding so much land that food prices increase, whereas a new generation of electric vehicles could allow the transportation sector to benefit increasingly from renewable electricity.

Although policymakers cannot manufacture beneficial shocks, they can foster conducive environments and prepare for future shocks. Investments in research and development increase the probability of the kind of technological momentum that allowed Denmark to become a pioneer in wind energy, and the high energy prices of recent years have been a key reason that renewable energy has become more competitive. Policymakers should invest in technological innovation under the assumption that the prices of fossil fuels may sharply increase in the future while preparing contingency plans for the rapid deployment of renewable energy technology, not only in electricity generation, but also in transportation and other sectors.

7.2 Creating Winners and Advocacy Coalitions

When renewable energy grows in the leading countries, it starts to perturb the prevailing equilibrium beyond the effects of the initial external shock. At this stage, the initial carbon lock-in is no longer guaranteed, and the opponents of renewable energy mobilize to undermine further growth of the new energy industry. Our case studies suggest that this is a common pattern. In the United States, a coalition of fossil-fuel producers and conservative ideologues was able to turn the tide against renewable energy during the last two decades of the 20th century. In Germany, the same interests also fought a bitter, if less successful, battle against a coalition of the Green Party, the broader environmental movement, and other renewable energy enthusiasts. Even in Denmark, where opposition to renewable energy was limited, political controversies and debates erupted when Anders Fogh Rasmussen's right-wing cabinet began to dismantle the FIT in 2002.

Renewable energy can survive politicization when the initial policies create a strong, robust constituency in favor of the continuation and expansion of renewable energy policy. Sometimes such a constituency is created through aggressive policies, although this strategy may prompt a backlash by the opponents of renewables. At other times, seemingly innocuous policies begin to create an advocacy coalition by stealth. One of the great puzzles in the political history of renewable energy is how the budding renewables industry can survive the onslaught of the fossil-fuel industry and its allies when the initial external shock wanes and enthusiasm for renewables should logically decrease. Building on theories of path dependence and positive reinforcement (Hathaway 1998; Pierson 2000; Michaelowa 2005; Aklin and Urpelainen 2013b), we have shown that the initial growth of renewable energy can, even within a relative short time span, build a politically powerful and influential advocacy coalition. This mechanism is a primary reason that renewables can grow beyond their modest initial levels even in a society under carbon lock-in. The growth of the renewable energy constituency provides the government with strong political incentives to continue investing in renewables, thwarting attacks of the renewable energy opponents.

The set of potential allies of renewable energy is diverse. Some actors, such as manufacturers and engineering associations, support renewable energy because it creates new markets for clean technology. Others, such as environmentalists, are interested in sustainable alternatives to fossil fuels. Yet others, such as farmers and foresters, are interested in policies that generate profits for the generation of electricity from wind, solar, and sustainable biomass. Municipalities may see renewable energy as a politically popular way to achieve energy independence and generate local employment, and agricultural cooperatives expect profits and self-reliance from renewable energy. Homeowners may expect subsidies for installing rooftop solar panels.

The case studies of Germany and Denmark offer a remarkable illustration of the diversity of the renewables advocacy coalition. In both countries, large segments of industry threw their weight on the side of renewables. In Germany, the national engineering community was decisive in helping the FIT to survive the challenge of the conservatives. Over time, wind and solar technologies became major employers and domains of industrial innovation. In Denmark, wind energy technology grew into one of the most important export industries. In both cases, renewable energy policies strengthened the advocacy coalition, and the growing political clout of the advocacy coalition allowed these policies to survive over time. More generally, studies of clean technology commercialization have found that domestic "lead markets" can be critical for the development and progress of a new technology (Beise 2004; Beise and Rennings 2005; Huber 2008), and renewable technology is no exception (Lewis and Wiser 2007). If domestic renewable energy policies allow a vibrant export industry to emerge, as they did in countries like Denmark, even a relatively small country can sustain a large renewable technology industry with political clout that produces sustained technological progress and generates support for new renewable energy policies.

A less obvious, yet politically critical, part of the advocacy coalition was agricultural producers. In Denmark, the agricultural cooperatives, which were formed already in the 19th century for reasons completely unrelated to renewable energy, turned out to be a natural ownership mechanism for wind power. Although large utilities did not show a lot of enthusiasm for wind power, local ownership of wind turbines grew exponentially over time. Because local ownership also meant that a large number of communities reaped direct benefits from the FIT, political controversies over renewables were kept at bay for decades. In Germany, the four major utilities owned only a small fraction of all renewable electricity capacity, whereas farmers, homeowners, and municipalities made major investments. Again, the spread of the generous benefits from the FIT to a large number of constituencies in different areas of Germany helped the government create a strong support base for renewables.

In the United States, a key obstacle to the growth of renewable energy despite initial success was the lack of a growing, politically influential advocacy coalition in favor of more generous policies and further growth of renewables. We found that the US renewable energy policies benefited the incumbent energy producers and technology providers, at the expense of new entrants that would have had a direct stake in renewable energy. An energy producer that is mostly dependent on coal or oil, for example, is hardly interested in aggressive renewable energy policies that would undermine the core business of fossil-fuel extraction and use. Even in California, where wind energy generation grew rapidly in the 1980s, the initial boom in renewable energy was driven by the economic rents made available to power generators through PURPA and other policies. The growth of wind energy did not allow the emergence of a wind power industry that would have been able to sustain high growth rates or produce political spillovers to other types of renewables. As documented in the previous chapter, today's situation is different—there is now a powerful renewable energy advocacy coalition in the United States that includes environmentalists, large clean technology companies, homeowners invested in solar, people working in the installation industry, and so on.

Approaches that target benefits to concentrated and politically organized sectors bring longevity and staying power to a nascent and generally vulnerable industry. This observation is important for a wide range of climate and energy policies. Although standard economic theory offers a seemingly unambiguous justification for emphasizing market-based instruments, such as emissions trading and carbon taxes (Stavins 1998), the argument for such instruments is weaker in a political setting. Political economists have found that concentrated and organized interests are politically influential due to their ability to influence electoral outcomes and lobby effectively (Olson 1968; Gawande, Krishna, and Olarreaga 2009), and renewable energy is no exception. Governments interested in promoting renewables must carefully weigh the economic costs of targeted benefits, such as subsidies and FITs, against the benefits of constituency creation.

At times, politics necessitates deviations from the economically optimal energy policies. Fast and steady growth of renewable energy depends on the continued growth of its advocacy coalition, and sometimes the policies needed for that growth are not efficient in the short run. Our case studies suggest that if governments are to promote the growth of renewable energy today, then they must implement policies that constitute an effective response to an external shock and also hold the potential for the continued growth of the advocacy coalition. It does not follow, however, that the political crafting of an advocacy coalition for renewables would necessarily cause large deadweight losses in the economy over time. If the advocacy coalition continues to grow, then certainty is brought to the market for renewable energy investment and the cost of relevant technologies due to learning effects is reduced. These benefits should be compared to the cost of building the advocacy coalition. If the cost–benefit ratio is acceptable, seemingly inefficient renewable energy policies can be the ideal approach when political barriers to alternatives are considered. From a dynamic perspective, the main challenge that governments face is to ensure that they provide sufficient support to the industry to prevent it from perishing in the "valley of death" that often stops nascent energy technologies from progressing (Grübler, Nakićenović, and Victor 1999). This is especially challenging to do while reducing these benefits over time and avoiding targeting resources to constituencies who do not really need them. This is not only important for fiscal reasons, but also because the legitimacy of renewable energy is undermined if too much taxpayer money is spent or consumer prices increase too much. Over time, aggressive and generous policies may backfire and prevent renewable energy from achieving its full potential.

The FIT is a great example of this general challenge. As documented earlier, both Denmark and Germany achieved rapid increases in their wind and solar generation capacity much earlier than other countries. As Mitchell, Bauknecht, and Connor (2006) argue, one of the key benefits of the FIT is that it significantly reduces the risk of investing in renewable energy by guaranteeing profits for years to come. However, this logic depends entirely on the continuation of political support for the feed-in policy. If the government is expected to repeal the FIT shortly after enactment under pressure to reduce electricity prices, forward-looking investors have no *ex ante* incentive to consider the future feed-in profits. The risk reduction from the FIT is contingent on the survival and thriving of the renewables advocacy coalition. The Danish and German FITs would not have generated a boom in renewable electricity generation without the expectation that they would remain in power for years to come.

This argument is not an encouragement to ignore the economic costs of renewable energy. The case of corn ethanol in the United States is a clear warning of the dangers of catering to special interests in the name of renewable energy. By 2007, the production of ethanol for transportation fuel had grown to more than 25 billion liters, most of it from corn (Tyner 2008). Much of this growth has been driven by generous agricultural subsidies by different states (Cotti and Skidmore 2010; Skidmore, Cotti, and Alm 2013). For example, Cotti and Skidmore (2010) report that Wisconsin gives 20 cents per gallon up to 15 million gallons, allowing large-scale production by major agribusinesses. However, it is difficult to justify ethanol subsidies as effective renewable energy policy. They carry a high cost, the technology is mature, and the energy balance of ethanol is weak because a lot of oil is required to grow and process the corn. The political power of the agricultural lobby is the only plausible explanation for the ethanol subsidies, as the environmental organizations oppose the ethanol policies. In the case of ethanol, the same political forces that have allowed solar and wind to prosper are causing wasteful spending on an energy form with little transformative potential.

Even Germany's success with the FIT constitutes, in some respects, a cautionary tale in recent years. Although Germany's FIT was a stunning success in promoting the rapid growth of renewables, the economic cost of *Die Energiewende* has grown rapidly with the number of beneficiaries. In a scathing critique, *The Economist* writes, "[a]n average household now pays an extra €260 ($355) a year to subsidise renewables: the total cost of renewable subsidies in 2013 was €16 billion."[8] The aggressive policy, which was necessary to enable investment in renewables in the early days, is now generously subsidizing producers that have already made the investments and are now enjoying rents. In this circumstance, it is not surprising that Chancellor Angela Merkel's bipartisan cabinet has chosen to reform the policy to constrain costs.

The main challenge for the government is to strike the right balance. If policies do not allow the advocacy coalition for renewable energy to grow, their seeming economic efficiency is irrelevant because the industry cannot survive the political backlash that its initial success generates. At the same time, renewable energy policies may generate a dangerous kind of lock-in if the advocacy coalition's main interest is in rent-seeking, instead of the growth of a healthy and vibrant renewable energy industry. Appropriate policies to support the growth of renewable energy should provide enough targeted benefits to political constituencies while ensuring that the cost of providing these benefits remains under control and, over time, competition in the industry grows.

In the case of the German turn toward renewables, the government is currently trying to deal with the second challenge, namely, cost containment. In April 2014, the Social Democrat Energy Minister Sigmar Gabriel pushed through a reform of renewable energy policy. Although Germany remains committed to increasing renewable electricity to 40% to 45% by 2025 and to 55% to 60% percent by 2035, the reform "will scale back green subsidies and upper limits will be placed on onshore wind power expansion (at 2.5 gigawatts in capacity per year), photovoltaic (2.5 GW per year) and offshore wind plants (6.5 GW to 2020)."[9] Beginning in 2017, renewable energy producers must also compete with conventional power generators for their electricity sales. Politically, an important move is to exempt certain heavy industrial users of electricity from the renewable energy surcharge, a cost that is carried by ordinary consumers.

Although it is too early to judge the success of Germany's policy reform, our analysis indicates that the direction is the correct one. Germany has reached such a high level of renewable electricity generation and created such a robust industrial ecosystem for renewables that the initial, aggressive approach is no longer warranted. To deal with the backlash against renewable energy, the German government should demonstrate the cost-effectiveness of renewable energy. In addition to protecting Germany's own policies over time, a demonstration of the cost-effectiveness of renewable energy may mitigate and undermine criticisms of renewable energy in other countries that have yet to match Germany and Denmark in renewable electricity penetration.

7.3 Managing the Lock-In of Renewables

Over time, the government's emphasis must shift from supporting the mobilization of the renewable advocacy coalition toward managing the impending or realized lock-in of renewables. Instead of trying to create a lock-in, the government must now ensure that the lock-in is economically and politically sustainable while serving the best interests of the society. This challenge is already reality in countries such as Denmark and Germany, and many other countries are on their way to a lock-in situation. This challenge is important for two main reasons. First, the management of lock-in determines whether renewable energy can actually solve the problems of society. In the worst case, renewables may prove ineffective or even outright counterproductive. Second, the cases of renewable lock-in almost inevitably draw international attention. Depending on the effects, costs, and benefits of renewable energy, other governments and stakeholders will update their beliefs and revise their preferences concerning the role of renewable energy. As renewables become a major contributor to a country's energy mix, the focus shifts from their growth to their performance. Domestic and international observers wonder about the cost, reliability, and future prospects of renewable energy.

The management of the lock-in of renewable energy presents the government with an altogether different problem. Once renewables have grown to play an important role in the energy economy, new obstacles will emerge. How to reduce the cost of renewables and allow them to grow without continuing government support in perpetuity? How to deal with the systemic challenges of renewable energy generation, such as the problems of intermittency in the electric grid? How to ensure continued experimentation in the renewable energy industry? How can the government make the transition from the aggressive support of an infant industry to a policy that facilitates cost reductions in renewable energy and enables a systemic transition?

The success of Germany's *Energiewende* depends on answers to these questions. The debate on the wisdom of phasing out nuclear notwithstanding, the widespread criticism of Germany's energy policy is largely driven by concerns about high electricity prices and the inequitable distribution of the burden. Although there is no question about the rapid expansion of renewable electricity generation in Germany (74% of the total demand on May 11, 2014, was met by renewable energy in the middle of the day[10]), the question is whether the federal government can reduce the cost of renewable electricity to consumers and taxpayers to such a level that the country can become a role model for the rest of the world. Germany now has a large advocacy coalition in favor of renewable energy, and the question is whether the government can steer renewable energy policy in a direction that benefits the entire society and provides an example for the rest of the world, as opposed to providing generous rents to renewable electricity generators.

From an environmental perspective, the sustainability of biomass use is also an important factor in judging the merits of the German *Energiewende*. Besides solar and wind, the growth of renewable electricity generation in Germany has been driven by the increased use of biomass. The environmental consequences of increased biomass use are potentially far more problematic than those of solar and wind. Because biomass generation requires land (with the exception of waste and residue), the rapid growth of biomass energy in Germany raises concerns about environmental sustainability.[11]

Land used for biomass is not available for agriculture, recreational use, or reforestation. As a result, the German government is now imposing limits on biomass use and focusing policy support on the development of waste and residue as sources of bioenergy.

In Denmark, major gains in wind electricity generation have already been achieved without much difficulty thanks in large part to the country's participation in the Nordic electricity market, Nordpool, in which access to abundant Norwegian hydroelectric power is guaranteed (Jacobsen and Zvingilaite 2010). Another advantage for Denmark is the large share of power production from combined heat and power plants, which can use excess wind electricity to provide heat if necessary. Finally, wind electricity generation in Denmark is highest in the cold winter months, when electricity prices are generally high because co-generation plants are geared toward heating. If Denmark is to further increase renewable electricity generation, however, variation in the supply of renewable electricity becomes a more important issue. Potential solutions include increased investment in flexible generation capacity, such as natural gas turbines, and "more interconnection capacity" both within and outside Denmark "to enlarge the market" (Jacobsen and Zvingilaite 2010). Perhaps because of Denmark's small size and favorable national conditions, renewable energy policy does not face the same kind of challenge as in the much larger and pivotal Germany.

From a political perspective, the key obstacle to effective renewable energy policy is now regulatory capture (Stigler 1971; Peltzman 1976). As the renewable energy industry grows increasingly powerful, it becomes a formidable interest group that can shape the government's political fortunes through campaign contributions and electoral mobilization. The tables are turned, and the government becomes dependent on the political support of the renewable energy industry. The danger is no longer that the renewable energy industry cannot grow; instead, there is a risk of harmful growth. For example, the advocacy coalition for renewables may be able to force the government to implement subsidies that have no economic or environmental merits, yet produce handsome profits for the advocacy coalition. The corn ethanol lobby in the United States is perhaps the clearest example of this phenomenon, and in Germany the generous FIT policies have been criticized on similar grounds.

The risk of regulatory capture presents an economic problem as well as a threat to the long-term viability of renewables. Even if regulatory capture may for a time allow renewables to grow at an explosive pace, in the end it will produce policies that endanger the future of renewables. The focus of policymaking under regulatory capture is on rent-seeking and profits, and there is no reason to believe that such policies will allow renewables to grow into a viable industry that can gradually become less and less dependent on government support. Although regulatory capture may initially appear to be good for the environment, it does not bode well for the long-term prospects of renewable energy.

Dealing with this threat requires a strong elite consensus, a critical but supportive public, and an emphasis on good results. For renewables to prosper in the long run, the government must prioritize the competitiveness of the industry over the rents available to special interest groups. Without a strong elite consensus in favor of renewable energy, the distributive politics of rent-seeking become pivotal for renewable energy policy. If the public is not both supportive and critical at the same time, then it is hard for the government to maintain the delicate balance between giving adequate support to renewables and relentlessly insisting on good results and improved performance. The debate on renewable energy must move beyond the question of "yes" versus "no" and instead focus on practical ways to ensure continued improvements in the functioning of the sector.

Of these three factors, emphasis on good results is particularly important for the future of renewable energy. A government willing to forgo short-run political gains to maximize the cost-effectiveness of renewable energy over time can strengthen the elite consensus in favor of renewables and maintain their public support. Although our case studies show that a single-minded focus on renewable energy deployment was initially a good idea, the evidence on the politicization and later lock-in of renewables shows that changed circumstances call for a different strategy. Renewables can now grow fast because their cost of power generation has decreased, but bringing renewable electricity generation to the next level will require a long-term strategy that contains costs.

In the German *Energiewende*, these themes are readily seen. The harshest criticisms of the government's past and current energy policies stem from centrist or conservative sources. In April 2014, for example, the prominent magazine *Der Spiegel* attacked renewable energy as an expensive, hopelessly naïve adventure that would be "romantic" for engineers.[12] Many commentators have also expressed concern about the cost of the renewable energy policy for poor residential consumers, worrying about the return of energy poverty to Germany.[13] Meanwhile, the government's efforts to reform the policy have been criticized, especially from the political left and by the Greens.[14] On May 10, 2014, as many as 20,000 people rallied in

Berlin to support renewable energy and criticize Energy Minister Gabriel's proposal to reduce support and slow down the deployment of renewable energy over time.

For the German government, this crossfire creates a difficult situation. On the one hand, steady progress of renewable energy is dependent on continued support by the state. At the same time, high costs and technical difficulties generate more criticism both within and outside Germany. If *Die Energiewende* is to promote renewable energy beyond Germany through a positive example, then the speed of renewable energy expansion is not the only, or even the primary, parameter of interest. Although some parts of the renewables advocacy coalition are primarily interested in the renewable energy subsidies that provide a material benefit to them, others are more interested in the environmental and social benefits of renewable energy. The government's challenge is, assuming that Germany's leaders want to promote renewable energy instead of only winning the next election, maintaining a sufficiently large advocacy coalition to thwart attacks against renewables while also ensuring that the costs of renewable energy can be contained.

The other major threat to renewables at this stage is the difficulty of systemic change. As discussed earlier, even the small Danish electricity market faces technical challenges as the share of renewable electricity increases over time. For a major electricity market such as Germany, the challenge is much greater. As a small electricity market, Denmark can largely rely on the stable supply of hydroelectric power from Norway to even out fluctuation in wind electricity supply. Because this is not possible for the large German electric market, a more significant systemic transformation is needed. The two primary solutions to this problem are enhanced regional integration and improved storage.

The challenge of dealing with renewables on a large scale is particularly relevant in the power sector due to the challenge that intermittency and storage needs pose to the electric grid. Both fossil fuels and nuclear power generate a relatively steady, predictable supply of electricity, making capacity assessments for the electric grid relatively easy. In contrast, many renewables are only intermittently available. The sun only shines during the day, whereas the strength of wind varies both within any day and over the year. A traditional grid that is designed for certain expected and peak capacities may not be able to accommodate fluctuations in the availability of renewable energy, and so either the grid needs to be made more flexible or energy storage facilities available. As a small economy capable of tapping into abundant hydroelectric power in the Nordic electricity market,

Denmark has managed to grow renewable electricity generation without major systemic challenges. In Germany, the need for a systemic transformation is already visible. In 2013, the German Parliament "adopted a law stipulating the need for 2800 km new high voltage power lines and the modification or extension of another 2900 km of existing power lines until 2022" (Steinbach 2013, 225). Moreover, the German regulatory framework for the authorization of grid construction has been revamped to facilitate expansion already in 2011. Local opposition to grid extension and the associated lack of democratic legitimacy and transparency are potential obstacles to the use of remote renewable energy sources, such as offshore wind. At the same time, the design and construction of a "smart grid" capable of dealing with the optimal storage and allocation of renewable electricity is a costly proposal. Although a smart grid, in conjunction with equally smart meters that can automatically regulate the use of electricity to exploit dynamic pricing over day and night, could improve the quality of power supply from renewables, such technologies are expensive and face a lot of resistance in Germany.[15] Solutions to these issues are necessary if Germany is to continue the expansion of renewable energy without unacceptable costs and serious technical problems.

The other solution to the intermittency problem is improved storage. Studies show that various storage technologies, some mature and others potential, exist (Ferreira et al. 2013). However, their use increases the cost of renewable energy, an issue that becomes increasingly critical as renewables penetrate the electricity market. In addition to technological solutions, the intermittency of renewables can be tackled with demand management. For example, Lund (2012) develops a model of converting renewable electricity into heat for storage in the Helsinki and Shanghai urban metropolitan areas, showing that such a conversion strategy can significantly improve the prospects of intermittent renewables. Another strategy is to use renewable electricity to power electric vehicles (Kempton and Tomić 2005). These and other unforeseen solutions can play a role in scaling up renewables over time.

If governments can overcome barriers to systemic change, the maintenance of a lock-in that has been achieved can have an important role in the diffusion and learning processes at the international level. Although renewables have made progress by leaps and bounds in several European countries, their global relevance is ultimately a question of their diffusion to emerging economies and, later, the least developed countries. As we showed in the previous chapter, such a shift is already occurring, and emerging economies, led by China, are increasingly central to the growth of renewable energy. However, the decarbonization of the emerging economies will require decades. Although China is a renewable energy leader in absolute terms, the share of modern renewables in the energy mix is still much lower than in, say, Germany or Denmark. As Chinese policymakers decide on investments and policies to reduce their dependence on coal, they can look at the experiences of industrialized frontrunners. Therefore, the image of renewables in other countries is at stake. If national renewable energy policies are widely criticized and ridiculed in other countries, their effect on a regional or global level may be negative despite national achievements. Suppose, for example, that the German renewables advocacy coalition remains strong and robust, pushing renewable electricity generation to ever higher levels. This may not be good for the global development of renewables unless the federal government succeeds in containing costs and avoiding rapid electricity price increases.

In summary, the central lesson from the analysis is that the government faces an important challenge in moving from promoting renewables and battling politicization to managing a new kind of lock-in. Over time, the government's focus must move from the obsessive expansion of the renewables advocacy coalition to the careful, prudent management of the renewables lock-in. Raising the share of renewables such that they become a serious competitor to fossil fuels and nuclear power requires paying attention to costs and systemic obstacles to growth.

7.4 Guiding International Cooperation

Many energy markets are now regional or global. When the deployment of renewable energy requires economic exchange across national borders, such as Chinese manufacturers exporting solar panels to the United States for installation, international governance is clearly needed. Indeed, none of the cases considered earlier was fully national. California's early leadership in wind power deployment in the United States depended on a Danish supply of reliable wind turbines, and Denmark's wind energy industry grew thanks to Governor Brown's policies in California. German renewable energy policies were driven by the Chernobyl nuclear accident and concerns about climate change. Today, India's solar policy is under a legal challenge by the United States in the World Trade Organization. China's solar subsidies have allowed solar installations to grow rapidly in Europe and the United States.

International cooperation is important to renewable energy for two reasons. First, international cooperation can help renewables overcome the challenge of a systemic transition. Although renewables are now in a secure position due to their improved competitiveness and political popularity, in most countries they have yet to trigger a systemic transition of the power sector. Regional efforts can facilitate such a systemic transition because the problem of intermittency is easier to deal with on a regional rather than a national scale.

To see the need for deeper regional integration as renewable energy use continues to grow, one only has to consider Germany's situation. As Europe's largest economy, the difference between Germany's domestic electricity generation and demand has major implications for the European electricity markets. Although much can be done through domestic infrastructure, such as the construction of a smart grid and the installation of smart meters for electricity consumption, another potentially less expensive route toward greater renewable energy penetration goes through greater integration with the broader regional markets. Although Germany's renewable electricity penetration continues to grow, the problem of intermittency and the low market prices for electricity promoted by renewable energy subsidies wreak havoc both within and outside German borders. In January 2014, the General Commission for Strategy and Forecasting of France published a report blaming Germany's renewable energy policies for problems with the European electricity market.[16] According to the report, generous renewable energy subsidies have driven the wholesale prices of electricity so low—often negative—that market prices are no longer any kind of a signal for the scarcity or abundance of electricity. Fluctuation in the supply of renewable electricity is hard to predict and has, at times, even created additional demand for coal to replace rapidly plummeting renewable electricity supply due to intermittency.

The problem is a lack of policy coordination and grid integration. Germany's renewable energy policies have always reflected domestic demand for subsidies by the renewables advocacy coalition and have not considered the effects of these policies on other countries. From a regional or global perspective, it hardly makes sense that Germany's renewable energy policies now drive down wholesale prices around Europe, encouraging electricity consumption and reducing the profitability of renewable energy outside German borders. At the same time, Germany fails to capitalize on renewable energy resources elsewhere. For example, although the Danish and North German wind electricity-generation profiles are relatively similar (Jacobsen and Zvingilaite 2010), the correlation over time, measured hourly, between the feed-in of renewable (solar, wind) power in Germany and Spain is only $r = 0.18$.[17] Given this low correlation, a system of policies to optimize the use of different power sources from Germany to Spain could greatly mitigate the problem of intermittency and stabilize electricity prices. Similarly, a regional grid integration could allow continental Europe to benefit from wind electricity generated in the favorable geographic conditions of the United Kingdom.

Although a European "supergrid" could produce major benefits for a regional renewable energy expansion, it faces political impediments.[18] Perhaps most important, the difficulty of bridging and reconciling the regulatory approaches of different countries is immense. Although most European countries have deregulated their power sectors over time, regulatory differences among countries remain large. Indeed, the European Union considers energy a field of "shared competence," meaning that policy formulation is still very much in the hands of member states (Braun 2011). This rule raises barriers to genuinely regional efforts to coordinate renewable energy policy. With 28 European Union members, the challenge is indeed a daunting one, and it may be more realistic to expect partial integration schemes among individual countries. For example, in January 2014, the president of France, François Hollande, announced increased French–German cooperation on renewable energy, smart grids, energy efficiency, and storage.[19] Although partial integration solutions would not reap the full gains from regional integration, they could gradually bring the European electric grid toward a better outcome.

Second, international cooperation can support the initiation of a renewable energy transition in developing countries. International institutions and ad hoc forms of cooperation can remove roadblocks to the spread of renewable energy across borders, from frontrunner countries to their followers. This is of particular import in the case of the least developed countries. Indeed, one key lesson from our case studies is the remarkable interdependence of both technological development and policy formulation. Denmark's initial success in wind technology development was largely dependent on demand for wind turbines in California because the Danish market was too small to support a large clean technology industry, and other European countries were not yet investing in wind power. The global triumph of the FIT over alternative renewable energy policies reflects positive early experiences and impressive results in Germany and Denmark. In light of these two observations, we focus our discussion of global policy coordination and international cooperation on renewable energy on learning and the spread of information.

The international governance of renewable energy remains inadequate, clearly behind national efforts to promote renewable energy. Although national policies to promote renewable energy have prompted annual investments of hundreds of billions of dollars, the role of international organizations and agreements is much less significant. Few international institutions invest in renewable energy, and their activities pale in comparison with national efforts. Technology agreements focused on renewable energy remain rare (de Coninck et al. 2008; Urpelainen 2013), and no major multilateral initiatives would channel significant funds into the renewable energy sector. Negotiations on permissive trade rules for clean technology have stalled in the World Trade Organization.

Yet the potential benefits from increased global cooperation are large. As governments and bureaucrats formulate and implement renewable energy policies, they face many difficult choices. Information about policy experiences and technological advances can be disseminated across borders faster if an international body collects, reviews, analyzes, simplifies, and communicates said information. Such an informational mechanism does not require large investments by national governments. A clear mandate, a competent staff, and the political support of a large number of national governments are enough.

Because renewable energy can provide great benefits to the poorest countries, building technical capacity in the least developed world should be a priority. Renewable energy can be deployed in a decentralized fashion even if the national grid is incomplete, and engineering studies show that, for remote rural communities, a decentralized solution, such as a community microgrid, is often less expensive than grid extension (Urpelainen 2014). However, the design of renewable energy systems for incomplete, poorly functioning power markets in the least developed world is a major policy challenge. In addition to suffering from limited technical and institutional capacity, least developed countries do not have access to a large body of experiences because most historical progress in renewable energy has been achieved in industrialized or at best emerging economies.

Urpelainen and Van de Graaf (2013) show that the International Renewable Energy Agency (IRENA), the leading multilateral organization for renewable energy, has had an excellent start in the promotion of renewables. IRENA was established in January 2009 and has headquarters in Abu Dhabi, United Arab Emirates. A small but effective organization, it "has a staff of about 70 people and an annual budget (2013) of almost USD 30 million" (Urpelainen and Van de Graaf 2013, 160). As they argue,

> By focusing on a narrowly defined set of goals related to the deployment of renewables, IRENA has, in spite of a small budget and the lack of a proven track record, established itself as a major provider of epistemic services to the least developed

countries. IRENA's mandate, which focuses on renewable energy, allows it to sidestep some of the political controversies that surround nuclear energy and climate change (Urpelainen and Van de Graaf 2013, 174).

According to the authors, IRENA's success in the provision of information and technical assistance to the least developed countries stems from a clearly defined and sharp focus. By remaining outside bodies intended to negotiate legally binding rules, IRENA avoids the politicization of information about renewable energy.

Examples of the benefits of technical assistance by IRENA abound. In one recent publication, for example, IRENA outlined a "roadmap" for doubling the share of renewable energy in the world by 2030 by aggregating data on national conditions in major economies (IRENA 2014b). Although most of the economies covered are industrialized or emerging, the publication also considers several of the least developed countries and smaller middle-income countries, including Ecuador, Indonesia, Morocco, Nigeria, and Tonga. The publication reviews available technologies and national challenges, offering policy recommendations for the international community. Another IRENA (2014a) publication offers guidelines for evaluating the design and effectiveness of renewable energy policy, including criteria such as "institutional feasibility" and "equity," which are of particular relevance to the least developed countries.

Indeed, one of IRENA's central achievements is the creation of comprehensive databases on renewable energy policies and resources across the world. In collaboration with the IEA, IRENA maintains a global database of renewable energy policies and measures.[20] A particularly notable feature of this database is that it contains a national profile for more than 100 countries, reaching significantly beyond the earlier data-collection efforts of the IEA among industrialized countries. IRENA has also launched a global portal for evaluating renewable energy potentials for wind, solar, biomass, hydroelectric, marine, and geothermal energy around the world.[21] The freely available portal evaluates the technical potential of different renewable energy sources for almost all countries in the world, allowing policymakers to focus their efforts on those energy sources that hold promise in any given national setting.

IRENA's explicitly nonpolitical, nonenvironmental stance has made the promotion of renewable energy easier. Although renewable energy is directly relevant to climate change mitigation, for example, IRENA has specifically avoided commenting on the role of renewables in climate policy. This, according to both Urpelainen and Van de Graaf (2013) and Meyer (2013), has shielded the agency from political controversies while allowing it to move forward with the important technical assistance, capacity building, and information dissemination functions that serve the least developed countries in particular.

To complement IRENA's technical agenda, Germany spearheaded the creation of the 10-country Renewables Club on June 1, 2013, in Berlin.[22] The Renewables Club is a smaller group of countries and expected to complement the work of IRENA through agenda setting and example. Although the performance of the Renewables Club remains to be evaluated in the future, it is a sensible strategy to separate the frontrunner approach from the more technical work of IRENA.

Another key issue for international cooperation is the development of global rules for trade and investment in renewable energy or perhaps in sustainable energy technologies more broadly. The renewable technology industry is global, and the cost of renewable energy deployment in a country depends to a large extent on the availability of imported technologies. At the same time, national renewable energy policies may shape trade flows by, for example, favoring domestic technology providers and investors. These cross-border externalities mean that the decisions of individual countries have important implications for the renewable energy futures of other countries through trade channels; therefore, guidelines and agreements on common rules would be useful.

The key challenge for the creation of trade and investment rules for renewables is to strike a fine balance between economics and politics. From a purely economic perspective, the ideal set of rules would impose a ban on policy discrimination against foreign players and remove tariffs on renewable energy technologies as soon as possible. From a political perspective, however, this does not appear wise. We found that when countries such as Germany, Denmark, and China built their renewable energy industries, they relied on industrial policies that promoted domestic generation. If national governments enact renewable energy policies in part to create and support domestic industries, then a blanket ban on discrimination against foreign countries could undermine political incentives to act. This is not to say that there should be no trade and investment rules at all, but it is important to recall that, because renewable energy generates many global benefits, policies that promote increased deployment should be encouraged. Indeed, China's solar manufacturing subsidies have, by bringing down the cost of solar photovoltaic modules, benefited other countries by giving them access to cheaper solar power.

Consider again the dispute over India's solar policies between India and the United States. In this case, the US Trade Representative's complaint pertains to rules such as local content requirements that prevent American technology providers from competing on a level playing field in the rapidly growing Indian solar market. This complaint is understandable from the perspective of a US trade representative, and we do not advocate a system that does not constrain protectionism. At the same time, it is important to consider the possibility that India's government might have not implemented such an ambitious national solar policy without any targeted benefits for domestic constituencies. Moreover, the continuation of India's national solar policies will, if our case studies are any indication, depend on the creation and expansion of a domestic advocacy coalition.

At the same time, tariffs on solar products may undermine the growth of the domestic energy service industry. In June 2014, India's newly appointed Minister for Power, Piyush Goyal, argued that India should not impose new tariffs on imports of solar products, saying, "[a]s things stand today, India doesn't have adequate manufacturing capacity to support the kind of thrust we want to give to solar."[23] With much of India's solar photovoltaic capacity being built on foreign solar panels, an increase in the price of imported solar panels could dramatically reduce the attractiveness of solar power in India, undermining Prime Minister Narendra Modi's plans to use solar to improve India's energy security and eradicate energy poverty. In this case, excessive trade protection could weaken the most robust part of India's solar industry, that is, installers and service providers.

One promising proposal for anticipating and preventing such disputes is the Sustainable Energy Trade Agreement by the International Centre for Trade and Sustainable Development (ICTSD 2011). The proposal would initially focus on trade barriers, leave a lot of flexibility in the agreement, and pay particular attention to the special and differential needs of less developed countries. This incremental approach, formulated in negotiations among governments, would allow the international community to reap gains from liberalizing energy technology trade without undermining the political foundations of renewable energy policy.

7.5 Promoting Renewables across Borders

Although the functions discussed previously are largely technical, the possibility of international cooperation to effect more tangible changes in national energy mixes is also worth considering. This approach would require more resources and a deeper commitment by a group of national governments, but it would also facilitate the deployment of renewable energy more directly than capacity building and technical assistance. For example, governments could agree on mutual increases in their national renewable energy investments through reciprocal policies. Because renewables generate transboundary benefits, each government would commit to increased investment in renewables, provided that other governments participating in the agreement also did so. This "tit for tat" strategy would allow renewable energy generation to increase.

In principle, standard channels of development finance can prove helpful here. Because renewable energy produces regional and global benefits, there is a strong rationale for funding by multilateral development agencies. In addition to the standard rationale for development assistance—in this case, improving the supply of energy to enable sustainable socioeconomic development—the positive externalities for countries outside the host of the renewable energy projects further increase the appeal of renewables for multilateral development lending, especially as the "global public good" rationale for multilateral financial assistance becomes increasingly salient (Kaul, Grunberg, and Stern 1999).

Consider, for example, the accomplishments of the GEF. Because the GEF specifically focuses on promoting projects that provide global environmental benefits, it is an ideal multilateral funding agency for looking at the idea that development funding can promote sustainable energy development. Dixon, Scheer, and Williams (2011) report GEF financing worth $1.3 billion between 1991 and 2010, with an estimated $7.6 billion catalyzed in cofinancing by recipient governments, the private sector, nongovernmental organizations, and other international funding agencies. Of this total, more than one-fourth has been invested in Africa, with a ratio of direct to cofinancing of 0.20. Although renewable energy has by now grown to become a big enough global business that the GEF's contributions to the total global investment are minuscule, the GEF and other development agencies can still provide critical funding to pilot and demonstrate new renewable energy technologies in countries with high potential but low penetration. Indeed, "GEF's investment approach to RE development and deployment is aimed directly at reducing these barriers through financial incentives, capacity building efforts, and support for policies and regulatory strategies that encourage local investment in renewable energy resources" (Dixon et al. 2011, 90).

Although little explicit international cooperation on renewable energy exists today besides development assistance, the potential scope of renewable energy agreements is wide (de Coninck et al. 2008). To illustrate, consider a simple deal whereby every member of a renewable energy agreement commits to investing in renewable energy. Each country's investment in renewable energy promotes technological progress, reduces carbon dioxide emissions, and mitigates the scarcity of fossil fuels. Therefore, each country benefits from the investments of others. A group of governments can agree to increase their investments, provided that none of the group members defects by not investing. This logic of reciprocity is similar to that proposed by Victor (2011) for climate mitigation commitments.

Why have countries not negotiated more of these agreements? Historically, governments have focused their efforts on national policy, which is both politically savvy and perhaps even necessary in the light of the difficulty of designing effective and robust renewable energy policy. Over time, however, the limits of purely national policy become increasingly apparent. If governments are committed to achieving even higher levels of renewable energy penetration, international cooperation offers one way forward. Of course successful cooperation depends on the formulation of credible, enforceable commitments. Governments cannot achieve success through reciprocal renewable energy deployment without considering their cooperation partners trustworthy with regard to their promises.

A related difficulty pertains to intellectual property rights. Because the global renewable energy technology market is growing, private companies have strong incentives to protect their innovations from competitors (Ockwell et al. 2010). For example, a bilateral renewable energy deal between the United States and China could be hampered by fears on the US side about the lack of enforcement of intellectual property rights in China. In May 2014, for example, US Attorney General Eric Holder said that Chinese hackers had attacked, among other targets, a West Coast solar technology company.[24] These concerns could make policymakers hesitant to agree on extensive renewable energy cooperation unless a clear understanding exists among the participants on knowledge sharing and the protection of intellectual property rights. More generally, the phenomenon of "techno-nationalism" (Ostry and Nelson 1995), although a possible driver behind national investment in renewable energy technology, can undermine cooperation. As Kim and Urpelainen (2013a) show, it is common for countries to refrain from environmental cooperation because they are worried about the loss of national competitiveness for their industry.

Given these problems, flexible approaches to renewable energy cooperation, perhaps on a bi- or minilateral basis, warrant attention. From intellectual property rights to technonationalism more generally, the nature and scope of the problem depends on the characteristics of the participating countries. The larger the group of countries negotiating a renewable energy agreement, the higher the likelihood of problems and conflicts. Although some forms of renewable energy cooperation, such as guidelines for trade and investment rules, clearly require a large number of countries to be effective—or even legal in the context of the multilateral trade regime—deployment agreements can be formulated on a bi- or minilateral basis. It is much easier for two countries to agree on a mutually profitable course of action than it is for a larger group of countries.

One form of international cooperation on renewable energy that warrants particular attention is North–South technology cooperation. For example, Urpelainen (2012) proposes North–South technology funds to facilitate technology transfer from wealthy to poor countries. In such funds, entrepreneurs from industrialized countries develop and commercialize technologies that are then sold to developing country markets, where the need for new energy infrastructure is apparent, with funding provided by the participating member state governments. Such an arrangement encourages entrepreneurs in industrialized countries to focus on renewable energy, whereas the governments of said industrialized countries reap political benefits from supporting industrial innovation and production. At the same time, less developed countries have clear incentives to participate because they gain access to new energy technologies at a subsidized rate, making it easier to deal with growing energy demands without excessive environmental deterioration or dependence on fossil-fuel imports.

The United States–India Partnership to Advance Clean Energy (PACE) is a possible example of how to engage in technology cooperation.[25] In November 2009, President Obama and Prime Minister Singh formed this new initiative:

> PACE focuses on accelerating the transition to high-performing, low emissions, and energy-secure economies. It aims to bolster joint efforts to demonstrate the viability of existing clean energy technologies as well as identify new technologies that can increase energy access and security. PACE also focuses on engaging the private sector, local governments, industries, and other stakeholders in sharing best practices on sustainable low carbon growth.

Although the size of the project, at approximately $100 million, is small, it is notable that both sides are committing funds to clean energy research and deployment. The project also uses public finance to leverage private contributions.

Similar to the national level, international efforts to promote renewable energy may provoke a political backlash. In fact, the problem may be even more difficult to control if the push for renewable energy comes from international cooperation. If cooperative efforts to increase renewable energy generation begin to have success in a country, they increase the level of renewable energy production beyond what the national government would have chosen in the absence of cooperation. After all, the goal of international cooperation on renewable energy is to reach beyond the status quo, that is, business as usual. This means that the government will have to deal with levels of politicization that it would not have been willing to accept in the absence of international cooperation. From a domestic political perspective, gains from international cooperation are artificial in that the government is accepting some domestic political backlash and additional economic costs in exchange for actions by other countries.

Under these conditions, a strategy for managing politicization is necessary. If a group of national governments are to achieve their joint goals for renewable energy, they must be prepared for a wave of politicization. Because the threat of politicization is common to the different governments, the international agreement on renewable energy can be designed considering the expected politicization. Institutional design can help governments deal with politicization by strengthening the renewables advocacy coalition and reducing the incentives of opponents to mobilize. The treaty should be designed to provide governments with the political and economic tools needed to mitigate and deal with politicization.

One strategy to create political support for compliance with the renewable energy agreement is "issue linkage" (Sebenius 1983). In addition to renewable energy provisions, the agreement among governments could contain articles on other issues. For example, the participating governments could agree to increase trade and investment in clean technologies as part of the agreement. Although such a provision could be warranted on substantive grounds alone, it could also be a useful instrument for enforcing the treaty. In the participating countries, domestic groups and actors that expect benefits from increased trade and investment would oppose any acts of noncompliance out of fear that said noncompliance undermines the agreement. International relations scholars have shown that domestic procompliance groups can play a key role in the enforcement of international agreements, and this strategy holds promise for renewable energy (Dai 2005).

The power of such linkages can be seen in the solar dispute between the United States and China. In 2012, the Obama administration imposed tariffs on Chinese solar panels to protect the domestic manufacturing industry. Solar manufacturers support the policy, whereas the much bigger installation and service industry has a different view.[26] In June 2014, Rhone Resch, the CEO of the Solar Energy Industries Association, said that import tariffs on solar panels "do a huge amount of damage to the installers and developers." Moreover, the solar installers are worried that China may retaliate by locking them out of the Chinese solar market, which is the largest in the world.

The mobilization of the prorenewables coalition should also be maximized through treaty design. Although the coalition always has a direct incentive to support renewable energy policy, the treaty can be designed to strengthen this incentive. Compliance with the treaty, for example, could be linked to access to joint funds for renewable energy investment. If a government fails to comply with the treaty, it no longer gains access to a pool of resources that the member states had initially created for joint investments. If this pool of resources is sufficiently large to be lucrative for renewable energy developers in all member states, their national governments will pay a high political cost for noncompliance. Similarly, the implementation of the renewable energy agreement could be based on policy instruments that furnish direct benefits to renewable energy producers, such as FITs or tenders. If a government does not comply with the treaty, then renewable energy investors and producers forgo these benefits. According to Urpelainen (2012), for example, a clean technology agreement can encourage industrialized countries to invest in R&D by setting aside funds for subsidized sales and procurement of new products in developing countries that need the new technology the most.

A third strategy to protect a renewable energy agreement from mobilization is a strong, unequivocal public commitment to it by member state governments. Guzman (2008, 72) maintains that reputational concerns drive states to abide by the rules, especially "in the multilateral context." If governments go public with their commitment to renewable energy investment and allow transparent monitoring by civil society, international organizations, and other governments, then they can tie their own hands through an increase in the reputational cost of noncompliance. Although this strategy neither empowers the renewables advocacy coalition nor directly mitigates the incentives of opponents to mobilize, it strengthens the national government's resolve. In turn, this strengthened resolve may indirectly deter the antirenewables from mobilizing because they believe that the government will not be responsive to their concerns.

After all, the government would pay a reputational price for any noncompliance induced by the antirenewables coalition. Although it is true that, "[l]ike all influences, compliance reputation operates at the margin"

(Guzman 2008, 118), it can, in conjunction with other measures, increase the likelihood of success in international cooperation on renewable energy.

Even simple pledges by national governments can help. The IRENA (2014b) roadmap for renewables until 2030, "REmap 2030," provides national roadmaps for 26 countries that are together responsible for three-fourths of all energy demand today. The roadmap proposes doubling the share of renewables in the global energy mix by 2030. The REmap 2030 plan is developed in collaboration with national energy experts, and every country with a roadmap has appointed at least one expert to participate in the project. National governments can use the IRENA plans as a benchmark for their own plans of action and, in so doing, enhance the credibility of their own plans and policies. After all, a strong public commitment to meeting the collaborative targets increases the reputational cost of failing to invest in renewable energy to a sufficient extent.

8 Conclusion

Renewable energy has weathered many storms. Only 15 years ago, the IEA predicted virtually no growth in the use of renewables in the coming decades. The prediction proved spectacularly incorrect. Every year, hundreds of billions of dollars are invested in renewable energy, due to both supportive government policy in key markets and rapidly declining costs of renewable energy generation. The first modern renewable source of energy that became commercially viable was onshore wind, and now solar photovoltaics are following its lead. Less mature technologies, such as offshore wind and tidal power, are waiting for their moment.

The success of renewable energy reflects, first and foremost, its political virtues. Since the story of modern renewables began in the 1970s, they have grown much faster than mainstream energy analysts have predicted. The pioneer countries invested in renewable energy for various reasons, and these reasons have changed over time. Although renewable energy policies have often proven expensive, governments have managed to build large and robust advocacy coalitions to support renewable energy generation. From farmers and homeowners to clean technology businesses and environmentalists, renewables have offered something for everyone. Their public image has also been positive, with large majorities of the population supporting increased renewable energy use across virtually every industrialized country. From a political economy perspective, renewables have been ideal for governments motivated by political survival. Similar to fossil fuels (Unruh 2000) and nuclear power (Cowan 1990), the political logic underpinning the growth of renewables has created a new kind of lock-in.

When the 1973 energy crisis opened a window of opportunity for renewable energy, few people anticipated a bright future for renewables, except in the long run. Enthusiasts expected rapid growth and a seamless transition from fossil fuels to renewables, whereas the much larger group of skeptics dismissed renewables as irrelevant tinkering. In reality, neither expectation has characterized the growth of renewables. As soon as renewables began to expand and show promise, they were politicized in the frontrunner countries. Over time, renewable energy enthusiasts won most of the political battles. Even where these battles were initially lost, such as in the United States, renewables made a comeback only a decade later as their economic performance improved. Today, renewable energy is a global growing industry, although until recently this growth has stemmed from government policy and political considerations. Only now, four decades later, are renewables beginning to compete with fossil fuels in important segments of the power sector.

Even today, renewables face many challenges. Although the economic cost of breaking the carbon lock-in has been high, renewables are still far from overwhelming fossil fuels in the global power and nonpower energy scenario. The relatively optimistic BNEF (2014b) prognosis would still leave fossil fuels generating most of the world's electricity in 2035, and the prognosis depends heavily on the massive deployment of rooftop solar technologies around the world, an assumption that remains unproven. In the power sector, the growth of renewable energy also creates a new set of systemic challenges. The more renewable energy generators feed into the grid, the more relevant the issue of intermittency becomes. High penetration rates require storage solutions and demand side management, smarter grid infrastructures, and regional integration. There is no fundamental reason that these barriers could not be overcome because the technical solutions either already exist or are being developed, but these new challenges require political strategies that are different from the earlier success of renewable energy. Governments can and should learn from history, but it is important to remember that renewables are now entering a new phase in their development.

Outside the power sector, the challenges to renewable energy are perhaps even more daunting. Although solar and wind power perform well in electricity generation, their potential for heating and transportation is more limited. Biomass holds potential for both heating and transportation, but the increased use of biomass puts upward pressure on food prices and may result in environmental degradation as forests are converted into plantations. Although the growth of modern renewables has largely depended on their limited social and environmental disadvantages, biomass use does not have this advantage. Efforts to increase the penetration of the use of renewable energy outside the electricity sector may prove difficult because the costs and disadvantages of expanding biomass use are more apparent than those of solar and wind power. A complementary approach would focus on increasing the electrification of the transportation sector, but this remains challenging due to the need for storage and the low cost of transportation based on the internal combustion engine.

This short concluding chapter reflects on the future of renewables. First, we discuss the main lessons of the book for understanding the past, present, and future of renewables. Our next step is to reach beyond renewables and discuss energy policy more generally. In a time of volatile energy prices, rapid economic growth in the global South, and pervasive planetary environmental change, the global energy system is in a state of turmoil. Most important, high energy prices and new production technologies have allowed the fossil fuel industry to exploit previously unavailable reserves of heavy oil and shale gas. Although the nuclear industry has been struggling, several advanced technologies on the horizon could increase the political and economic feasibility of atomic power. Energy conservation through advanced technology is an environmentally beneficial alternative to increased energy generation.

For the broader academic audience in comparative and international political economy, our study also offers some general lessons. The themes of external shocks, politicization, and lock-in play an important role in the study of public policy across specific fields (R. Keohane and Nye 1977; Hathaway 1998; Pierson 2000; Torvanger and Meadowcroft 2011). Our results shed new light on how external shocks trigger policy change, when and how policy may break path dependence, and how political struggles surrounding new technologies and innovations play out.

We conclude this book with a discussion of renewables as a force for positive change. Although our research has revealed many barriers to a more widespread use of renewables, there is reason for cautious optimism about their future as a source of clean energy for the world. This final discussion offers some concrete suggestions for how renewable energy advocates can promote their cause and how governments can use renewables to achieve their socioeconomic goals.

8.1 Renewables and Global Energy Futures

External shocks have historically played a central role in spurring the rapid growth of renewables. In 1973 and 1979, oil price spikes created two panics that prompted previously skeptical or uninterested governments to experiment with renewable energy. The difficulties with nuclear energy, such as the 1979 Three Mile Island and the 1986 Chernobyl accidents, strengthened the appeal of renewables among many political constituencies. In countries such as Germany, new scientific evidence about the deleterious effects of climate change was yet another effective argument for renewable energy. Even later, when renewables had already begun to grow, the Fukushima nuclear accident in Japan gave renewables another push.

Experimentation, stealth politics, and low levels of distributional conflict all played an important role in the early years of renewable energy. Although much of the literature on renewable energy begins with Unruh's (2000) premise of carbon lock-in, we have found that, during and after an external shock, there is space for modest renewable energy policies. In the study of new energy technologies, the concept of "niche markets" plays an important role (Lopolito, Morone, and Taylor 2013; Smith et al. 2014). According to theories that emphasize niche markets, new energy technologies can grow and become increasingly competitive in specific niches or, as Smith et al. (2014) put it, "in protective spaces." For example, solar technologies could, even under a general carbon lock-in, prove economically attractive in remote locations, such as military bases or scientific research stations. In practice, our findings suggest that the construction of protective spaces need not—at least initially—be difficult. If an external shock creates social demand for new energy innovations, the government is impelled to respond, and as long as renewables do not constitute a credible threat to the status quo, resistance is also limited. Although the ideas and assumptions behind energy policy have always been contested (Laird 2001), our evidence suggests that this kind of conflict is one of low intensity and does not reach high levels of politicization in the early years.

The politicization of renewable energy began in earnest with President Reagan's 1981 inauguration in the United States. The evidence from the United States, Germany, and, a few decades later, Denmark is largely consistent with the hypothesis that the initial success of renewables provokes a backlash when the supporters of fossil fuels and nuclear power begin to see renewables as a credible threat to the dominant modes of energy production and consumption. We found evidence of politicization in all three cases of frontrunner countries, but the central political struggles always came at least a decade after renewable energy policy began. At that time, the opponents of renewables had recognized them as a real concern and a possible threat to their interests and ideology.

What is particularly striking about the case studies of politicization is the variation in the outcome. In the United States, an account emphasizing left-right partisan politics, along with a lackluster public opinion and a strong antirenewables coalition, can explain why the growth of renewables came to an abrupt halt in 1981 and only recovered more than two decades later. In Germany, partisan politics also played an important role, but ultimately the power of public opinion and strong pro-Green interest groups—both economic and ideological—tilted the scale in favor of renewables. In Denmark, the role of partisan politics was much less pronounced, whereas institutional capacity in the form of strong agricultural cooperatives and a large, vocal advocacy coalition supported by the public were decisive. These comparisons show that no parsimonious model, including the one we presented in our earlier work (Aklin and Urpelainen 2013b), can fully capture the complexities of any given case.

Over time, renewables achieved their own kind of lock-in and began to spread across the globe. During the first decade of the 21st century, renewables, led by onshore wind, achieved record growth rates and started to displace significant quantities of fossil fuel and nuclear generation capacity in key countries. Although Denmark stands out as the leader, many other countries have also moved forward in leaps and bounds. The combination of high fossil-fuel prices, politically unpalatable nuclear power, and rapid improvements in renewable energy technologies has brought renewables to the forefront of energy policy; climate change has also recently drawn a great deal of attention to renewables. These clean energy sources now play a decisive role in China's response to environmental and energy security concerns, and in Europe they have a realistic chance of overtaking fossil fuels in the power sector in the next two decades.

At this time, the challenges and barriers to renewable energy have also begun to change. Especially in Germany, cost containment has become the central issue in renewable energy policy. Due to the success of renewable energy policy, the total cost of subsidies and other policies has risen rapidly. This development has both threatened the positive public image of renewables and invigorated opposition to renewables by the right wing and heavy industry. As the United States and major emerging economies continue to catch up with Europe, opposition to generous renewable energy policies cannot be fully avoided. Although renewables have many features that make them politically expedient, a government's failure to contain the costs in the form of public spending and higher electricity prices could endanger the future development of renewables. The kind of lock-in that renewables have achieved so far has not been purely economic because the political virtues of renewable energy have been a major factor in their rapid growth. This alone is not a reason to discount renewables in the future, but rising total costs could undermine the political clout of the advocacy coalition for renewables. Unless renewables achieve a decisive economic advantage at the same time, any loss of political advantage could prove fatal to them.

Another cautionary note about renewables concerns the prospects for systemic change. As Baker et al. (2013) show, contemporary cost estimates for renewable energy may prove too optimistic because they do not account for the systemic changes required to accommodate intermittency and decentralized power generation. We expect this to be politically significant as well. If the political success of renewables has reflected the large size of their advocacy coalition, large investments in upgrading the electricity grid, higher residential and industrial electricity prices, and technical difficulties with storage will create a host of new political challenges that renewables supporters must consider. The primary lesson from our research for managing these challenges is that governments should specifically design their policies to create a large advocacy coalition. Because systemic change is required, such a coalition must be even larger than the one supporting renewables in earlier years. As the minimal required size of the coalition grows, cost-effectiveness and the provision of public goods become increasingly important.

An important lesson from the study pertains to the relationship between national policy and regional or international cooperation. We have found that renewable energy policies and generation capacity are now growing rapidly outside the initial frontrunners, and this development opens new windows of opportunity for cooperation. As the number of countries with expanding renewable energy industries grows, it will be in their mutual interest to develop a supranational strategy for integrating higher levels of renewable energy generation into the energy sector. However, as Gullberg, Ohlhorst, and Schreurs (2014) have shown, cooperation on renewables is difficult because it requires resolving disputes between a large and heterogeneous group of interest groups in different countries. Again, the key challenge is to create a sufficiently large advocacy coalition in each pivotal country so that the pivotal national governments have incentives to support renewable energy cooperation.

The final two comments in this section are about the role of renewables in climate policy and eradicating energy poverty. In the case of climate policy, renewables have a lot of mitigation potential because they generate electricity with little carbon dioxide emissions. The growing importance of renewables in the power sector has already reduced global carbon dioxide emissions, and their role may be expected to become more and more important in the future. However, critics may argue that renewables are but a partial approach to reducing these emissions. In the absence of actual constraints on carbon emissions, the increased use of renewables in one country puts downward pressure on fossil-fuel prices in other countries. This "carbon leakage" (Babiker 2005) reduces the net effect of renewables on carbon dioxide emissions. However, given the political clout of renewables, they may play an indirect role in creating public support for climate mitigation. If renewable energy policy creates interest groups that benefit from climate change mitigation, then these interest groups may support additional policies that constrain carbon emissions.

Another politically important virtue of renewables is that they can be used to combat energy poverty. Many renewable sources, such as solar, are not only carbon-free but also suited to electricity generation in remote rural communities that are not connected to the electric grid (Hiremath et al. 2009; Brass et al. 2012). To the extent that the North–South distributional conflict in climate policy raises barriers to global climate cooperation, an issue linkage that connects decarbonization and energy access among the rural poor could help negotiators break the "global warming gridlock" (Victor 2011). Renewable energy is not a panacea for solving the problem of energy poverty, but public support for creating access to energy through renewables is an effective and compassionate strategy that also helps countries bridge their political differences and create real synergies between the environment and development.

8.2 Implications for Energy

Although energy has historically seen few real breakthroughs (Grübler, Nakićenović, and Victor 1999), incremental change is the norm. New energy technologies come and go, with profound effects on the broader society. As Smil (2010) has shown, energy systems change slowly, and governments cannot expect to manage these transitions in full. Yet the case of renewable energy shows that ambitious and sustained government policy can produce large effects over time. According to scholars of carbon lock-in (Unruh 2000; Sandén and Azar 2005), it should be hard, perhaps even impossible, for renewables to penetrate the energy systems of industrialized countries. In practice, government policy has already enabled this penetration in many countries.

To illustrate the broader applications of our theory for energy, consider the case of natural gas fracking in North America. As Davis (2012, 179) notes, "the recent upsurge in its use was prompted by the discovery of large new reserves of coal- or shale-bound gas throughout the United States and by technological improvements such as combining fracking with horizontal drilling techniques adopted from deep-water oil and gas wells operating in the Gulf of Mexico." Between 2000 and 2012, the share of natural gas in US electricity generation almost doubled from 16% to 30% (EIA 2014a). Although the increase is much smaller than that in renewable energy in proportional terms, it is larger in absolute terms. These discoveries and technological advances were, in turn, driven by high fossil-fuel prices—external pressure—in the first decade of the 21st century. The main difference to the case of renewable energy is that fracking is an activity promoted by private energy companies. Under high natural gas prices, fracking became commercially viable almost immediately, and some commentators are talking about the "shale gas revolution" (Deutch 2011). Although the external shock played a key role in this case as well, the technological breakthrough was supported by an existing infrastructure for fossil fuel extraction and allowed a rapid increase in production.

The technology of fracking has been politicized. Besides environmentalists worried about fossil fuels and methane emissions, the coalition opposing fracking comprises local residents worried about water pollution, air quality, and noise. Davis (2012) offers a comparison of the politics of fracking in Texas and Colorado. In Texas, which is the frontrunner state for fracking in the United States, no major legislative proposals for the regulation of fracking have been made. The traditional dominance of oil and gas interests has kept the advocates of new regulation at bay, and local campaigns against fracking have had limited success at best. In this case, the key factor supporting fracking appears to be the dominance of the fossil fuel industry in Texas. In contrast, the political clout of the oil and gas industry in Colorado is both less prominent than in Texas and partially offset by the large environmentalist community. Moreover, rural landowners in Colorado—ranchers, retirees, tourism business owners—are a politically powerful constituency with a strong interest in protecting natural landscapes and wildlife. As a result, Colorado has created a rigorous regulatory framework that specifically addresses concerns about fracking. Similar to the case of renewable energy, the political fortunes of fracking have depended on the relative power of supporting and opposing advocacy coalitions.

Could fracking achieve the same kind of lock-in that renewables have achieved? In the United States, it already has, given the rapidly growing production. Although the geographic concentration of fracking remains high, with no action in states such as California and New York, it is clear that the growth of fracking has been the primary driver of the rapid reduction in the use of coal for electricity generation over the past decade. Outside North

America, however, the diffusion process has not begun. In July 2014, for example, Germany imposed a moratorium on fracking until 2021.[1] France had already imposed a ban on fracking earlier.[2]

Besides fracking, many other emergent energy technologies are on the horizon. The combination of the rapid growth of energy demand in the developing world and the environmental imperative of mitigating climate change is putting tremendous pressure on energy technology innovation, and it is not possible to say yet what kinds of technologies will emerge in the coming decades. From electric vehicles to advanced nuclear reactors as well as carbon capture and storage, a number of possibilities could have major effects on the global energy system. If the case of renewable energy is any indication, the fortunes of these energy technologies will depend on a series of political struggles that pit the advocates and opponents of said technologies against each other.

The management of side effects will be key to understanding the political fortunes of different energy technologies. The success of renewable energy reflects a combination of concentrated benefits to key constituencies and a generally positive public image. Fracking has the former advantage, and it may be even more fortuitous because fracking benefits established fossil-fuel industries. However, the fracking industry is vulnerable to local opposition campaigns. Compared with local opposition to windmills, fracking has clearly provoked more vocal campaigns. What is more, solar power does not really provoke any local opposition, except in the case of the astroturf campaigns mobilized by the fossil-fuel industry or electric utilities.

In the short to medium run, renewables cannot be the exclusive solution to decarbonization, even if one accepts that renewables could technically provide a comprehensive energy solution for the planet, as Jacobson and Delucchi (2011) have argued. Because today's global energy sector is heavily dependent on fossil fuels, a sustainable and decarbonizing energy trajectory will feature gradual reductions in energy consumption and the carbon intensity of the remainder. For example, carbon dioxide emissions can be reduced by substituting natural gas or nuclear power for coal. Energy conservation also reduces carbon dioxide emissions, provided it displaces fossil fuels instead of, say, renewables or nuclear power.

If a mix of technologies is necessary, then interactions between them become crucial. We have yet to explore this theme in any detail, but we can illustrate the concept by considering the relationship between renewables and natural gas. Because both solar and wind power suffer from intermittency, their ability to meet power demand is seasonal and cyclical. On the contrary, natural gas turbines can be easily turned on and off depending on power demand. Because natural gas produces less than one-half of the carbon dioxide from coal in electricity generation per unit of power, an optimal combination of renewables and natural gas could allow rapid decreases in carbon dioxide emissions even in the absence of smart grid technologies, improved storage, and new techniques for demand side management in the power sector. Such a system would require that the advocates of renewable energy acknowledge the priority of reducing coal use, as opposed to all fossil fuels, and that natural gas can be produced in an environmentally sustainable and socially acceptable fashion.

Energy conservation adds another dimension to the policy design problem. In expectation, efforts to reduce energy consumption in different countries will cause a decrease in the demand of a variety of different energy sources. From the perspective of climate change mitigation and environmental protection, it would be ideal if this decrease focused on coal power plants. The risk is that the reduction in energy demand brings down the demand for new renewable electricity generation capacity, as previously built power plants that rely on fossil fuels continue to operate because their capital costs have already been sunk in. To avoid this possibility, energy conservation measures should be coordinated policy mechanisms that support renewable energy and put a price on carbon dioxide emissions. Such coordination will create a new set of political conflicts.

The past four decades of renewable energy provide interesting insights into the logic of energy transitions more generally. Historically, human societies have undergone many important transitions, beginning with the replacement of wood with coal and continuing with the growing dominance of liquid fossil fuels at the expense of coal (Fouquet 2008; Smil 2010). Despite rapid technological advances and generous policy support, only after three decades of blood, sweat, and toil did renewables begin to surge. Even now, they provide a small percentage of world electricity. Even under optimistic scenarios, a world powered largely by renewables is many decades away. Technological, economic, and political factors conspire to lengthen the timespan of energy transitions. If we use the year 1970 as the origin to measure the length of the ongoing renewable energy transition, we are already into its fifth decade and are expecting another three to five decades at least. Combining these numbers, the total length of such a transition does not fall short of an entire century. Even if the construction of power plants running on fossil fuel were to cease completely in the next two decades, a life span of 40 to 60 years would mean that, in 2070, an entire generation of conventional power plants would continue to be in operation. In the absence of expensive dismantling of already operational power plants, a fully renewable global power sector will only become reality decades after the last fossil-fuel power plant is already built.

8.3 Implications for International and Comparative Political Economy

For a long time, external shocks have played an important role in international and comparative political economy. In their study of "complex interdependence," Keohane and Nye (1977) argued that, although all countries are to some extent "sensitive" to external shocks and pressure, there is considerable variation in their "vulnerability." Some countries have the political and economic capacities to adjust to changing circumstances, whereas others do not. Since then, political economists have analyzed national policy responses to a variety of external shocks, ranging from the oil crisis (Mendershausen 1976; Ikenberry 1986) to financial distress (Rosas 2006) and changes in commodity prices (Finlayson and Zacher 1988). The "second image reversed" framework proposed by Gourevitch (1978) lays the analytical foundation for many of these studies because it focuses on evaluating the effect of changes in the international system on domestic politics and societies. According to Krasner (1985), the international strategy of developing countries during the Cold War with respect to the "global commons" focused on collective management exactly because of the vulnerability and weakness of the governments of these countries. Indeed, the 1973 oil crisis was a defining moment for the study of international political economy because it created a global economic recession that, for the first time since the Second World War, required extensive policy coordination and cast doubt on the economic strategies of industrialized countries (R. Keohane 1984).

The 1973 and 1979 oil crises left a positive legacy in the form of renewable energy. Although many scholars have recognized the connection between energy prices and government policy to support renewables (Agnolucci 2007; Lipp 2007; Cheon and Urpelainen 2012; Aklin and Urpelainen 2013b), this book has filled several gaps in the account. The early investments in renewable energy were modest and did not contribute to solving the immediate problem of an energy crisis in any of the countries under study, but they created an opening for modest growth and, more important, impressive cost reductions in renewable energy. Mainstream scholarship in international political economy has largely ignored renewables as an outcome of the oil crises, focusing instead on other economic and political developments. This omission probably reflects the initial invisibility of renewable energy and the long delay between the initial external shocks and today's boom in renewable energy. Our findings highlight the importance of a long view, as the growth of renewables resembles that of a rolling snowball, with modest beginnings resulting in rapid expansion much later.

Our case studies show that dependence on fossil-fuel imports played an important role in the early stages of renewable energy development, consistent with the model built by R. Keohane and Nye (1977). However, this state of affairs did not last. As renewables grew and became politicized, dependence on fossil fuels no longer provided a compelling explanation for variation in policy. Reagan's efforts to undercut renewables began at a time of high fossil-fuel prices and the United States' continued dependence on fossil fuels, whereas Germany and Denmark began to move to the direction of increased renewable energy deployment without much change in their situation. If anything, the Danish case shows that the growth of wind energy occurred concurrently with greatly improved prospects of natural gas exploration in the North Sea. The most powerful Danish renewable energy policies were implemented at a time when Denmark's dependence on fossil-fuel imports was decreasing. These observations are not consistent with the kind of stimulus-response models that scholars of international political economy apply to understand external shocks.

These dynamics provide a possible solution to the problem of time inconsistency (Kydland and Prescott 1977). The time-inconsistency trap is avoidable under low levels of opposition, as the political clout of the renewables advocacy coalition provides the government with incentives to continue supporting renewables over long periods of time. Stated differently, time inconsistency disappears because the endogenous growth of the renewables advocacy coalition keeps the government's emphasis on continued policy support. A similar analysis is found in Jacobs (2011), who focuses his argument concerning reforms that produce benefits over the long run on inter alia electoral incentives. Our argument encompasses the electoral incentives, as well as interest groups and coalitions. The strong path dependency that favored the carbon lock-in could be broken when policymaker kept supporting renewables over the long run. In turn, they did so when domestic interests were strong enough to counterbalance the supporters of the old energy regime. Thus, gradual—although not linear—technological change was possible over the past forty years.

Although vulnerability provides an explanation for why countries act initially, path dependence and long temporal lags ensure that other structural and institutional factors begin to play an important role over time. The stimulus-response model can adequately capture variation in policy responses in the beginning, but it is not a good framework for understanding the dynamic evolution of policies and outcomes. At the politicization stage, policies and outcomes inevitably become entangled in domestic political conflicts, and the role of vulnerability as a causal factor begins to diminish.

Similar to Hathaway (1998), who showed that trade liberalization in the United States has been largely driven by the decreasing size and political clout of previously influential import-competing industries, our results emphasize the importance of constituency creation and the expansion of the advocacy coalition for renewable energy. Although trade policies are notorious for generating rents to preexisting constituencies, the case of renewables is different in that governments had to create these constituencies through policy formulation. In the early stages of renewable energy development, few power generators or clean technology manufacturers would have reaped large benefits from renewable energy policy. These interest groups were created through government policy; as policy spurred their growth, they began to demand more ambitious policy. This positive reinforcement was the decisive factor in allowing renewables to break the carbon lock-in, first in Denmark and Germany and later in other countries.

Another important result is the initial irrelevance of cost-effectiveness as a criterion for policy formulation. Renewable energy policies were, by and large, a response to the demands of the rapidly growing advocacy coalition and reflected the need to allocate resources to the advocates while maintaining support among the broader public and appeasing opponents. This result is largely consistent with those studies of political economy that focus on competition between interest groups as a source of public policy (Truman 1951; Chubb 1983; Grossman and Helpman 1994; N. Keohane, Revesz, and Stavins 1998; Cheon and Urpelainen 2013) because they also emphasize that governments are often ready to depart from economic criteria to allocate resources to politically powerful constituencies. However, our case studies add two new components. First, much of the lack of interest in cost-effectiveness has developed over time due to the positive reinforcement dynamics between renewable energy policy and the clout of the advocacy coalition. Second, cost-effectiveness has over time become an increasingly critical factor in the politics of renewable energy.

8.4 Powering the Future With Renewables

During the past four decades, renewable energy has grown from a fringe pursuit into an important, dynamic element of the power sector of most countries in the world. The outlook for modern renewables is bright.

Although many obstacles to continued expansion remain, widespread consensus now exists about the excellent growth prospect for renewable energy. At a time of rapid climate change, the key question is no longer whether renewables will grow in importance. Rather, the question is how fast they can transform the world energy economy. In the long run, renewables hold the potential for running the entire world economy, although the transition to this steady state may require many decades or even centuries. Such a transformation is crucial to the implementation of the December 2015 Paris agreement on climate change, which lays the foundation for dynamic bottom-up climate policy over the coming decades.

Given the environmental merits of renewables and four decades of successful growth, increased investment into renewable energy is a proposition worth consideration. In addition to other measures, such as energy conservation, a transition toward renewable energy can support climate change mitigation, play a role in eradicating energy poverty, and enhance the energy security of nations. For climate change mitigation, renewables are advantageous because they generate no carbon at all. Although the future use of fossil fuels, including natural gas, is only compatible with climate mitigation targets in combination with heavy investment in carbon capture and storage, renewables are foolproof because they do not generate carbon dioxide. Although efforts to reduce energy poverty are not about the kind of fuel used to generate electricity and heat, renewables hold the potential for providing relief to remote communities that cannot realistically expand electric grid coverage any time soon. Because renewable energy resources are inexhaustible and local, their positive effect on energy security is clear.

Smart policies can promote renewables by creating an advocacy coalition in their support, facilitating international cooperation, dismantling systemic obstacles to increased renewable energy use, and emphasizing cost containment. Although past policies were right to focus on breaking the carbon lock-in and focusing on the expansion of renewables, future policies must achieve more than that. The changing nature of renewable energy politics indicates that new policies must appeal to a larger group of constituencies, forge a robust consensus on renewable energy, and relentlessly pursue cost reductions. In addition to increased deployment, funding for technology innovation and basic energy engineering is also required. The burden on solar photovoltaics and onshore wind is greatly relieved if innovation spurs major progress in other renewables, such as tidal power and offshore wind. The history of today's mature renewable energy technologies shows that such progress is feasible.

Aggressive investment in renewable energy may even enable more comprehensive climate policy at the global level. Although we support efforts to invest in renewable energy, the cost of mitigating climate change through renewable energy alone would be much higher than necessary. Other technologies from energy conservation to carbon capture and storage can reduce the cost of climate mitigation. However, renewable energy can create coalitions for climate action by creating economic opportunities for clean technology. If renewables continue their breakthrough and investments in clean energy technology continue to grow, the number of interest groups with a direct economic interest in reduced carbon dioxide emissions will grow. As a result, climate negotiations will become easier. If such a trajectory were to materialize, then renewables would exert a much larger positive influence than their direct environmental benefits.

Notes

Chapter 1: Introduction

1. The data from the United States should be interpreted with some caution. The sharp increase in 1989–1990 is due to later revisions by the EIA. These revisions took into account nonutility electric plant data but did not cover the period before 1989. That said, this change amounts to a one-time increase and does not affect the general trend observed in the United States. See "What's New?" for the year 2003, http://www.eia.gov/totalenergy/data/monthly/whatsnew.cfm (accessed September 20, 2015).

2. See https://www.weforum.org/press/2016/06/after-paris-climate-agreement-a-world-of-renewable-energy-is-emerging/ (accessed November 3, 2016).

3. Indeed, recent scholarship suggests that resources such as oil encourage "revolutionary" regimes to engage in costly conflict (Colgan 2013).

4. For the full text of the Paris Agreement, see https://unfccc.int/resource/docs/2015/cop21/eng/l09r01.pdf.

5. An FIT is a policy that requires that utilities purchase electricity generated from renewables at a price premium.

6. To be sure, governments across the world also subsidize the production and consumption of fossil fuels (IEA 2014a).

7. "German Farmers Reap Benefits of Harvesting Renewable Energy," *Financial Times*, December 2, 2013.

8. See http://www.iea.org/topics/renewables/ (accessed January 10, 2014).

9. See http://www.eia.gov/energyexplained/index.cfm?page=renewable_home (accessed January 10, 2014).

10. In individual countries, variation in this pattern occurs. In Brazil, for example, biofuels have been a key focus.

11. Fossil fuels were used before the industrial revolution, but in a limited fashion. For instance, oil was sometimes used for lighting (Sørensen 1991).

12. Corn ethanol, for example, requires a lot of energy to produce and results in land-use change, so the total carbon dioxide emissions from it are high (Searchinger et al. 2008). More generally, when biomass use results in forest clearing, substituting biomass for gasoline or other fossil fuels can increase the rate of climate change. In contrast, GHG emissions from solar and wind power are low.

13. The experiences of individual countries vary, but we maintain that globally the prospect of decarbonizing the power sector through wind and solar power is the most important development, especially when the climate mitigation imperative is considered.

14. For estimates of nuclear GHG emissions over the entire life cycle, see Warner and Heath (2012). They find that even the highest estimates of nuclear life-cycle emissions fall below those from fossil fuels.

15. See chapter 4 for evidence of the lack of fundamental concerns about fossil fuels and the carbon lock-in before the 1973 oil crisis.

16. Other considerations, such as the perceptions and leadership of individual organizations and political leaders, also prove important. For the purposes of our analytical framework, however, we focus on a parsimonious set of factors that we consider central.

17. In chapter 3, we explain and justify our case-selection strategy.

18. "Opponents Dislike Dodd-Frank Because It Works," *Boston Globe*, January 28, 2016, https://www.bostonglobe.com/opinion/editorials/2016/01/28/opponents-dislike-dodd-frank-because-works/YKhpct94gg83In2BxqzdLJ/story.html (accessed February 20, 2016).

Chapter 2: Renewable Energy

1. "How to lose half a trillion euros," *The Economist*, October 12, 2013, http://www.economist.com/news/briefing/21587782-europes-electricity-providers-face-existential-threat-how-lose-half-trillion-euros/ (accessed June 28, 2014).

2. "Swiss hydro power under threat from German energy shift," *Reuters*, April 29, 2014, http://in.reuters.com/article/2014/04/29/alpiq-hldg-germany-power-idINL6N0NL3KG20140429 (accessed November 4, 2014).

3. Reversals are more likely when we look at different fuels (instead of entire energy systems). For instance, some cities in the United States experienced a back and forth between coal and natural gas depending on their relative price (Tarr and Clay 2014).

4. This is not to say that nuclear power did not face obstacles. In our reading, however, these obstacles were related to issues such as public opinion, safety concerns, and high constructions costs—as opposed to the systemic barriers that renewables faced and still face today.

5. See http://www.windpowermonthly.com/article/1138562/close---e126-worlds-biggest-turbine/ (accessed December 4, 2013).

6. See http://www.euronuclear.org/info/encyclopedia/n/nuclear-power-plant-europe.htm (accessed December 4, 2013).

7. "Sunny uplands." *The Economist*, November 21, 2012, http://www.economist.com/news/21566414-alternative-energy-will-no-longer-be-alternative-sunny-uplands (accessed June 26, 2017).

8. "A look at wind and solar, Part 2: Is there an upper limit to variable renewables?" The Energy Collective, May 28, 2015, http://www.theenergycollective.com/jessejenkins/2233311/look-wind-and-solar-part-2-there-upper-limit-intermittent-renewables (accessed August 25, 2015).

9. This number was computed as follows. Natural gas (which includes shale gas) generated about 814,752 million kWh in 2007 (Table 7.2b Electricity Net Generation: Electric Power Sector, Monthly Energy Review). Additional EIA data tell us that shale gas represented about 8% of all the natural gas extracted in the United States that year (Natural Gas Gross Withdrawals and Production, EIA, http://www.eia.gov/dnav/ng/ng_prod_sum_dcu_NUS_a.htm). Therefore, about 65,743 million kWh can be attributed to shale gas. The same calculation can be done for 2013. Shale extractions were about 40% of total gas extractions (11,896,204 million cubic feet out of 30,005,254). Because natural gas generated about 1,028,949 million kWh, this means that about 407,948 million kWh came from shale gas.

10. Prices would later climb even higher after a few years of stability. In 1979, they doubled to about $100 per barrel, a record for the 20th century and a price that would not be matched before 2008.

11. "Table 7.2a Electricity Net Generation: Total (All Sectors)," EIA, http://www.eia.gov/totalenergy/data/monthly/pdf/sec7_5.pdf (accessed February 2, 2014).

12. The concept of energy security predates the 1973 oil crisis. Before the First World War, Winston Churchill decided to impose the use of oil instead of coal on the British Navy to make the ships faster. Churchill clearly realized the tradeoff between using domestic coal and being dependent on foreign oil, and his solution to this dilemma was to diversify the oil supply (Yergin 1988).

13. http://www.mhivestasoffshore.com/new-24-hour-record/ (accessed July 15, 2017).

14. The decrease can be explained by the replacement of a large number of old turbines by more efficient ones.

15. The data for renewable energy generation in the United States referred to here and below are from the EIA, "Net generation by state by type of producer by energy source (EIA-906, EIA-920, and EIA-923)," http://www.eia.gov/electricity/data/state/ (accessed January 30, 2014).

16. Data from "Stamdataregister for vindmøller," Danish Energy Agency, http://www.ens.dk/info/tal-kort/statistik-noegletal/oversigt-energisektoren/stamdataregister-vindmoller (accessed February 2, 2014).

17. For instance, see "Google, KKR invest $400 million in six U.S. solar power plants," *Reuters*, November 14, 2013 (accessed June 26, 2017).

18. "Physical progress (achievements)," Ministry of New and Renewable Energy, http://www.mnre.gov.in/ (accessed June 26, 2017).

19. "China solar panel exports to ease on EU curbs" *Reuters*, August 9, 2013; "The next battle in our trade war with China" *The New Republic*, January 21, 2014.

20. See, for instance, "An ill wind blows for Denmark's green energy revolution," *The Telegraph*, September 12, 2010, http://www.telegraph.co.uk/news/worldnews/europe/denmark/7996606/An-ill-wind-blows-for-Denmarks-green-energy-revolution.html (accessed August 17, 2015).

21. See also "Solar panels could destroy U.S. utilities, according to U.S. utilities," *Grist*, April 10, 2013, http://grist.org/climate-energy/solar-panels-could-destroy-u-s-utilities-according-to-u-s-utilities/ (accessed November 6, 2014).

22. See, for instance, "Tesla moves into batteries that store energy for homes, businesses," *Reuters*, May 1, 2015; "Bill Gates backs renewable battery startup," NASDAQ, January 31, 2014 (accessed June 26, 2017).

Chapter 3: Policy Responses to External Shocks

1. An early exception to this pattern is Brazil. Because Brazil's investments in biofuels after the 1973 oil crisis reflected unusually favorable geographic conditions, they have not provoked the kind of global shift that early investments in solar and wind did elsewhere over time.

2. However, a shock caused by another government's activities would qualify as external. To the extent that the 1973 oil crisis was caused by the Arab oil embargo, it would still be an external shock for the United States and European countries.

3. US Energy Information Administration Annual Energy Review, Tables 1.3, 10.1, and E1.

4. Consider, for example, the situation in October 1973, at the time of the first oil crisis. Although natural resources had not run out by then, contemporary predictions

raised concerns (Meadows et al. 1972). These predictions were later proven wrong, but this outcome was not obvious to everyone at the time. Similarly, the prospect of a "peak oil" seemed plausible at the time of record high oil prices in late 2007, but only a few years later oil companies reported a massive expansion in nonconventional oil reserves, such as shale oil.

5. Here it is important to consider the possibility that other industries could have an interest in renewable energy or other experimental policies (e.g., Hughes and Urpelainen 2015). In such circumstances, it is possible that interest group politics plays a more important role already at the time of the external shock. Such cases are rare, however, when the policy in focus is risky and uncertain and thus does not promise profits in the short or even medium run.

6. However, later shocks at the politicization stage need not follow this logic. If renewables have already been politicized, then additional external shocks provoke political conflicts.

7. "Energy Bill Adopted by House Requires Utilities to Use Renewable Power Sources," *New York Times* (accessed December 22, 2013).

8. For these events, see "White House Solar Panels Installed This Week," *Washington Post*, August 15, 2013.

9. "Americans Want More Emphasis on Solar, Wind, Natural Gas," *Gallup*, March 27, 2013, http://www.gallup.com/poll/161519/americans-emphasis-solar-wind-natural-gas.aspx (accessed February 3, 2014).

10. "Keystone Backed in Poll by 56% of Americans as Security," *Bloomberg*, December 12, 2013.

11. "German Farmers Reap Benefits of Harvesting Renewable Energy," *Financial Times*, December 2, 2013.

12. Indeed, the case of Finland shows that a proportional electoral system need not cause the success of Green Parties or the development of renewable energy.

13. Unruh (2000) considers nuclear power a component of the carbon lock-in because the infrastructure for nuclear electricity generation is, thanks to economies of scale, favorable to fossil fuels. That said, we must acknowledge that nuclear power is a low-carbon source.

14. We do not include comparisons of left- and right-wing parties because these parties are difficult to compare systematically across countries, and our hypothesis is centered on the timing of renewable energy within a country.

15. It must be acknowledged, however, that Brazil's support for the ethanol industry has been wavering recently.

Chapter 4: External Shocks

1. Transcript of a televised speech by President Jimmy Carter on April 18, 1977. Available at http://www.pbs.org/wgbh/americanexperience/features/primary-resources/carter-energy/ (accessed February 21, 2014).

2. Oil scarcity had occasionally caused worries before 1973. Take the United States, for instance. In 1908, US scientists already feared that "[a]ll large [oil] fields had been discovered" (R. Stern 2016, 222). After World War I, it was politically induced scarcity that worried US officials because oil appeared to be mostly in British hands. Later, fears about shortages were renewed during World War II, and oil prices increased steeply after the war ended, as supplies lagged booming demand (Yergin 2011a). New supply from the Middle East would soon satisfy the postwar needs.

3. Between 1973 and 2012, consumption in non-OECD countries tripled and is currently about the same as in the OECD.

4. At the time of this writing, a car's mileage is expected to reach about 54 miles per gallon by 2025. See "The 1973 Arab Oil Embargo: The Old Rules No Longer Apply," NPR, http://www.npr.org/blogs/parallels/2013/10/15/234771573/the-1973-arab-oil-embargo-the-old-rules-no-longer-apply.

5. We normalize the number of renewables patents by total patents to account for the relative lower number of filings in earlier periods.

6. "Politically Safe Oil Stocks Win Investors' Esteem," *New York Times*, October 28, 1973.

7. "Saudi Oil Is Cut Off," *New York Times*, October 21, 1973; "Highlights of the Week," *New York Times*, October 28, 1973.

8. "The Arab Oil Weapon Comes Into Play," *New York Times*, October 21, 1973.

9. In the summer of 1980, the Carter administration also called for the creation and funding of a Synthetic Fuels Corporation in the Energy Security Act (Tugwell 1988). In practice, the effort failed, and the corporation was closed by 1985 (see Cohen and Noll 1991).

10. "Status of Nuclear Power: United States," American Physical Society, 2014, http://www.aps.org/policy/reports/popa-reports/energy/fission.cfm (accessed November 19, 2014).

11. According to the original Act, the formal name of the commission was State Energy Resources Conservation and Development Commission (CEC 2016).

12. As Heymann (1998) shows, however, the modest investments in wind energy R&D produce fewer gains.

13. Among the American public, lack of awareness about oil imports in the first place contributed to conspiracy theories (M. Jacobs 2016). As late as in November 1979, *The New York Times* reported that "[m]ore than half the respondents in the latest New York Times/CPS News Poll said they thought the energy shortage had been fabricated" (see "Poll Shows Doubts Over Energy Shortage," Section Business & Finance, D6). However, such conspiracy theories did not characterize elite beliefs. US policy elites were aware of the real causes of oil scarcity and the resulting record prices.

14. See "National Energy Plan: Address to the Nation," http://www.presidency.ucsb.edu/ws/?pid=6904 (accessed September 6, 2016).

15. See "Address to the Nation About National Energy Policy," http://www.presidency.ucsb.edu/ws/?pid=4051 (accessed September 6, 2016).

16. See "Executive Order 11814: Activation of the Energy Resources Council," http://www.presidency.ucsb.edu/ws/?pid=23901 (accessed September 6, 2016).

17. See "Address to the Nation on Energy and Economic Program," http://www.presidency.ucsb.edu/ws/?pid=4916 (accessed January 13, 1975).

18. See "Solar Energy Remarks Announcing Administration Proposals," http://www.presidency.ucsb.edu/ws/?pid=32500 (accessed September 6, 2016).

19. "Ölwechsel für die BRD," *Die Zeit*, October 26, 1973.

20. The share of imports from OPEC was 49.6% in 1973. See Table 5.7 in the EIA's Annual Energy Review, http://www.eia.gov/totalenergy/data/annual/showtext.cfm?t=ptb0507 (accessed September 30, 2015).

21. "Die Russen sind da," *Die Zeit*, October 5, 1973.

22. "Die Furcht vor einem 'dunklen und kalten Winter,"' "Der Ölkrieg findet nicht statt," *Die Zeit*, October 26, 1973.

23. "Die Erpresser machen Ernst," *Die Zeit*, November 9, 1973.

24. For instance, *Die Zeit* wrote, "OPEC General Secretary Khene suggested a solution [to the oil shortages]: limit consumption and curb individual mobility. But he did not reveal how to do this" ("OPEC-Generalsekretär Khene weist den Ausweg: Schränkt den Verbrauch ein, drosselt den Individualverkehr, so lautet sein Rat. Nur verrät er nicht, wie man das macht"); "Planspiele in Öl," *Die Zeit*, November 2, 1973.

25. "Sparzwang—nein danke!" *Die Zeit*, July 4, 1980.

26. "Vorrang für Ölsparen und neue Technologien," *Süddeutsche Zeitung*, July 2, 1979.

27. http://energytransition.de/category/book/chapter4/.

28. Population and output data from the Penn World Tables (Heston, Summers, and Aten 2012). Unemployment data from "International Comparison of Labor Force Statistics, 1970-2012," BLS, June 17, 2013, http://www.bls.gov/fls/flscomparelf/lfcom pendium.pdf (accessed March 11, 2014).

29. "Ce choix a été débattu dans les enceintes compétentes, et les meilleurs ingénieurs du pays y ont été associés. Il a été validé scientifiquement et politiquement –y compris par le Parti communiste, qui se situait pourtant alors dans l'opposition. Depuis, il n'a jamais été remis en cause par les gouvernements successifs –notamment pendant les deux septennats du président Mitterrand– parce qu'il n'existe pas d'alternative!" Quoted in "Comment la France est devenue nucléaire (et nucléocrate)," *Slate*, April 6, 2011.

30. As of 2011, France had 58 operational reactors. "Nuclear Power Plants, World-Wide," European Nuclear Society, http://www.euronuclear.org/info/encyclopedia/n /nuclear-power-plant-world-wide.htm (accessed March 11, 2014).

31. "Le nuage s'est arrêté à la frontière." This sentence was misattributed to the head of the national nuclear public health safety agency, Pierre Pelerin, by a journalist. "Tchernobyl: quand le nuage s'est (presque) arrêté à la frontière," *Le Nouvel Obs*, September 7, 2011.

32. This slogan, invented by EDF in 1975, can literally be translated as "all electricity, all nuclear."

33. "Tasavallan Presidentin puhe Tampereen Paasikivi-Seurassa," December 19, 1973.

34. "Kekkonen runnoi maakaasun Suomeen," *Tekniikka ja Talous*, January 15, 2004.

35. Note, however, that another nuclear reactor was constructed in Olkiluoto, Finland, by a Swedish company in December 1972 with private capital from Teollisuuden Voima Oy, a corporation owned by major industrial interests (Sunell 2001). This development shows that the Soviet connection is not a sufficient explanation for Finland's strong interest in nuclear power.

36. "Päätös ydinvoiman lisärakentamisesta sinänsä ei liene ollut vaikea, koska asiantuntijat olivat lisärakentamisen kannalla."

37. "Esimerkiksi Energiapolitiikan neuvottelukunnan kanta ydinvoiman lisärakentamiseen oli selvä."

38. "Hiilivoiman ja vähitellen myös öljyn varassa teollistuvassa Länsi-Euroopassa ydinvoima nähtiin pelastavana energiamuotona, mutta myös Euroopasta 'kolmatta voimaa' rakentavana poliittisena tekijänä."

39. For example, the board emphasized the importance of large facility size and the opportunities that inexpensive electricity would create for heavy industry: "Atomivoimassa ei tunnettu kuin suuria laitoksia. Takana olivat kuvitelmat

puunjalostustehtaiden omista paikallisista reaktoreista. Mittakaavaedun lisäksi polttoaineen tuonti hidastui, varmuusvarastoinnin kustannuksia voitiin alentaa, epäedullisen vesivoiman rakentaminen saatiin loppumaan ja syntyi edellytyksiä uuden energiaintensiivisen teollisuuden luomiselle."

40. "Tasavallan Presidentin *Göteborgs-Postenille* myöntämä haastattelu," May 12, 1980.

Chapter 5: Politicization

1. "The Koch Attack on Solar Energy," *New York Times*, April 27, 2014, http://www.nytimes.com/2014/04/27/opinion/sunday/the-koch-attack-on-solar-energy.html (accessed August 13, 2014).

2. Ronald Reagan/Jimmy Carter Presidential Debate, October 28, 1980.

3. The raw data are in 2002 constant prices and available at http://rael.berkeley.edu/old_drupal/sites/default/files/old-site-files/2006/RandD2006.html (accessed February 4, 2017).

4. Measured in 2002 constant prices.

5. Another state that warrants a brief digression is Texas. The Texas state government enacted a renewable portfolio standard based on a market for tradeable certificates in the late 1990s, but the initial target was modest, and few people in the state expected the policy to have much of an effect (Rabe 2004; Hurlbut 2008). In the 2000s, however, the enhanced competiveness of wind power resulted in a "Texas wind rush" and made this deeply conservative state an unlikely renewable energy powerhouse (Galbraith and Price 2013).

6. "Bad Faith on Ecology Laid to Reagan," *New York Times*, April 1, 1982.

7. "Exxon Foresees 'Real' Energy Price Up By 50% By 2000," *New York Times*, December 19, 1980.

8. "Our and Our Environment's Future Demand Nuclear Power," *New York Times*, July 12, 1983.

9. "Solar Research Now Leans on Big Oil Money," *New York Times*, October 16, 1983.

10. "Coal Mining: Long-Term Contribution Trends," https://www.opensecrets.org/industries/totals.php?cycle=2016&ind=E1210 (accessed September 29, 2016).

11. "Oil & Gas: Long-Term Contribution Trends," https://www.opensecrets.org/industries/totals.php?ind=E01++ (accessed September 29, 2016). To be sure, the clean technology industry was not very active either. In 1990, the average contribution by this sector to Senators was a paltry $370 for Democrats and $111 for Republicans. See "Alternative Energy Production & Services," https://www.opensecrets.org/industries/indus.php?ind=E1500 (accessed September 15, 2016).

12. The renewable energy sector, again, remained almost entirely inactive. In the 1990s, average contributions to Democratic or Republican Senators, for example, did not reach $1,000 in election. See "Alternative Energy Production & Services," https://www.opensecrets.org/industries/indus.php?ind=E1500 (accessed September 15, 2016).

13. On oil and gas: "Oil & Gas: Long-Term Contribution Trends," https://www.opensecrets.org/industries/totals.php?ind=E01++. On coal: "Coal Mining: Long-Term Contribution Trends," https://www.opensecrets.org/industries/totals.php?ind=E1210 (accessed September 30, 2016).

14. See https://www.opensecrets.org/lobby/top.php?showYear=1998&indexType=i (accessed October 6, 2015).

15. See http://www.world-nuclear.org/info/country-profiles/countries-g-n/germany/ (accessed April 20, 2014).

16. For example, "Säure-Regen: Da Liegt was in der Luft," *Der Spiegel*, November 20, 1981.

17. The EEG underwent a number of reforms since its inception. In its most recent avatar (after the 2014 reform), renewables producers do not receive a fixed price anymore but instead have to sell their electricity on the market; they do, however, receive a market premium. To give an example, as of 2014, solar panels with a capacity up to 1 megawatt received a premium of 11.49 euro cent per KWh. See EEG 2014, § 48 Solare Strahlungsenergie, http://www.gesetze-im-internet.de/eeg_2014/__48.html (accessed October 9, 2015). Before these reforms, the fixed price varied markedly over time, by type of energy source, and by size of the installation.

18. "1973 Full List," *Forbes*, http://archive.fortune.com/magazines/fortune/fortune500_archive/full/1973/ (accessed October 4, 2016).

19. See the database on contributions above €20,000 at Unklarheiten, http://www.parteispenden.unklarheiten.de, and the Bundestag's own database ("Parteispenden über 50.000 C"), http://www.bundestag.de/bundestag/parteienfinanzierung/fundstellen50000/2002 (accessed October 15, 2015).

20. Renewable energy companies were identified based on their names and authors' information of the German renewable energy sector.

21. The European Court of Justice would side with renewable energy producers in a ruling published in March 2001. See "PreussenElektra AG v Schhleswag AG," European Court of Justice, March 13, 2001, http://eur-lex.europa.eu/legal-content/EN/TXT/PDF/?uri=CELEX:61998CJ0379&from=EN (accessed October 10, 2015).

22. Numbers reported in constant 1981 dollars. On January 5, 1981, the exchange rate was 7.14 kronors for one dollar. See https://research.stlouisfed.org/fred2/series/AEXDNUS (accessed October 7, 2015).

23. Numbers reported in 2001 prices. On January 5, 2001, the exchange rate was 8.33 kronors for one dollar. See https://research.stlouisfed.org/fred2/series/AEXD NUS (accessed October 7, 2015).

24. "Denmark's 50 Percent Wind Commitment and a Path to Fully Renewable Power," *Ars Technica*, March 30, 2012, http://arstechnica.com/tech-policy/2012/03 /denmarks-50-percent-wind-commitment-is-only-the-beginning/ (accessed November 26, 2014).

25. See http://www.statbank.dk/tabsel/146519 (accessed October 9, 2015).

26. Another notable Danish oil and gas producer is the state-owned DONG Energy. It is much smaller than Maersk Oil in oil/gas production but a major electric utility in Denmark.

27. See http://www.maerskoil.com/operations/denmark/pages/oil-and-gas-production .aspx (accessed October 7, 2016).

28. http://world-nuclear.org/NuclearDatabase/Default.aspx?id=27232 (accessed May 5, 2014).

29. These numbers are estimated based on the table on p. 27 of Department of Energy and Climate Change (2013).

30. Letter to Prime Minister Matti Vanhanen and four other ministers, Helsinki, Finland, June 20, 2005. See http://www.sll.fi/ajankohtaista/tiedotteet/2005/minis terikirje.pdf (accessed November 26, 2014).

31. "Suomeen kaksi uutta ydinvoimalaa." *MTV3 Uutiset* July 1, 2010. See http:// www.mtv.fi/uutiset/kotimaa/artikkeli/suomeen-kaksi-uutta-ydinvoimalaa-/1851174 (accessed April 29, 2014).

32. Energiaennakko 2009, Tilastokeskus, http://www.stat.fi/til/ehkh/2009/04/ehkh _2009_04_2010-03-24_tau_001.xls (accessed April 29, 2014).

Chapter 6: Lock-In

1. "State Renewable Portfolio Standards and Goals," National Conference of State Legislatures, July 1, 2015, http://www.ncsl.org/research/energy/renewable-portfolio -standards.aspx (accessed October 10, 2015).

2. For the latest figures, see EIA, "Table 7.2b Electricity Net Generation: Electric Power Sector," http://www.eia.gov/totalenergy/data/monthly/pdf/sec7_6.pdf (accessed February 4, 2017).

3. "Bilan des énergies renouvelables en France en 2013," Ministère de l'Écologie, du Développement Durable et de l'Énergie, http://www.statistiques.developpement-durable .gouv.fr/fileadmin/documents/Themes/Energies_et_climat/Les_differentes_energies

/Energies_renouvelables/2013/bilan-enr-2013-prod-conso.xls (accessed August 20, 2014).

4. "Production d'électricité renouvelable en France jusqu'en 2012 (metropole et DOM)," http://www.statistiques.developpement-durable.gouv.fr/energie-climat/r/energies-renouvelables.html?tx_ttnews[tt_news]=20647&cHash= 470a0ab6e6d4aad-97459c455743b12bf (accessed May 30, 2014).

5. By the early 2010s, China added about 90 to 100 GW of capacity per year; as solar grew up by about 12.9 GW, this means that the contribution of solar to total capacity growth is more than 10%.

6. The actual target has changed over time. See the Ministry of New and Renewable Energy's description, http://www.mnre.gov.in/solar-mission/jnnsm/introduction-2/ (accessed September 20, 2014).

7. Total investments were about $39.5 billion; this includes $1.7 billion devoted to small-scale hydro. Thus, excluding all forms of hydropower, we estimate total investments to be about $38 billion.

8. The LCOE of an electricity project, such as a wind farm, measures the cost at which electricity must be sold to cover all the costs over the lifespan of that project. It is calculated as the time-discounted costs (investments, operations, fuel) divided by the amount of electricity generated. LCOE varies by energy source and regional characteristics. A low LCOE implies that the feasibility of a project is higher because it is more likely to be profitable in the long run.

9. "Asia to outpace Europe in renewables investment says BNEF," *Power Engineering International*, July 3, 2014, http://www.powerengineeringint.com/articles/2014/07/asia-to-outpace-europe-in-renewables-investment-says-bnef.html (accessed July 3, 2014).

10. "Brazil's Ethanol Sector, Once Thriving, Is Being Buffeted by Forces Both Manmade, Natural," *Washington Post*, January 1, 2014, https://www.washingtonpost.com/world/brazils-ethanol-sector-once-thriving-is-being-buffeted-by-forces-both-man-made-natural/2014/01/01/9587b416-56d7-11e3-bdbf-097ab2a3dc2b_story.html (accessed October 10, 2015).

11. "Germany's Green Energy Is an Expensive Success," *Bloomberg*, September 22, 2014, http://www.bloombergview.com/articles/2014-09-22/germany-s-green-energy-is-an-expensive-success (accessed October 15, 2015); and "Germany's Expensive Gamble on Renewable Energy," *Wall Street Journal*, August 26, 2014, http://www.wsj.com/articles/germanys-expensive-gamble-on-renewable-energy-1409106602 (accessed October 15, 2015).

12. "Are Denmark's Renewable Energy Goals Wishful Thinking?", *BBC News*, http://www.bbc.co.uk/news/science-environment-17628146 (accessed June 27, 2014).

13. "Facts about Anholt Offshore Wind Farm," DONG Energy, http://www.dongenergy.com/en (accessed June 27, 2014).

14. "Dong to Spend $795 Million Turning Fossil Plants to Wood," *Bloomberg News*, http://www.bloomberg.com/news/2012-04-11/dong-energy-to-spend-794-million-turning-fossil-plants-to-wood.html (accessed June 27, 2014).

15. Besides Denmark, the countries were Germany, the Netherlands, Spain, Sweden, Switzerland, and the United States.

16. "Public Opinion," Danish Wind Power Association, http://www.windpower.org/en/policy/public_opinion.html (accessed October 15, 2016).

17. "Obstacles to Danish Wind Power," *New York Times*, January 22, 2012, http://www.nytimes.com/2012/01/23/business/global/obstacles-to-danish-wind-power.html?_r=0 (accessed June 28, 2014).

18. "How Angela Merkel Became Germany's Unlikely Green Energy Champion," *Guardian*, May 9, 2011, http://www.theguardian.com/environment/2011/may/09/angela-merkel-green-energy (accessed June 28, 2014).

19. The quotes by Merkel and Röttgen were reported by Christian Schwägerl, a German journalist for the Spiegel, in "Germany's Unlikely Champion of a Radical Green Energy Path," Yale E360, May 9, 2011, http://e360.yale.edu/feature/germanys_unlikely_champion_of_a_radical_green_energy_path/2401/ (accessed December 1, 2015).

20. "Last Decade of German Nuclear Power," *World Nuclear News*, May 31, 2011, http://www.world-nuclear-news.org/NP_Last_decade_of_German_nuclear_power_3105111.html (accessed June 28, 2014).

21. "Germany Ushers in Renewable Energy Reform," *Reuters*, April 8, 2014, http://www.reuters.com/article/2014/04/08/germany-energy-idUSL6N0N011P20140408 (accessed June 28, 2014).

22. "Die Energiewende sei jedoch unumkehrbar, betonte Müller. Es geht nicht mehr darum, ob, sondern wie sie umgesetzt werden wird," in "Strombranche: Neues Marktdesign für Energiewende," VDI, April 27, 2012, http://www.vdi-nachrichten.com/Technik-Wirtschaft/Strombranche-Neues-Marktdesign-fuer-Energiewende (accessed June 28, 2014).

23. "Coal Resurgence Darkens Germans' Green Image," *Financial Times*, October 13, 2015, http://www.ft.com/cms/s/0/719ea15e-68fa-11e5-a57f-21b88f7d973f.html#axzz3pxi2gMkU (accessed October 15, 2015).

24. "Germany's Renewables Electricity Generation Grows in 2015, But Coal Still Dominant," EIA, May 24, 2016, http://www.eia.gov/todayinenergy/detail.php?id=26372 (accessed October 15, 2016).

25. "Coal Returns to German Utilities Replacing Lost Nuclear," *Bloomberg News*, April 15, 2014, http://www.bloomberg.com/news/2014-04-14/coal-rises-vampire-like-as-german-utilities-seek-survival.html (accessed June 28, 2014).

26. "How to Lose Half a Trillion Euros," *The Economist*, October 12, 2013, http://www.economist.com/news/briefing/21587782-europes-electricity-providers-face-existential-threat-how-lose-half-trillion-euros (accessed June 28, 2014).

27. Data from RWE (2015, 4). RWE indicates that renewables represent 7.5% of its electricity capacity, the vast majority of which comes from wind. To obtain the figure of 5.25 GW, we added the share of off- and on-shore wind: 13% and 57%, respectively. Biomass only represents 7% of all renewables in RWE's portfolio.

28. "E.ON Climate & Renewables closes financing for Panther Creek Wind Farm I&II, LLC," *PRNewswire*, December 12, 2013, http://www.prnewswire.com/news-releases/eon-climate—renewables-closes-financing-for-panther-creek-wind-farm-iii-llc-235613801.html (accessed June 26, 2017).

29. "E.ON and E.OUT," *The Economist*, December 6, 2014, http://www.economist.com/news/business/21635503-german-power-producer-breaking-itself-up-face-future-eon-and-eout (accessed December 6, 2014).

30. "Germany's New Coal Plants Push Power Glut to 4-Year High," *Bloomberg News*, June 26, 2014, http://www.bloomberg.com/news/articles/2014-06-26/germany-s-new-coal-plants-push-power-glut-to-4-year-high (accessed October 15, 2015).

31. See the DSIRE database of the US Department of Energy, http://programs.dsireusa.org/system/program/tables (accessed June 27, 2017).

32. "Renewables Portfolio Standard," DSIRE, http://programs.dsireusa.org/system/program/detail/840 (accessed June 27, 2017).

33. "Renewables Portfolio Standard," DSIRE, http://programs.dsireusa.org/system/program/detail/2898 (accessed June 27, 2017).

34. "Smarter Electricity in New York," *New York Times*, May 12, 2014, http://www.nytimes.com/2014/05/13/opinion/smarter-electricity-in-new-york.html (accessed June 28, 2014).

35. Iowa, Kansas, Minnesota, Missouri, Nebraska, North Dakota, and South Dakota.

36. Data from May 2014, "Electric Power Monthly," http://www.eia.gov/electricity/monthly/ (accessed June 28, 2014).

37. "Berkshire's Renewable Energy Investment to Hit $15 Billion," *The Street*, March 1, 2014, http://www.thestreet.com/story/12462501/1/berkshires-renewable-energy-investment-to-hit-15-billion.html (accessed June 28, 2014).

38. "Google Makes Huge Investment in Clean Energy," CNBC, February 16, 2014, http://www.cnbc.com/id/101417698 (accessed June 28, 2014).

39. Table 8.4. Average Power Plant Operating Expenses for Major U.S. Investor-Owned Electric Utilities, 2003 through 2013 (Mills per Kilowatthour), *EIA*, http://www.eia.gov/electricity/annual/html/epa_08_04.html (accessed October 15, 2015).

40. "Koch Brothers and Big Utilities Campaign to Unplug Solar Power," *Los Angeles Times*, April 23, 2014, http://www.latimes.com/opinion/topoftheticket/la-na-tt-koch-brothers-and-solar-power-20140422-story.html (accessed June 28, 2014).

41. "Wind Energy Facts at a Glance," http://www.awea.org/Resources/Content.aspx?ItemNumber=5059 (accessed May 13, 2014).

42. See "Texas Hits New Wind Peak Output," Energy Information Administration, http://www.eia.gov/todayinenergy/detail.cfm?id=16811 (accessed December 5, 2014).

43. "Technology, Not Regulation, Will Kill Coal Fired Power," *ABC Australia*, June 4, 2014, http://www.abc.net.au/news/2014-06-04/technology-not-regulation-will-kill-coal-fired-power/5500356 (accessed June 28, 2014).

44. "Production d'électricité renouvelable en France jusqu'en 2012 (metropole et DOM)," http://www.statistiques.developpement-durable.gouv.fr/energie-climat/r/energies-renouvelables.html?tx_ttnews[tt_news]=20647&cHash=470a0ab6e6d4aad97459c455743b12bf (accessed May 30, 2014).

45. "The Simple Guide to the Renewable Energy Tariffs," http://www.fitariffs.co.uk/library/info/Ownergys_Simple_Guide_to_the_Renewable_Energy_Tariffs.pdf (accessed June 28, 2014).

46. "Conservatives to End Onshore Wind Farm Subsidies," *Reuters*, April 24, 2014, http://uk.reuters.com/article/2014/04/24/uk-britain-energy-windfarms-idUKBREA3N0KV20140424 (accessed June 28, 2014).

47. "Eolien en mer: GDF Suez remporte deux appels d'offres importants," *Le Monde*, May 7, 2014, http://www.lemonde.fr/economie/article/2014/05/07/eolien-en-mer-gdf-suez-remporte-les-appels-d-offres-au-treport-et-a-noirmoutier_4413044_3234.html (accessed May 8, 2014); and "Total garde le cap dans l'industrie solaire, *Le Figaro*, May 6, 2014, http://bourse.lefigaro.fr/devises-matieres-premieres/actu-conseils/total-garde-le-cap-dans-l-industrie-solaire-200259 (accessed May 8, 2014).

48. "Conservatives to End Onshore Wind Farm Subsidies," *Reuters*.

49. We searched the Electoral Commission's contribution database between 2001 and 2015. Searching for keywords such as "electricity," "energy," or "power," we only found a handful of contribution to any of the mainstream parties. See the Electoral Commission's contribution database at http://search.electoralcommission.org.uk (accessed October 15, 2015).

50. Notice that these are unweighted predictions: we do not claim that the world-wide expected share of renewables has gone up that much. Instead, this figure is the expected value for a random country.

51. In comparison, the United States generated 234 GWh from renewable sources in that year and had passed 136 GWh as recently as 2009.

52. "China's Solar Growth Facing Funding Shortfall, Reports Reuters," *PV Magazine*, April 30, 2015, http://www.pv-magazine.com/news/details/beitrag/chinas-solar-growth-facing-funding-shortfall—reports-reuters_100019294/#axzz3fsuBtOco (accessed October 15, 2015).

53. In general, however, distributed renewable electricity generation has grown only slowly in China (Liang 2014).

54. For a large collection of relevant statements and speeches, see http://en.ccchina.gov.cn/list.aspx?clmId=98 (accessed December 7, 2014).

55. See, for example, "Paris Climate Conference: China," National Resources Defense Council, https://www.nrdc.org/sites/default/files/paris-climate-conference-China-IB.pdf (accessed October 18, 2016).

56. Note, however, that the economic benefits from household electrification appear modest based on an impact evaluation by Burlig and Preonas (2016).

57. "India Confronts Mountain of Coal Problems," *Washington Post*, December 22, 2012, http://www.washingtonpost.com/world/india-confronts-mountain-of-coal-problems/2012/12/22/dd631a9e-327c-11e2-bfd5-e202b6d7b501_story.html (accessed December 8, 2014).

58. "India Blackouts Leave 700 Million without Power," *The Guardian*, July 31, 2012, http://www.theguardian.com/world/2012/jul/31/india-blackout-electricity-power-cuts (accessed December 9, 2014).

59. "Despite Having One of the World's Largest Reserves of Coal, India Is Unable to Dig It up Fast Enough," *Economic Times*, June 14, 2012, http://articles.economictimes.indiatimes.com/2012-06-14/news/32236139_1_coal-mines-coal-india-underground-mining (accessed December 10, 2014).

60. "Modi Govt Doubles UPA's Solar Mission Targets," *Business Standard*, September 19, 2014, http://www.business-standard.com/article/economy-policy/modi-govt-doubles-upa-s-solar-mission-targets-114091801225_1.html (accessed December 8, 2014).

61. "Solar, Wind to Beat Coal on Costs in China, India by 2020," *Reneweconomy*, July 10, 2014, http://reneweconomy.com.au/2014/solar-wind-beat-coal-costs-china-india-2020 (accessed October 15, 2015).

62. "At WTO, India May Oppose USA's Stance on Jawaharlal Nehru Solar Mission," *Economic Times*, April 28, 2014, http://economictimes.indiatimes.com/news/economy/policy/at-wto-india-may-oppose-usas-stance-on-jawaharlal-nehru-solar-mission/articleshow/34304534.cms (accessed June 28, 2014).

63. "BJP Manifesto: Lok Sabha Elections 2009," BJP Party Manifesto for 2009, http://www.bjp.org/documents/manifesto/manifesto-lok-sabha-election-2009 (accessed December 10, 2014).

64. See EIA analysis, http://www.eia.gov/todayinenergy/detail.cfm?id=22652 (accessed October 18, 2015).

65. "Solar Boom Boosts South Africa Salaries With 25% Jobless: Energy," *Bloomberg News*, December 11, 2013, http://www.businessweek.com/news/2013-12-10/solar-boom-boosts-south-africa-salaries-with-25-percent-jobless-energy (accessed August 18, 2014).

66. "South Africa Doubles Plans for Renewable Energy on Program Wins," *Bloomberg News*, December 12, 2014, http://www.bloomberg.com/news/2014-12-12/south-africa-doubles-plans-for-renewable-energy-on-program-wins.html (accessed December 13, 2014).

67. "In a Decade Brazil Must Increase Power Generating Capacity by 56%," *MercoPress*, January 5, 2012, available at http://en.mercopress.com/2012/01/05/in-a-decade-brazil-must-increase-power-generating-capacity-by-56 (accessed December 14, 2014).

68. "Solar Powers Brazil's World Cup," *PV Magazine*, June 12, 2014, http://www.pv-magazine.com/news/details/beitrag/solar-powers-brazils-world-cup_100015391/ (accessed December 14, 2014).

69. "In Brazil, the Wind Is Blowing in a New Era of Renewable Energy," *Washington Post*, October 30, 2013, https://www.washingtonpost.com/world/in-brazil-the-wind-is-blowing-in-a-new-era-of-renewable-energy/2013/10/30/8111b7e8-2ae0-11e3-b141-298f46539716_story.html (accessed December 16, 2014).

70. See the World Resource Institute's "A Closer Look at Brazil's New Climate Plan (INDC)," http://www.wri.org/blog/2015/09/closer-look-brazils-new-climate-plan-indc (accessed October 18, 2016).

71. "Special Report Africa: Kenya," *PV Magazine*, November 20, 2013, http://www.pv-magazine.com/news/details/beitrag/special-report-africa—kenya_100013508/ (accessed December 16, 2014).

72. "Vestas Gets Biggest Wind Order as Africa Market Accelerates," *Bloomberg News*, December 12, 2014, http://www.bloomberg.com/news/2014-12-12/vestas-gets-biggest-wind-turbine-order-as-africa-market-erupts.html (accessed December 16, 2014).

73. "Kenya Power to Invest $1.8 bln in 5 Years to Expand Network" *Reuters*, October 11, 2013, http://www.reuters.com/article/2013/10/11/kenya-power-idUSL6N0I11RQ20131011 (accessed October 15, 2015).

74. "Africa Has to Start Deliberating Energy Policy—Raila Odinga," *Engineering News*, September 2, 2014, http://www.engineeringnews.co.za/article/africa-has-to-start-deliberating-energy-policy-raila-odinga-2014-09-02 (accessed December 16, 2014).

75. "Kenya's Trade Minister Moses Wetangula Woos Danish Investors," *Today Financial News*, http://todayfinancialnews.com/trade-and-investment/1485-kenyas-trade-minister-moses-wetangula-woos-danish-investors (accessed December 16, 2014).

76. "Kenya Eyes Domestic Gas, Coal for New Power Plants," *Reuters*, June 25, 2015, http://af.reuters.com/article/investingNews/idAFKBN0P51U720150625 (accessed October 20, 2015).

Chapter 7: Improving National Policy

1. "Fossil Fuels Remain at the Forefront, Energy Expert Says," *Boston Globe*, February 22, 2014, https://www.bostonglobe.com/business/2014/02/22/fossil-fuels-will-continue-dominate-energy-mix-says-daniel-yergin/bXFiJqIyBocTTRiLsxsxcL/story.html (accessed May 11, 2014).

2. See "Renewable Portfolio Standard Policies, March 2013," http://www.dsireusa.org/documents/summarymaps/RPS_map.pdf (accessed May 11, 2014).

3. "The Fallout from Fukushima," *New Scientist*, http://www.newscientist.com/special/fukushima-crisis (accessed May 12, 2014).

4. "Japan Unveils Draft Energy Policy in Wake of Fukushima," *Guardian*, February 25, 2014.

5. "Fukushima Pledges to Go 100% Renewable," *CleanTechnica*, February 6, 2014. http://cleantechnica.com/2014/02/06/fukushima-100-renewable-energy/ (accessed May 12, 2014).

6. "AEO2014 Early Release Overview," US Energy Information Administration, December 16, 2013. http://www.eia.gov/forecasts/aeo/er/early_elecgen.cfm (accessed May 12, 2014).

7. "Industry Sees Costly Rules After Obama's Climate Report," *Bloomberg*, May 7, 2014. http://www.bloomberg.com/news/2014-05-07/industry-sees-costly-rules-after-obama-s-climate-report.html (accessed May 12, 2014).

8. "Germany's Energy Transition: Sunny, Windy, Costly and Dirty," *Economist*, January 18, 2014.

9. "Germany Ushers in Renewable Energy Reform." *Reuters*, April 8, 2014. http://www.reuters.com/article/2014/04/08/us-germany-energy-idUSBREA3716I20140408 (accessed May 13, 2014).

10. "Renewables Surged to 74% of German Demand Last Sunday," *Greentech Media*, May 16, 2014. http://www.greentechmedia.com/articles/read/Renewables-Surged-to-74-of-German-Demand-Last-Sunday (accessed May 16, 2014).

11. "The Future of Power from Biomass in Germany," *Renewables International* January 24, 2014. http://www.renewablesinternational.net/the-future-of-power-from-biomass-in-germany/150/537/76305/ (accessed June 25, 2014).

12. "Reform der Energiewende: Teure Ingenieursromantik," *Spiegel Online*, April 8, 2014. http://www.spiegel.de/wirtschaft/soziales/energiewende-die-folgen-der-oekostrom-reform-a-963094.html (accessed May 18, 2014).

13. "Germany's Effort at Clean Energy Proves Complex," *New York Times*, September 18, 2013.

14. "Germany Plans to Curb Energy Transition." *Deutsche Welle*, November 30, 2013. http://www.dw.de/germany-plans-to-curb-energy-transition/a-17263482 (accessed May 18, 2014).

15. "Smart Grid Not Sexy Enough for German Households," *Renewables International*, April 14, 2014. http://www.renewablesinternational.net/smart-grid-not-sexy-enough-for-german-households/150/537/78186/ (accessed May 18, 2014).

16. "French Body Blames Renewables for EU Power Market Failures." *EurActive*, January 31, 2014. http://www.euractiv.com/energy/eu-electric-system-failure-news-533145 (accessed May 19, 2014).

17. "EU vs. Regional Electricity Markets: Don't Think Too Small," *Bruegel*, October 8, 2013. http://www.bruegel.org/nc/blog/detail/article/1163-eu-vs-regional-electricity-markets-dont-think-too-small/ (accessed May 19, 2014).

18. "An Energy Supergrid for Europe Faces Big Obstacles," *New York Times*, January 16, 2012.

19. "Franco-German Energy Plan to Focus on Renewables, not Mergers," *Reuters*, January 15, 2014. http://uk.reuters.com/article/2014/01/15/france-germany-energy-idUKL5N0KP1UX20140115 (accessed January 15, 2014).

20. See http://www.iea.org/policiesandmeasures/renewableenergy/ (accessed May 19, 2014).

21. See https://www.irena.org/potential_studies/index.aspx (accessed May 19, 2014).

22. "Renewables Club Founded by 10 Countries," *CleanTechnica*, June 1, 2013. http://cleantechnica.com/2013/06/01/renewables-club-founded-by-10-countries/ (accessed June 1, 2013).

23. "India Power Ministry Seeks to Reverse Solar Duties," *Bloomberg* June 23, 2014. http://www.bloomberg.com/news/2014-06-23/india-power-ministry-seeks-to-reverse-solar-duties.html (accessed June 23, 2014).

24. "U.S. Says China Hacked West Coast Solar Firm, Nuclear Power and Metals Industries," *San Jose Mercury News*, May 19, 2014.

25. See "U.S.-India Energy Cooperation," http://energy.gov/ia/initiatives/us-india-energy-cooperation (accessed June 30, 2014).

26. "US Solar Industry Opposes Import Tariffs on Chinese Panels," *Clean Energy Authority*, June 11, 2014. http://www.cleanenergyauthority.com/solar-energy-news/us-solar-industry-opposes-import-tariffs-061114 (accessed June 30, 2014).

Chapter 8: Conclusion

1. "German Proposal Seeks to Sharply Curtail Fracking," *New York Times*, July 4, 2014, https://www.nytimes.com/2014/07/05/business/international/german-proposal-to-curtail-fracking.html (accessed August 24, 2014).

2. "Gaz de schiste: un engouement mondial, mais beaucoups de doutes," *Le Monde*, December 12, 2012. http://www.lemonde.fr/planete/article/2012/12/21/gaz-de-schiste-un-engouement-mondial-mais-beaucoup-de-doutes_1809052_3244.html (accessed August 24, 2014).

References

Acemoglu, Daron, and James A. Robinson. 2008. Persistence of Power, Elites, and Institutions. *American Economic Review* 98 (1): 267–293.

Acemoglu, Daron, Philippe Aghion, Leonardo Bursztyn, and David Hemous. 2012. The Environment and Directed Technical Change. *American Economic Review* 102 (1): 131–166.

Achen, Christopher H., and Duncan Snidal. 1989. Rational Deterrence Theory and Comparative Case Studies. *World Politics* 41 (2): 143–169.

Acker, Richard H, and Daniel M. Kammen. 1996. The Quiet (Energy) Revolution: Analysing the Dissemination of Photovoltaic Power Systems in Kenya. *Energy Policy* 24 (1): 81–111.

Ackermann, Thomas, and Lennart Söder. 2002. An Overview of Wind Energy-Status 2002. *Renewable & Sustainable Energy Reviews* 6 (1–2): 67–127.

ADEME. 2012. *Maîtrise de l'Énergie et Développement des Énergies Renouvelables*. [*Mastery of Energy and Renewable Energy Development*.] Angers, France: Stratégie & Études.

AGEE. 2013. *Erneuerbare Energien in Zahlen*. [*Renewable Energy in Numbers*.] Berlin: Arbeitsgruppe Erneuerbare Energien-Statistik.

AGEE. 2016. *Erneuerbare Energien in Zahlen: Nationale und Internationale Entwicklung im Jahr 2015*. [*Renewable Energy in Numbers: National and International Development in 2015*.] Berlin: Arbeitsgruppe Erneuerbare Energien-Statistik.

Agnolucci, Paolo. 2007. Wind Electricity in Denmark: A Survey of Policies, Their Effectiveness and Factors Motivating Their Introduction. *Renewable & Sustainable Energy Reviews* 11 (5): 951–963.

Agrawala, Shardul, and Steinar Andresen. 1999. Indispensability and Indefensibility? The United States in the Climate Treaty Negotiations. *Global Governance* 5 (4): 457–482.

Åhman, Max. 2006. Government Policy and the Development of Electric Vehicles in Japan. *Energy Policy* 34 (4): 433–443.

Aidt, Toke S. 1998. Political Internalization of Economic Externalities and Environmental Policy. *Journal of Public Economics* 69 (1): 1–16.

Aiken, Philip. 2002. *BHP Billiton Petroleum and Growth*. Melbourne, Australia: Analyst Briefing.

Ailawadi, V. S., and Subhes C. Bhattacharyya. 2006. Access to Energy Services by the Poor in India: Current Situation and Need for Alternative Strategies. *Natural Resources Forum* 30 (1): 2–14.

Aklin, Michaël, and Johannes Urpelainen. 2013a. Debating Clean Energy: Frames, Counter Frames, and Audiences. *Global Environmental Change* 23 (5): 1225–1232.

Aklin, Michaël, and Johannes Urpelainen. 2013b. Political Competition, Path Dependence, and the Strategy of Sustainable Energy Transitions. *American Journal of Political Science* 57 (3): 643–658.

Aklin, Michaël, and Johannes Urpelainen. 2014. The Global Spread of Environmental Ministries: Domestic-International Interactions. *International Studies Quarterly* 58 (4): 764–780. doi:10.1111/isqu.12119.

Aklin, Michaël, Patrick Bayer, S. P. Harish, and Johannes Urpelainen. 2015. Quantifying Slum Electrification in India and Explaining Local Variation. *Energy* 80 (1): 203–212.

Anderson, C. Lindsay, and Judith B. Cardell. 2008. Reducing the Variability of Wind Power Generation for Participation in Day Ahead Electricity Markets. Paper presented at the 41st Hawaii International Conference on System Sciences, University of Hawaii, Honolulu, HI, January 7–10.

Anderson, Soren T., Ian W. H. Parry, James M. Sallee, and Carolyn Fischer. 2011. Automobile Fuel Economy Standards: Impacts, Efficiency, and Alternatives. *Review of Environmental Economics and Policy* 5 (1): 89–108.

Andrews, Thomas G. 2008. *Killing for Coal: America's Deadliest Labor War*. Cambridge, MA: Harvard University Press.

Aoki, Masahiko. 2001. *Toward a Comparative Institutional Analysis*. Cambridge, MA: MIT Press.

Arnett, J. C., L. A. Schaffer, J. P. Rumberg, and R. E. L. Tolbert. 1984. Design, Installation and Performance of the ARCO Solar One-Megawatt Power Plant. In *5th Photovoltaic Solar Energy Conference*, edited by Willeke Palz and F. Fittipaldi, pp. 314–320. Dordrecht, the Netherlands: Springer.

Arrow, Kenneth. 1962. The Economic Implications of Learning by Doing. *Review of Economic Studies* 29 (3): 155–173.

Arthur, W. Brian. 1989. Competing Technologies, Increasing Returns and Lock-in by Historical Events. *Economic Journal (London)* 99 (1): 106–131.

Asif, M., and T. Muneer. 2007. Energy Supply, Its Demand and Security Issues for Developed and Emerging Economies. *Renewable & Sustainable Energy Reviews* 11 (7): 1388–1413.

Atkeson, Andrew, and Patrick J. Kehoe. 2007. Modeling the Transition to a New Economy: Lessons from Two Technological Revolutions. *American Economic Review* 97 (1): 64–88.

Auken, Svend. 2002. Answers in the Wind: How Denmark Became a World Pioneer in Wind Power. *Fletcher Forum of World Affairs* 26:149–157.

Babiker, Mustafa H. 2005. Climate Change Policy, Market Structure, and Carbon Leakage. *Journal of International Economics* 65 (2): 421–445.

Backwell, Ben. 2014. *Wind Power: The Struggle for Control of a New Global Industry*. New York: Routledge.

Baker, Erin, Meredith Fowlie, Derek Lemoine, and Stanley S. Reynolds. 2013. The Economics of Solar Electricity. Energy Institute at Haas, University of California, Berkeley, Working Paper 240.

Baker, Lucy. 2011. Governing Electricity in South Africa: Wind, Coal and Power Struggles. Governance of Clean Development Series, Working Paper 15, University of East Anglia.

Balogh, Brian. 1991. *Chain Reaction: Expert Debate and Public Participation in American Commercial Nuclear Power, 1945–1975*. New York: Cambridge University Press.

Barnes, Douglas F. 2007. *The Challenge of Rural Electrification: Strategies for Developing Countries*. Washington, DC: Resources for the Future and Energy Sector Management Assistance Program.

Barrett, Scott. 2003. *Environment and Statecraft: The Strategy of Environmental Treaty-Making*. Oxford: Oxford University Press.

Barroso, Luiz. 2012. Renewable Energy Auctions: The Brazilian Experience. Workshop on Energy Tariff-based Mechanisms, IRENA.

Baumgartner, Frank R. 1989. Independent and Politicized Policy Communities: Education and Nuclear Energy in France and in the United States. *Governance: An International Journal of Policy, Administration and Institutions* 2 (1): 42–66.

Bauwens, Thomas, Boris Gotchev, and Lars Holstenkamp. 2016. What Drives the Development of Community Energy in Europe? The Case of Wind Power Cooperatives. *Energy Research & Social Science* 13:136–147.

Bayer, Patrick, and Johannes Urpelainen. 2016. It's All About Political Incentives: Democracy and the Renewable Feed-In Tariff. *Journal of Politics* 78 (2): 603–619.

BDEW. 2014. *Erneuerbare Energien und das EEG: Zahlen, Fakten, Grafiken*. [*Renewable Energy and the EEG: Numbers, Facts, Graphics*.] Berlin: Bundesverband der Energie- und Wasserwirtschaft.

Beaudin, Marc, Hamidreza Zareipour, Anthony Schellenberglabe, and William Rosehart. 2010. Energy Storage for Mitigating the Variability of Renewable Electricity Sources: An Updated Review. *Energy for Sustainable Development* 14 (4): 302–314.

Bechberger, Mischa, and Danyel Reiche. 2004. Renewable Energy Policy in Germany: Pioneering and Exemplary Regulations. *Energy for Sustainable Development* 8 (1): 47–57.

Becker, Gary S. 1983. A Theory of Competition among Pressure Groups for Political Influence. *Quarterly Journal of Economics* 98 (3): 371–400.

Beise, Marian. 2004. Lead Markets: Country-Specific Drivers of the Global Diffusion of Innovations. *Research Policy* 33 (6–7): 997–1018.

Beise, Marian, and Klaus Rennings. 2005. Lead Markets and Regulation: A Framework for Analyzing the International Diffusion of Environmental Innovations. *Ecological Economics* 52 (1): 5–17.

Beland, Daniel. 2007. Ideas and Institutional Change in Social Security: Conversion, Layering, and Policy Drift. *Social Science Quarterly* 88 (1): 20–38.

Benoit, Kenneth, and Michael Laver. 2006. *Party Policy in Modern Democracies*. London: Routledge.

Bergek, Anna, and Staffan Jacobsson. 2003. The Emergence of a Growth Industry: A Comparative Analysis of the German, Dutch, and Swedish Wind Turbine Industries. In *Change, Transformation, and Development*, edited by J. Stan Metcalfe and Uwe Cantner, 197–227. Heidelberg, Germany: Physica Verlag.

Bertani, Ruggero. 2005. World Geothermal Power Generation in the Period 2001–2005. *Geothermics* 34 (6): 651–690.

Bettencourt, Luis M. A., Jessika E. Trancik, and Jasleen Kaur. 2013. Determinants of the Pace of Global Innovation in Energy Technologies. *PLoS One* 8 (10): 1–6.

Bharatiya Janata Party. 2014. *Election Manifesto 2014*.

Bhattacharya, Saugata, and Urjit R. Patel. 2007. The Power Sector in India: An Inquiry into the Efficacy of the Reform Process. *India Policy Forum* 4 (1): 211–283.

Bhattacharyya, Subhes C. 2007. Power Sector Reform in South Asia: Why Slow and Limited so Far? *Energy Policy* 35 (1): 317–332.

Billon, Philippe Le. 2001. The Political Ecology of War: Natural Resources and Armed Conflicts. *Political Geography* 20 (5): 561–584.

Bird, L., B. Parsons, T. Gagliano, M. Brown, R. Wiser, and M. Bolinger. 2003. Policies and Market Factors Driving Wind Power Development in the United States. National Renewable Energy Laboratory, NREL/TP-620–34599.

Blanchard, Olivier J., and Jordi Gali. 2007. The Macroeconomic Effects of Oil Shocks: Why Are the 2000s So Different from the 1970s? NBER Working Paper 13368.

Blinder, Alan S. 1997. Is Government Too Political? *Foreign Affairs* 76 (6): 115–126.

Bloomberg. 2014a. Green Bonds Market Outlook 2014. Bloomberg New Energy Finance.

Bloomberg. 2014b. H1 2014 Levelized Cost of Electricity—PV. Bloomberg New Energy Finance.

BMWi. 2015. *Erneuerbare Energien in Zahlen: Nationale und Internationale Entwicklung im Jahr 2014.* [*Renewable Energy in Numbers: National and International Development in 2014.*] Berlin: BMWi.

BNEF. 2012. US Solar: Re-Imagining US Solar Financing. White Paper.

BNEF. 2014a. *2030 Market Outlook: Global Overview*. Bloomberg New Energy Finance; Accessed July 7, 2014.

BNEF. 2014b. *Global Trends in Renewable Energy Investment 2014*. Frankfurt, Germany: Frankfurt School-UNEP Centre and Bloomberg New Energy Finance.

BNEF. 2015. *Global Trends in Renewable Energy Investment 2015*. Frankfurt, Germany: Frankfurt School-UNEP Centre and Bloomberg New Energy Finance.

BNEF. 2016. Off-Grid Solar Market Trends Report 2016. Bloomberg New Energy Finance and Lighting Global. https://www.lightingglobal.org/launch-of-off-grid-solar-market-trends.

Boix, Carles. 1998. *Political Parties, Growth and Equality: Conservative and Social Democratic Economic Strategies in the World Economy*. New York: Cambridge University Press.

Bolinger, Mark. 2001. *Community Wind Power Ownership Schemes in Europe and their Relevance to the United States*. Berkeley, CA: Ernest Orlando Lawrence Berkeley National Laboratory.

Bolinger, Mark A. 2005. Making European-Style Community Wind Power Development Work in the {US}. *Renewable & Sustainable Energy Reviews* 9 (6): 556–575.

Bolsen, Toby, and Fay Lomax Cook. 2008. The Polls-Trends: Public Opinion on Energy Policy, 1974-2006. *Public Opinion Quarterly* 72 (2): 364–388.

BP. 2013. *BP Statistical Review of World Energy June 2013*. London: BP.

BP. 2014. *BP Energy Outlook 2035*. London: BP.

BP. 2015. *BP Statistical Review of World Energy June 2015*. London: BP.

Brass, Jennifer N., Sanya Carley, Lauren M. MacLean, and Elizabeth Baldwin. 2012. Power for Development: A Review of Distributed Generation Projects in the Developing World. *Annual Review of Environment and Resources* 37:107–136.

Braun, Jan Frederik. 2011. EU Energy Policy under the Treaty of Lisbon Rules: Between a New Policy and Business as Usual. European Policy Institutes Network, Working Paper 31.

Brechin, Steven R. 2003. Comparative Public Opinion and Knowledge of Global Climatic Change and the Kyoto Protocol: The US versus the World? *International Journal of Sociology and Social Policy* 23 (10): 106–134.

Bruns, Elke, Dörte Ohlhorst, Bernd Wenzel, and Johann Köppel. 2011. *Renewable Energies in Germany's Electricity Market*. Berlin: Springer.

Bugaje, I. M. 2006. Renewable Energy for Sustainable Development in Africa: A Review. *Renewable & Sustainable Energy Reviews* 10 (6): 603–612.

Burke, Marshall B., Edward Miguel, Shanker Satyanath, John A. Dykema, and David B. Lobell. 2009. Warming Increases the Risk of Civil War in Africa. *Proceedings of the National Academy of Sciences of the United States of America* 106 (49): 20670–20674.

Burlig, Fiona, and Louis Preonas. 2016. Out of the Darkness and Into the Light? Development Effects of Rural Electrification in India. Energy Institute at Haas, University of California, Berkeley, Working Paper 268.

Busch, Per-Olof, and Helge Jörgens. 2005. The International Sources of Policy Convergence: Explaining the Spread of Environmental Policy Innovations. *Journal of European Public Policy* 12 (5): 1–25.

Calder, Kent E. 2012. *The New Continentalism: Energy and Twenty-First-Century Eurasian Geopolitics*. New Haven: Yale University Press.

Campbell, Scott. 1996. Green Cities, Growing Cities, Just Cities?: Urban Planning and the Contradictions of Sustainable Development. *Journal of the American Planning Association* 62 (3): 296–312.

Carlin, P. W., A. S. Laxson, and E. B. Miljadi. 2003. The History and State of the Art of Variable-Speed Wind Turbine Technology. *Wind Energy* 6:129–159.

Carolin, Mabel, and Eugene Fernandez. 2008. Growth and Future Trends of Wind Energy in India. *Renewable & Sustainable Energy Reviews* 12 (6): 1745–1757.

Carson, Rachel. 2002. *Silent Spring: 40th Anniversary Edition*. New York: Mariner Books. (Originally published in 1962).

Carvalho, Georgia O. 2006. Environmental Resistance and the Politics of Energy Development in the Brazilian Amazon. *Journal of Environment & Development* 15 (3): 245–268.

CEC. 2016. Warren-Alquist Act. CEC-140–2016–002. http://www.energy.ca.gov/reports/Warren-Alquist_Act/.

Chakrabarti, Snigdha, and Subhendu Chakrabarti. 2002. Rural Electrification Programme with Solar Energy in Remote Region—A Case Study in an Island. *Energy Policy* 30 (1): 33–42.

Chalvatzis, Konstantinos J., and Elizabeth Hooper. 2009. Energy Security vs. Climate Change: Theoretical Framework Development and Experience in Selected EU Electricity Markets. *Renewable & Sustainable Energy Reviews* 13 (9): 2703–2709.

Chen, Dingding, Chao-yo Cheng, and Johannes Urpelainen. 2014. Support for Renewable Energy among the Chinese Middle Class: A Survey Experiment. Working Paper, Columbia University.

Chen, Yuyu, Avraham Ebenstein, Michael Greenstone, and Hongbin Li. 2013. Evidence on the Impact of Sustained Exposure to Air Pollution on Life Expectancy from China's Huai River Policy. *Proceedings of the National Academy of Sciences of the United States of America* 110 (32): 12936–12941.

Chenoweth, Jonathan, and Eral Feitelson. 2005. Neo-Malthusians and Cornucopians Put to Test: *Global 2000* and *The Resourceful Earth* Revisited. *Futures* 37 (1): 51–72.

Cheon, Andrew, and Johannes Urpelainen. 2012. Oil Prices and Energy Technology Innovation: An Empirical Analysis. *Global Environmental Change* 22 (2): 407–417.

Cheon, Andrew, and Johannes Urpelainen. 2013. How Do Competing Interest Groups Influence Environmental Policy? The Case of Renewable Electricity in Industrialized Democracies, 1989–2007. *Political Studies* 61 (4): 874–897.

Cherni, Judith A., and Joanna Kentish. 2007. Renewable Energy Policy and Electricity Market Reforms in China. *Energy Policy* 35 (7): 3616–3629.

Cherrington, R., V. Goodship, A. Longfield, and K. Kirwan. 2013. The Feed-In Tariff in the UK: A Case Study Focus on Domestic Photovoltaic Systems. *Renewable Energy* 50:421–426.

China. 2012. Full Text: China's Energy Policy 2012. Information Office of the State Council, People's Republic of China. http://www.gov.cn/english/official/2012-10/24/content_2250497.htm.

Chu, Steven, and Arun Majumdar. 2012. Opportunities and Challenges for a Sustainable Energy Future. *Nature* 488 (7411): 294–303.

Chubb, John E. 1983. *Interest Groups and the Bureaucracy: The Politics of Energy*. Stanford, CA: Stanford University Press.

Cipolla, Carlo M. 1994. *Before the Industrial Revolution: European Society and Economy, 1000–1700*. New York: Norton.

Clark, Wilson. 1974. *Energy for Survival: The Alternative to Extinction*. Garden City, NJ: Anchor Press.

Cochet, Yves. 2000. Stratégie et Moyens de Développement de l'Efficacité Energétique et des Sources d'Énergie Renouvelables en France. [Strategy and Means of Developing Energy Efficiency and Sources of Renewable Energy in France.] Technical report for the Prime Minister.

Cohen, Linda R., and Roger G. Noll. 1991. Synthetic Fuels from Coal. In *The Technology Pork Barrel*, edited by Linda R. Cohen and Roger G. Noll, 259–320. Washington, DC: Brookings Institution Press.

Colgan, Jeff D. 2013. *Petro-Aggression: When Oil Causes War.* New York: Cambridge University Press.

Collier, David. 2011. Understanding Process Tracing. *PS, Political Science & Politics* 44 (4): 823–830.

Collier, Paul, and Anthony J. Venables. 2012. Greening Africa? Technologies, Endowments and the Latecomer Effect. *Energy Economics* 34 (S1): S75–S84.

Collingridge, D. 1984. Lessons of Nuclear Power French "Success" and the Breeder. *Energy Policy* 12 (2): 189–200.

Commission de Régulation de l'Énergie. 2014. Coûts et Rentabilité des Énergies Renouvelables en France Métropolitaine. [Costs and Profitability of Renewable Energies in Metropolitan France.] Analysis.

Congressional Research Service. 2011. *U.S. Wind Turbine Manufacturing: Federal Support for an Emerging Industry*. Washington, DC: Federal Publications.

Cornot-Gandolphe, Sylvie. 2014. China's Coal Market: Can Beijing Tame "King Coal"? Oxford Institute for Energy Studies, OIES Paper CL-1.

Cotti, Chad, and Mark Skidmore. 2010. The Impact of State Government Subsidies and Tax Credits in an Emerging Industry: Ethanol Production 1980-2007. *Southern Economic Journal* 76 (4): 1076–1093.

Council on Foreign Relations. 2012. *Public Opinion on Global Issues*. New York: Council on Foreign Relations.

Cowan, Robin. 1990. Nuclear Power Reactors: A Study in Technological Lock-In. *Journal of Economic History* 50 (3): 541–567.

Cowan, Robin, and Staffan Hultén. 1996. Escaping Lock-In: The Case of the Electric Vehicle. *Technological Forecasting and Social Change* 53 (1): 61–79.

CTI. 2014. *Unburnable Carbon: Are The World's Financial Markets Carrying a Carbon Bubble?* London: Carbon Tracker Initiative.

Dai, Xinyuan. 2005. Why Comply? The Domestic Constituency Mechanism. *International Organization* 59 (2): 363–398.

Dalton, Russell J. 1994. *The Green Rainbow: Environmental Groups in Western Europe*. New Haven, CT: Yale University Press.

DAMVAD Danmark. 2015. Branchestatistik for Vindmølleindustrien. May 2015, Copenhagen.

Dangerman, A. T. C. Jérôme, and Hans Joachim Schellnhuber. 2013. Energy Systems Transformation. *Proceedings of the National Academy of Sciences of the United States of America* 110 (7): E549–E558.

Davis, Charles. 2012. The Politics of "Fracking": Regulating Natural Gas Drilling Practices in Colorado and Texas. *Review of Policy Research* 29 (2): 177–191.

de Carmoy, Guy. 1978. The USA Faces the Energy Challenge. *Energy Policy* 6 (1): 36–52.

de Carmoy, Guy. 1982. The New French Energy Policy. *Energy Policy* 10 (3): 181–188.

de Coninck, Heleen, Carolyn Fischer, Richard G. Newell, and Takahiro Ueno. 2008. International Technology-Oriented Agreements to Address Climate Change. *Energy Policy* 36 (1): 335–356.

DEA. 2012. *Green Production in Denmark*. Copenhagen: Danish Energy Agency.

DEA. 2013. *Oil and Gas Production in Denmark 2013*. Copenhagen: Danish Energy Agency.

DEA. 2014. *Energy Statistics 2012*. Copenhagen: Danish Energy Agency.

Deichmann, Uwe, Craig Meisner, Siobhan Murray, and David Wheeler. 2011. The Economics of Renewable Energy Expansion in Rural Sub-Saharan Africa. *Energy Policy* 39 (1): 215–227.

del Río, Pablo, and Gregory Unruh. 2007. Overcoming the Lock-Out of Renewable Energy Technologies in Spain: The Cases of Wind and Solar Electricity. *Renewable & Sustainable Energy Reviews* 11 (7): 1498–1513.

Delmas, Magali A., and Maria J. Montes-Sancho. 2011. U.S. State Policies for Renewable Energy: Context and Effectiveness. *Energy Policy* 39 (5): 2273–2288.

Department of Energy and Climate Change. 2013. *UK Energy in Brief 2013*. London: Government of the United Kingdom.

Department of Minerals and Energy. 1998. White Paper on Energy Policy for South Africa. Report.

Department of Minerals and Energy. 2002. White Paper on the Promotion of Renewable Energy and Clean Energy Development. Report.

Deshmukh, Ranjit, Ashwin Gambhir, and Girish Sant. 2010. Need to Realign India's National Solar Mission. *Economic and Political Weekly* 45 (12): 41–50.

Deutch, John. 2011. The Good News About Gas: The Natural Gas Revolution and Its Consequences. *Foreign Affairs* 90 (1): 82–93.

Devine, Warren D., Jr. 1983. From Shafts to Wires: Historical Perspective on Electrification. *Journal of Economic History* 43 (2): 347–372.

Dincer, Furkan. 2011. The Analysis on Photovoltaic Electricity Generation Status, Potential and Policies of the Leading Countries in Solar Energy. *Renewable & Sustainable Energy Reviews* 15 (1): 713–720.

Dincer, Ibrahim. 2000. Renewable Energy and Sustainable Development: A Crucial Review. *Renewable & Sustainable Energy Reviews* 4 (2): 157–175.

Dixon, Robert K., Richard M. Scheer, and Gareth T. Williams. 2011. Sustainable Energy Investments: Contributions of the Global Environment Facility. *Mitigation and Adaptation Strategies for Global Change* 16 (1): 83–102.

Dobbin, Frank, Beth Simmons, and Geoffrey Garrett. 2007. The Global Diffusion of Public Policies: Social Construction, Coercion, Competition, or Learning. *Annual Review of Sociology* 33:449–472.

Drazen, Allan, and Vittorio Grilli. 1993. The Benefit of Crises for Economic Reforms. *American Economic Review* 83 (3): 598–607.

Dryzek, John S., David Downes, Christian Hunold, David Schlosberg, and Kans-Kristian Hernes. 2003. *Green States and Social Movements*. Oxford: Oxford University Press.

Dubash, Navroz K., and Sudhir Chella Rajan. 2001. Power Politics: Process of Power Sector Reform in India. *Economic and Political Weekly* 36:3367–3390.

Dunlap, Riley E., and Angela G. Mertig. 1992. *American Environmentalism: The US Environmental Movement, 1970–1990*. New York: Routledge.

Earth Policy Institute. 2014. *PV Cell Module Production Data*. Washington, DC: GTM Research Electronic Database.

Edelman Significa. 2014. *Brazilian Elections: 2014*. 2nd edition. São Paulo, Brazil: Edelman Significa Public Affairs Team.

Edenhofer, Ottmar, Lion Hirth, Brigitte Knopf, Michael Pahle, Steffen Schlömer, Eva Schmid, and Falko Ueckerdt. 2013. On the Economics of Renewable Energy Sources. *Energy Economics* 40 (Supplement 1): S12–S23.

EIA. 2012. *Electric Power Monthly*. Washington, DC: US Energy Information Administration.

EIA. 2013a. *Electric Power Monthly*. Washington, DC: US Energy Information Administration.

EIA. 2013b. International Energy Statistics. Accessed December 1, 2013. http://www.eia.gov/countries/data.cfm.

EIA. 2014a. *Annual Energy Outlook 2014*. Washington, DC: US Energy Information Administration.

EIA. 2014b. *Electric Power Monthly*. Washington, DC: US Energy Information Administration.

EIA. 2014c. Levelized Cost and Levelized Avoided Cost of New Generation Resources in the Annual Energy Outlook 2014. Accessed May 11, 2014. http://www.eia.gov/forecasts/aeo/electricity_generation.cfm.

EIA. 2017a. International Energy Statistics. Accessed January 24, 2017. http://www.eia.gov/countries/data.cfm.

EIA. 2017b. *Monthly Energy Review*. Washington, DC: US Energy Information Administration.

EPA. 2014. *Renewable Energy Cost Database*. Washington, DC: Online Database.

EPIA. 2013. *Global Market Outlook For Photovoltaics 2013–2017*. Brussels, Belgium: European Photovoltaic Industry Association.

Ericsson, Karin, Suvi Huttunen, Lars J. Nilsson, and Per Svenningsson. 2004. Bioenergy Policy and Market Development in Finland and Sweden. *Energy Policy* 32 (15): 1707–1721.

Ernst & Young. 2014. RECAI: Renewable Energy Country Attractiveness Index.

Eurobarometer. 2008. Europeans' Attitudes Towards Climate Change. Report.

Eurobarometer. 2013. Public Opinion in the European Union. *Standard Eurobarometer* 79.

Eurobarometer. 2015. Climate Change. Special Eurobarometer Report 435.

Evans, Peter B. 2002. *Livable Cities? Urban Struggles for Livelihood and Sustainability*. Berkeley: University of California Press.

Farhar, Barbara C. 1994a. Trends in US Public Perceptions and Preferences on Energy and Environmental Policy. *Annual Review of Energy and the Environment* 19:211–239.

Farhar, Barbara C. 1994b. Trends: Public Opinion About Energy. *Public Opinion Quarterly* 58 (4): 603–632.

Ferreira, Helder Lopes, Raquel Garde, Gianluca Fulli, Wil Kling, and Joao Pecas Lopes. 2013. Characterisation of Electrical Energy Storage Technologies. *Energy* 53:288–298.

Fiedler, Martin, and Howard Gospel. 2010. The Top 100 Largest Employers in UK and Germany in the Twentieth Century. Cologne Economic History Paper No 3.

Finlayson, Jock A., and Mark W. Zacher. 1988. *Managing International Markets: Developing Countries and the Commodity Trade Regime*. New York: Columbia University Press.

Finnis, John, Germain Grisez, and Joseph Boyle. 1988. *Nuclear Deterrence, Morality and Realism*. Oxford: Oxford University Press.

Fischer, David. 1997. *History of the International Atomic Energy Agency: The First Forty Years*. Vienna, Austria: IAEA.

Fitzgerald, Joan. 2010. *Emerald Cities: Urban Sustainability and Economic Development*. New York: Oxford University Press.

Forsyth, P. J., and J. A. Kay. 1980. The Economic Implications of North Sea Oil Revenues. *Fiscal Studies* 1 (3): 1–28.

Fouquet, Roger. 2008. *Heat, Power and Light: Revolutions in Energy Services*. Northampton, UK: Edward Elgar.

Foxon, T. J., R. Gross, A. Chase, J. Howes, A. Arnall, and D. Anderson. 2005. UK Innovation Systems for New and Renewable Energy Technologies: Drivers, Barriers and Systems Failures. *Energy Policy* 33 (16): 2123–2137.

Frankel, Eugene. 1986. Technology, Politics and Ideology: The Vicissitudes of Federal Solar Energy Policy, 1974–1983. In *The Politics of Energy Research and Development*, edited by John Byrne and Daniel Rich. New Brunswick, NJ: Transaction Books.

Frantzeskaki, Niki, and Derk Loorbach. 2010. Towards Governing Infrasystem Transitions: Reinforcing Lock-in or Facilitating Change? *Technological Forecasting and Social Change* 77 (8): 1292–1301.

Fredriksson, Per G., Herman R. J. Vollebergh, and Elbert Dijkgraaf. 2004. Corruption and Energy Efficiency in OECD Countries: Theory and Evidence. *Journal of Environmental Economics and Management* 47 (2): 207–231.

Frondel, Manuel, Nolan Ritter, Christoph M. Schmidt, and Colin Vance. 2010. Economic Impacts from the Promotion of Renewable Energy Technologies: The German Experience. *Energy Policy* 38 (8): 4048–4056.

Galbraith, Kate, and Asher Price. 2013. *The Great Texas Wind Rush: How George Bush, Ann Richards, and a Bunch of Tinkerers Helped the Oil and Gas State Win the Race to Wind Power*. Austin: University of Texas Press.

Gambhir, Ashwin, Vishal Toro, and Mahalakshmi Ganapathy. 2012. Decentralised Renewable Energy (DRE) Micro-Grids in India. Prayas Report.

Gamson, William A., and Andre Modigliani. 1989. Media Discourse and Public Opinion on Nuclear Power: A Constructionist Approach. *American Journal of Sociology* 95 (1): 1–37.

Garrett, Geoffrey. 1998. *Partisan Politics in the Global Economy*. New York: Cambridge University Press.

Gawande, Kishore, Pravin Krishna, and Marcelo Olarreaga. 2009. What Governments Maximize and Why: The View from Trade. *International Organization* 63 (3): 491–532.

Geller, Howard, Philip Harrington, Arthur H. Rosenfeld, Satoshi Tanishima, and Fridtjof Unander. 2006. Polices for Increasing Energy Efficiency: Thirty Years of Experience in OECD Countries. *Energy Policy* 34 (5): 556–573.

Geller, Howard, Roberto Schaeffer, Alexandre Szklo, and Mauricio Tolmasquim. 2004. Policies for Advancing Energy Efficiency and Renewable Energy Use in Brazil. *Energy Policy* 32 (12): 1437–1450.

George, Alexander L. 1979. Case Studies and Theory Development: The Method of Structured, Focused Comparison. In *Diplomacy: New Approaches in History, Theory, and Policy*, edited by Paul Gordon Lauren. New York: Free Press.

Gerring, John. 2004. What Is a Case Study and What Is It Good For? *American Political Science Review* 98 (2): 341–354.

Gersick, Connie J. G. 1991. Revolutionary Change Theories: A Multilevel Exploration of the Punctuated Equilibrium Paradigm. *Academy of Management Review* 1 (1): 10–36.

Gipe, Paul. 1991. Wind Energy Comes of Age: California and Denmark. *Energy Policy* 19 (8): 756–767.

Gipe, Paul. 1995. *Wind Energy Comes of Age*. New York: Wiley.

Glaser, Peter E. 1968. Power from the Sun: Its Future. *Science* 162 (3856): 857–861.

Global Wind Energy Council. 2013a. Global Wind Report. Annual Market Update 2013.

Global Wind Energy Council. 2013b. *Global Wind Statistics 2012*. Brussels: GWEC.

Gourevitch, Peter A. 1978. The Second Image Reversed: The International Sources of Domestic Politics. *International Organization* 32 (4): 881–912.

Government of India. 2011. 2011 Census Report, Houselisting and Housing Census Data Highlights. Official Website. http://www.censusindia.gov.in/2011census/hlo/hlo_highlights.html.

Government of Kenya. 2012. *Summary of Vision 2030 Investment Projects*. KenInvest Promotion Investments in Kenya.

Goyal, Richa, and Marcus Wiemann. 2015. *The India Off-Grid Electricity Market: Policy Framework, Players and Business Opportunities*. New Delhi, India: European Business Technology Centre and Alliance for Rural Electrification.

Graetz, Michael J. 2011. *The End of Energy: The Unmaking of America's Environment, Security and Independence*. Cambridge, MA: MIT Press.

Greening, Lorna A., David L. Greene, and Carmen Difiglio. 2000. Energy Efficiency and Consumption—The Rebound Effect—A Survey. *Energy Policy* 28 (6–7): 389–401.

Grossman, Gene M., and Elhanan Helpman. 1994. Protection for Sale. *American Economic Review* 84 (4): 833–850.

Grossman, Gene M., and Elhanan Helpman. 1996. Electoral Competition and Special Interest Politics. *Review of Economic Studies* 63 (2): 265–286.

Grübler, Arnulf. 2012. Energy Transitions Research: Insights and Cautionary Tales. *Energy Policy* 50:8–16.

Grübler, Arnulf, Nebojša Nakićenović, and David G. Victor. 1999. Dynamics of Energy Technologies and Global Change. *Energy Policy* 27 (5): 247–280.

Gullberg, Anne Therese, Dörte Ohlhorst, and Miranda Schreurs. 2014. Towards a Low Carbon Energy Future: Renewable Energy Cooperation between Germany and Norway. *Renewable Energy* 68:216–222.

Guzman, Andrew T. 2008. *How International Law Works: A Rational Choice Theory*. New York: Oxford University Press.

Haas, Peter M. 1989. Do Regimes Matter? Epistemic Communities and Mediterranean Pollution Control. *International Organization* 43 (3): 377–403.

Haas, Peter M. 1992. Introduction: Epistemic Communities and International Policy Coordination. *International Organization* 46 (1): 1–35.

Haas, Reinhard, Christian Panzer, Gustav Resch, Mario Ragwitz, Gemma Reece, and Anne Held. 2011. A Historical Review of Promotion Strategies for Electricity from Renewable Energy Sources in EU Countries. *Renewable & Sustainable Energy Reviews* 15 (2): 1003–1034.

Hacker, Jacob S. 2004. Privatizing Risk without Privatizing the Welfare State: The Hidden Politics of Social Policy Retrenchment in the United States. *American Political Science Review* 98 (2): 243–260.

Hacker, Jacob S., and Paul Pierson. 2010. Winner-Take-All Politics: Public Policy, Political Organization, and the Precipitous Rise of Top Incomes in the United States. *Politics & Society* 38 (2): 152–204.

Hadjilambrinos, Constantine. 2000. Understanding Technology Choice in Electricity Industries: A Comparative Study of France and Denmark. *Energy Policy* 28 (15): 1111–1126.

Hale, Thomas, and Johannes Urpelainen. 2015. When and How Can Unilateral Policies Promote the International Diffusion of Environmental Policies and Clean Technology? *Journal of Theoretical Politics* 27 (2): 177–205.

Hamilton, James D. 1983. Oil and the Macroeconomy Since World War II. *Journal of Political Economy* 91 (2): 228–248.

Harborne, Paul, and Chris Hendry. 2009. Pathways to Commercial Wind Power in the US, Europe and Japan: The Role of Demonstration Projects and Field Trials in the Innovation Process. *Energy Policy* 37 (9): 3580–3595.

Harish, Santosh M., and Shuba V. Raghavan. 2011. Redesigning the National Solar Mission for Rural India. *Economic and Political Weekly* 76 (23): 51–58.

Hathaway, Oona A. 1998. Positive Feedback: The Impact of Trade Liberalization on Industry Demands for Protection. *International Organization* 52 (3): 575–612.

Hecht, Gabrielle. 2009. *The Radiance of France*. Cambridge, MA: MIT Press.

Heede, Richard. 2014. Tracing Anthropogenic Carbon Dioxide and Methane Emissions to Fossil Fuel and Cement Producers, 1854-2010. *Climatic Change* 122 (1–2): 229–241.

Heide, Dominik, Lueder von Bremen, Martin Greiner, Clemens Hoffmann, Markus Speckmann, and Stefan Bofinger. 2010. Seasonal Optimal Mix of Wind and Solar Power in a Future, Highly Renewable Europe. *Renewable Energy* 35 (11): 2483–2489.

Heide, Dominik, Martin Greiner, Lüder von Bremen, and Clemens Hoffmann. 2011. Reduced Storage and Balancing Needs in a Fully Renewable European Power System with Excess Wind and Solar Power Generation. *Renewable Energy* 36 (9): 2515–2523.

Heinrich Böll Stiftung. 2014. The German Coal Conundrum. Report.

Helynen, Satu. 2004. Bioenergy Policy in Finland. *Energy for Sustainable Development* 8 (1): 36–46.

Herring, Horace. 2006. Energy Efficiency: A Critical View. *Energy* 31 (1): 10–20.

Heston, Alan, Robert Summers, and Bettina Aten. 2012. Penn World Tables 7.1. Center for International Comparison of Production, Income, and Price at the University of Pennsylvania.

Heymann, Matthias. 1998. Signs of Hubris: The Shaping of Wind Technology Styles in Germany, Denmark, and the United States, 1940–1990. *Technology and Culture* 39 (4): 641–670.

Hiltzik, Michael. 2011. *The New Deal: A Modern History*. New York: Free Press.

Hira, Anil, and Luiz Guilherme de Oliveira. 2009. No Substitute for Oil? How Brazil Developed Its Ethanol Industry. *Energy Policy* 37 (6): 2450–2456.

Hiremath, Rahul B., Bimlesh Kumar, P. Balachandra, N. H. Ravindranath, and B. N. Raghunandan. 2009. Decentralised Renewable Energy: Scope, Relevance and Applications in the Indian Context. *Energy for Sustainable Development* 13 (1): 4–10.

Hirschi, Bernd. 2008. *Erneuerbare Energien-Politik: Eine Multi-Level Policy-Analyse mit Fokus auf den deutschen Strommarkt.* [*Renewable Energy Policy: A Multi-Level Policy Analysis Focusing on the German Electricity Market.*] Wiesbaden, Germany: VS Research.

Hirschman, Albert O. 1970. *Exit, Voice, and Loyalty: Response to Decline in Firms, Organizations, and States*. Cambridge, MA: Harvard University Press.

Hirsh, Richard F. 1999. *Power Loss: The Origins of Deregulation and Restructuring in the American Electric Utility System*. Cambridge, MA: MIT Press.

Hirth, Lion. 2013. The Market Value of Variable Renewables: The Effect of Solar Wind Power Variability on their Relative Price. *Energy Economics* 38:218–236.

Hisnanick, John J., and Kern O. Kymn. 1999. Modeling Economies of Scale: The Case of US Electric Power Companies. *Energy Economics* 21 (3): 225–237.

Horowitz, Roberrt Britt. 1989. *The Irony of Regulatory Reform: The Deregulation of American Telecommunications*. New York: Oxford University Press.

Huber, Joseph. 2008. Pioneer Countries and the Global Diffusion of Environmental Innovations: Theses From the Viewpoint of Ecological Modernisation Theory. *Global Environmental Change* 18 (3): 360–367.

Hughes, Llewelyn, and Johannes Urpelainen. 2015. Interests, Institutions, and Climate Policy: Explaining the Choice of Policy Instruments for the Energy Sector. *Environmental Science & Policy* 54:52–53.

Hurlbut, David. 2008. A Look Behind the Texas Renewable Portfolio Standard: A Case Study. *Natural Resources Journal* 48:129–161.

Hustedt, Michaele. 1998. Windkraft: Made in Germany. In *Windiger Protest: Konflikte um das Zukunftspotential der Windkraft* [*Windy Protest: Conflicts about the Future Potential of Wind Power*], edited by Franz Alt, Jürgen Claus, and Hermann Scheer. Bochum, Germany: Ponte Press.

Huttunen, Suvi. 2009. Ecological Modernisation and Discourses on Rural Non-Wood Bioenergy Production in Finland from 1980 to 2005. *Journal of Rural Studies* 25 (2): 239–247.

Hvelplund, Frede. 2013. Innovative Democracy, Political Economy, and the Transition to Renewable Energy: A Full-Scale Experiment in Denmark 1976-2013. *Environmental Research, Engineering and Management* 66 (4): 5–21.

ICTSD. 2011. *Fostering Low Carbon Growth: The Case for a Sustainable Energy Trade Agreement*. Geneva: International Centre for Trade and Sustainable Development.

IEA. 2000. *World Energy Outlook*. Paris: International Energy Agency.

IEA. 2011. *Energy Policies of IEA Countries: Denmark 2011 Review*. Paris: International Energy Agency.

IEA. 2013. *World Energy Outlook*. Paris: International Energy Agency.

IEA. 2014a. *World Energy Outlook*. Paris: International Energy Agency.

IEA. 2014b. *Energy Supply Security: Emergency Response of IEA Countries*. Paris: International Energy Agency.

IEA. 2014c. WDS. Accessed February 1, 2014. http://wds.iea.org/WDS/TableViewer/tableView.aspx.

IEA. 2014d. *World Energy Investment Outlook*. Paris: International Energy Agency.

IEA. 2015. *India Energy Outlook*. Paris: International Energy Agency.

Ikenberry, G. John. 1986. The Irony of State Strength: Comparative Responses to the Oil Shocks in the 1970s. *International Organization* 40 (1): 105–137.

Indian National Congress. 2009. Lok Sabha Elections 2009 Manifesto.

Indian National Congress. 2014. Lok Sabha Elections 2014 Manifesto.

IPCC. 2007. *IPCC Fourth Assessment Report*. Geneva: IPCC.

IPCC. 2011. *IPCC Special Report on Renewable Energy Sources and Climate Change Mitigation*. New York: Cambridge University Press.

IPCC. 2013. Contribution of Working Group I to the Fifth Assessment Report of the Intergovernmental Panel on Climate Change. [Summary for Policymakers.] *Climate Change 2013: The Physical Science Basis*. http://www.ipcc.ch/pdf/assessment-report/ar5/wg1/WG1AR5_SPM_FINAL.pdf.

IRENA. 2012. Wind Power. *Renewable Energy Technologies: Cost Analysis Series* 1 (5). IRENA Working Paper. https://www.irena.org/DocumentDownloads/Publications/RE_Technologies_Cost_Analysis-WIND_POWER.pdf.

IRENA. 2014a. *Evaluating Renewable Energy Policy: A Review of Criteria and Indicators for Assessment*. Abu Dhabi: IRENA. https://www.irena.org/documentdownloads/publications/evaluating_re_policy.pdf.

IRENA. 2014b. *REmap 2030: A Renewable Energy Roadmap*. Abu Dhabi: IRENA. https://www.irena.org/publications/2014/Jun/REmap-2030-Full-Report.

IRENA. 2014c. *Renewable Energy and Jobs: Annual Review 2014*. Abu Dhabi: IRENA. https://www.irena.org/Publications/rejobs-annual-review-2014.pdf.

IRENA. 2015. *Renewable Energy and Jobs: Annual Review 2015*. Abu Dhabi: IRENA http://www.irena.org/-/media/Files/IRENA/Agency/Publication/2015/IRENA_RE_Jobs_Annual_Review_2015.pdf.

IRENA/GWEC. 2012. *30 Years of Policies for Wind Energy: Lessons from 12 Wind Energy Markets*. Abu Dhabi: IRENA. http://www.irena.org/DocumentDownloads/Publications/GWEC_WindReport_All_web%20display.pdf.

Jacobs, Alan M. 2011. *Governing for the Long Term: Democracy and the Politics of Investment*. New York: Cambridge University Press.

Jacobs, Meg. 2016. *Panic at the Pump: The Energy Crisis and the Transformation of American Politics in the 1970s*. New York: Hill and Wang.

Jacobsen, Henrik Klinge, and Erika Zvingilaite. 2010. Reducing the Market Impact of Large Shares of Intermittent Energy in Denmark. *Energy Policy* 38 (7): 3403–3413.

Jacobson, Arne. 2007. Connective Power: Solar Electrification and Social Change in Kenya. *World Development* 35 (1): 144–162.

Jacobson, Mark Z., and Mark A. Delucchi. 2011. Providing All Global Energy with Wind, Water, and Solar Power, Part I: Technologies, Energy Resources, Quantities and Areas of Infrastructure, and Materials. *Energy Policy* 39 (3): 1154–1169.

Jacobsson, Staffan, and Anna Bergek. 2004. Transforming the Energy Sector: The Evolution of Technological Systems in Renewable Energy Technology. *Industrial and Corporate Change* 13 (5): 815–849.

Jacobsson, Staffan, and Anna Johnson. 2000. The Diffusion of Renewable Energy Technology: An Analytical Framework and Key Issues for Research. *Energy Policy* 28 (9): 625–640.

Jacobsson, Staffan, Björn Sandén, and Lennart Bångens. 2004. Transforming the Energy System: The Evolution of the German Technological System for Solar Cells. *Technology Analysis and Strategic Management* 16 (1): 3–30.

Jacobsson, Staffan, and Volkmar Lauber. 2006. The Politics and Policy of Energy System Transformation: Explaining the German Diffusion of Renewable Energy Technology. *Energy Policy* 34 (3): 256–276.

Jahn, Detlef. 1992. Nuclear Power, Energy Policy and New Politics in Sweden and Germany. *Environmental Politics* 1 (3): 383–417.

Jamison, Andrew. 1978. Democratizing Technology. *Environment* 20 (1): 25–28.

Jamison, Andrew, Ron Eyerman, and Jacqueline Cramer. 1991. *The Making of the New Environmental Consciousness: A Comparative Study of Environmental Movements in Sweden, Denmark and the Netherlands*. Edinburgh, UK: Edinburgh University Press.

Jasper, James M. 1992. Gods, Titans and Mortals: Patterns of State Involvement in Nuclear Development. *Energy Policy* 20 (7): 653–659.

Jones, Bryan D., and Frank R. Baumgartner. 2005. *The Politics of Attention: How Government Prioritizes Problems*. Chicago: University of Chicago Press.

Joppke, Christian. 1990. Nuclear Power Struggles after Chernobyl: The Case of West Germany. *West European Politics* 13 (2): 178–191.

Joppke, Christian. 1993. *Mobilizing against Nuclear Energy*. Berkeley: University of California Press.

Joskow, Paul L. 2011. Comparing the Costs of Intermittent and Dispatchable Electricity Generating Technologies. *American Economic Review* 101 (3): 238–241.

Jubilee. 2013. *Transforming Kenya: Securing Kenya's Prosperity.* Manifesto.

Kahneman, Daniel, and Amos Tversky. 1979. Prospect Theory: An Analysis of Decision under Risk. *Econometrica* 47 (2): 263–292.

Kanase-Patil, A. B., R. P. Saini, and M. P. Sharma. 2010. Integrated Renewable Energy Systems for Off Grid Rural Electrification of Remote Area. *Renewable Energy* 35 (6): 1342–1349.

Karapin, Roger. 2012. Explaining Success and Failure in Climate Policies: Developing Theory through German Case Studies. *Comparative Politics* 45 (1): 46–68.

Karapin, Roger. 2016. *Political Opportunities for Climate Policy: California, New York, and the Federal Government.* New York: Cambridge University Press.

Karekezi, Stephen 1994. Disseminating Renewable Energy Technologies in Sub-Saharan Africa. *Annual Review of Energy and the Environment* 19:387–421.

Karl, Terry Lynn. 1997. *The Paradox of Plenty: Oil Booms and Petro-States.* Berkeley: University of California Press.

Kaul, Inge, Isabelle Grunberg, and Marc A. Stern. 1999. *Global Public Goods: International Cooperation in the 21st Century.* New York: Oxford University Press.

Kavasseri, Rajesh G., and Krithika Seetharaman. 2009. Day-Ahead Wind Speed Forecasting Using f-ARIMA Models. *Renewable Energy* 34 (5): 1388–1393.

Kempton, Willett, and Janša Tomić. 2005. Vehicle-to-Grid Power Implementation: From Stabilizing the Grid to Supporting Large-Scale Renewable Energy. *Journal of Power Sources* 144 (1): 280–294.

Keohane, Nathaniel O., Richard L. Revesz, and Robert N. Stavins. 1998. The Choice of Regulatory Instruments in Environmental Policy. *Harvard Environmental Law Review* 22 (2): 313–367.

Keohane, Robert O. 1984. *After Hegemony: Cooperation and Discord in the World Political Economy.* Princeton, NJ: Princeton University Press.

Keohane, Robert O., and Joseph S. Nye. 1977. *Power and Interdependence: World Politics in Transition.* Boston: Little, Brown.

Ketterer, Janina C. 2014. The Impact of Wind Power Generation on the Electricity Price in Germany. *Energy Economics* 44:270–280.

Khandker, Shahidur R., Hussain A. Samad, Zubair K. M. Sadeque, Mohammed Asaduzzaman, Mohammad Yunus, and A. K. Enamul Haque. 2014. *Surge in Solar-Powered Homes: Experience in Off-Grid Rural Bangladesh.* Washington, DC: International Bank for Reconstruction and Development.

Kilian, Lutz. 2008. The Economic Effects of Energy Price Shocks. *Journal of Economic Literature* 46 (4): 871–909.

Kim, Sung Eun, and Johannes Urpelainen. 2013a. Technology Competition and International Co-operation: Friends or Foes? *British Journal of Political Science* 44 (3): 545–574.

Kim, Sung Eun, and Johannes Urpelainen. 2013b. When and How Can Advocacy Groups Promote New Technologies? Conditions and Strategies for Effectiveness. *Journal of Public Policy* 33 (3): 259–293.

Kim, Sung Eun, Joonseok Yang, and Johannes Urpelainen. 2016. Does Power Sector Deregulation Promote or Discourage Renewable Energy Policy? Evidence from the States, 1991-2012. *Review of Policy Research* 33 (1): 22–50.

King, Gary, Robert O. Keohane, and Sidney Verba. 1994. *Designing Social Inquiry: Scientific Inference in Qualitative Research*. Princeton, NJ: Princeton University Press.

Kirby, Eric G. 1995. An Evaluation of the Effectiveness of US CAFE Policy. *Energy Policy* 23 (2): 107–109.

Kitschelt, Herbert P. 1986. Political Opportunity Structures and Political Protest: Anti-Nuclear Movements in Four Democracies. *British Journal of Political Science* 16 (1): 57–85.

Kivimaa, Paula, and Per Mickwitz. 2011. Public Policy as a Part of Transforming Energy Systems: Framing Bioenergy in Finnish Energy Policy. *Journal of Cleaner Production* 19 (16): 1812–1821.

Klick, Holly, and Eric R. A. N. Smith. 2010. Public Understanding of and Support for Wind Power in the United States. *Renewable Energy* 35 (7): 1585–1591.

Könnölä, Totti, Gregory C. Unruh, and Javier Carrillo-Hermosilla. 2006. Prospective Voluntary Agreements for Escaping Techno-Institutional Lock-In. *Ecological Economics* 57 (2): 239–252.

Krasner, Stephen D. 1985. *Structural Conflict: The Third World Against Global Liberalism*. Berkeley: University of California Press.

Krause, Florentin, Hartmut Bossel, and Karl-Friedrich Müller-Reißmann. 1980. *Energiewende: Wachstum und Wohlstand ohne Erdöl und Uran*. [Energy Transition: Growth and Prosperity without Oil and Uranium.] Frankfurt: S. Fischer Verlag.

Krohn, Søren, and Steffen Damborg. 1999. On Public Attitudes towards Wind Power. *Renewable Energy* 16 (1–4): 954–960.

Krupa, Joel, and Sarah Burch. 2011. A New Energy Future for South Africa: The Political Ecology of South African Renewable Energy. *Energy Policy* 39 (10): 6254–6261.

Krupnick, Alan, Zhongmin Wang, and Yushuang Wang. 2013. Sectoral Effects of the Shale Gas Revolution in the United States. Resources for the Future, Discussion Paper 13–20.

KTM. 2005. *Lähiajan energia- ja ilmastopolitiikan linjauksia: Kansallinen strategia Kioton pöytäkirjan toimeenpanemiseksi.* [*The Future of Energy and Climate Policy: A National Strategy for the Implementation of the Kyoto Protocol.*] Report prepared for Parliament of Finland by Energy Department, Ministry of Trade and Industry.

Kunnas, Jan, and Timo Myllyntaus. 2009. Postponed Leap in Carbon Dioxide Emissions: Impacts of Energy Efficiency, Fuel Choices and Industrial Structure on the Finnish Energy Economy, 1800–2005. *Global Environment* 3:154–189.

Kydland, Finn E., and Edward C. Prescott. 1977. Rules Rather Than Discretion: The Inconsistency of Optimal Plans. *Journal of Political Economy* 85 (3): 473–492.

Laird, Frank N. 1990. Technocracy Revisited: Knowledge, Power and the Crisis in Energy Decision Making. *Organization & Environment* 4 (1): 49–61.

Laird, Frank N. 2001. *Solar Energy, Technology Policy, and Institutional Values*. New York: Cambridge University Press.

Laird, Frank N. 2003. Constructing the Future: Advocating Energy Technologies in the Cold War. *Technology and Culture* 44 (1): 27–49.

Laird, Frank N., and Christoph Stefes. 2009. The Diverging Paths of German and United States Policies for Renewable Energy: Sources of Difference. *Energy Policy* 37 (7): 2619–2629.

Lake, David A. 2009. Open Economy Politics: A Critical Review. *Review of International Organizations* 4 (3): 219–244.

Landry, Marc. 2012. Water as White Coal. *RCC Perspectives* 2:7–11.

Langniss, Ole, and Ryan Wiser. 2003. The Renewables Portfolio Standard in Texas: An Early Assessment. *Energy Policy* 31 (6): 527–535.

Lauber, Volkmar. 2002. The Different Concepts of Promoting Res-Electricity and their Political Careers. In *Proceedings of the 2001 Berlin Conference on the Human Dimensions of Global Environmental Change*, edited by Frank Biermann, Rainer Brohm, and Klaus Dingwerth, 296–304. Potsdam, Germany: Potsdam Institute for Climate Impact Research.

Lauber, Volkmar, and Lutz Mez. 2004. Three Decades of Renewable Electricity Policies in Germany. *Energy & Environment* 15 (4): 599–623.

Lee, April, Owen Zinaman, and Jeffrey Logan. 2012. *Opportunities for Synergy between Natural Gas and Renewable Energy in the Electric Power and Transportation Sectors.* National Renewable Energy Laboratory, NREL/TP-6A50–56324.

Leiserowitz, Anthony, Edward Maibach, Connie Roser-Renouf, Geoff Feinberg, Jennifer Marlon, and Peter Howe. 2013. Public Support for Climate and Energy Policies in April 2013. Yale University and George Mason University, Project on Climate Change Communication.

Leuphana University. 2013. Definition und Marktanalyse von Bürgerenergie in Deutschland. [Definition and Market Analysis of Citizen Energy in Germany.] Report for Die Wende: Energie in Bürgerhand [The Turning Point: Energy in the Hands of Citizens].

Levi, Michael A. 2013. *The Power Surge: Energy, Opportunity, and the Battle for America's Future*. New York: Oxford University Press.

Levy, Brian, and Pablo T. Spiller. 1994. The Institutional Foundations of Regulatory Commitment: A Comparative Analysis of Telecommunications Regulation. *Journal of Law Economics and Organization* 10 (2): 201–246.

Levy, David L., and Ans Kolk. 2002. Strategic Responses to Global Climate Change: Conflicting Pressures on Multinationals in the Oil Industry. *Business and Politics* 4 (3): 275–300.

Levy, Jack S. 2008. Case Studies: Types, Designs, and Logics of Inference. *Conflict Management and Peace Science* 25 (1): 1–18.

Lewis, Joanna I., and Ryan H. Wiser. 2007. Fostering a Renewable Energy Technology Industry: An International Comparison of Wind Industry Policy Support Mechanisms. *Energy Policy* 35 (3): 1844–1857.

Li, Xianguo. 2005. Diversification and Localization of Energy Systems for Sustainable Development and Energy Security. *Energy Policy* 33 (17): 2237–2243.

Liang, Xiupei. 2014. Lost in Transmission: Distributed Solar Generation in China. Briefing Paper, Wilson Center, China Environment Forum.

Lipp, Judith. 2007. Lessons for Effective Renewable Electricity Policy from Denmark, Germany and the United Kingdom. *Energy Policy* 35 (11): 5481–5495.

List, John A., and Daniel M. Sturm. 2006. How Elections Matter: Theory and Evidence from Environmental Policy. *Quarterly Journal of Economics* 121 (4): 1249–1281.

Litmanen, Tapio, and Matti Kojo. 2011. Not Excluding Nuclear Power: The Dynamics and Stability of Nuclear Power Policy Arrangements in Finland. *Journal of Integrative Environmental Sciences* 8 (3): 171–194.

Loiter, Jeffrey M, and Vicki Norberg-Bohm. 1999. Technology Policy and Renewable Energy: Public Roles in the Development of New Energy Technologies. *Energy Policy* 27 (2): 85–97.

Lopolito, A., P. Morone, and R. Taylor. 2013. Emerging Innovation Niches: An Agent Based Model. *Research Policy* 42 (6–7): 1225–1238.

Løvdal, Nicolai, and Frank Neumann. 2011. Internationalization as a Strategy to Overcome Industry Barriers: An Assessment of the Marine Energy Industry. *Energy Policy* 39:1093–1100.

Lovins, Amory B. 1976. Energy Strategy: The Road Not Taken? *Foreign Affairs* 55:65–96.

Lowry, William R. 2008. Disentangling Energy Policy from Environmental Policy. *Social Science Quarterly* 89 (5): 1195–1211.

Lucas, Robert E. 2003. Macroeconomic Priorities. *American Economic Review* 93 (1): 1–14.

Lund, Henrik. 2000. Choice Awareness: The Development of Technological and Institutional Choice in the Public Debate of Danish Energy Planning. *Journal of Environmental Policy and Planning* 2 (3): 249–259.

Lund, Peter D. 2007. The Link between Political Decision-Making and Energy Options: Assessing Future Role of Renewable Energy and Energy Efficiency in Finland. *Energy* 32 (12): 2271–2281.

Lund, Peter D. 2011. Boosting New Renewable Technologies towards Grid Parity: Economic and Policy Aspects. *Renewable Energy* 36 (11): 2776–2784.

Lund, Peter D. 2012. Large-Scale Urban Renewable Electricity Schemes: Integration and Interfacing Aspects. *Energy Conversion and Management* 63:162–172.

Luukkanen, Jyrki. 2003. Green Paper with Green Electricity? Greening Strategies of Nordic Pulp and Paper Industry. *Energy Policy* 31 (7): 641–655.

Madrigal, Alexis. 2011. *Powering the Dream: The History and Promise of Green Technology*. Cambridge, MA: Da Capo Press.

Maersk Oil. 2012. *Maersk Oil: Under the Surface*. Electronic book. http://www.e-pages.dk/maerskoil/14/.

Markham, William T. 2011. *Environmental Organizations in Modern Germany*. Oxford: Berhahn Books.

Markussen, Peter, and Gert Tinggaard Svendsen. 2005. Industry Lobbying and the Political Economy of GHG Trade in the European Union. *Energy Policy* 33 (2): 245–255.

Martinot, Eric, Akanksha Chaurey, Debra Lew, José Roberto Moreira, and Njeri Wamukonya. 2002. Renewable Energy Markets in Developing Countries. *Annual Review of Energy and the Environment* 27 (1): 309–348.

Martinot, Eric, and Junfeng Li. 2007. *Powering China's Development: The Role of Renewable Energy*. Washington, DC: Worldwatch Institute.

McCright, Aaron M., Chenyang Xiao, and Riley E. Dunlap. 2014. Political Polarization on Support for Government Spending on Environmental Protection in the USA, 1974–2012. *Social Science Research* 48:251–260.

McCright, Aaron M., and Riley E. Dunlap. 2003. Defeating Kyoto: The Conservative Movement's Impact on U.S. Climate Change Policy. *Social Problems* 50 (3): 348–373.

McCright, Aaron M., and Riley E. Dunlap. 2011. The Politicization of Climate Change and Polarization in the American Public's Views of Global Warming, 2001–2010. *Sociological Quarterly* 52 (2): 155–194.

McFarland, Andrew S. 1984. Energy Lobbies. *Annual Review of Energy* 9:501–527.

McVeigh, James, Dallas Burtraw, Joel Darmstadter, and Karen Palmer. 2000. Winner, Loser, or Innocent Victim? Has Renewable Energy Performed as Expected? *Solar Energy* 68 (3): 237–255.

Meadows, Donella H., Jorgen Randers, Dennis L. Meadows, and William W. Behrens. 1972. *The Limits to Growth*. New York: Universe Books.

Mendershausen, Horst. 1976. *Coping with the Oil Crisis: French and German Experiences*. Baltimore: Johns Hopkins University Press.

Mendonça, Miguel. 2007. *Feed-In Tariffs: Accelerating the Deployment of Renewable Energy*. London: Earthscan.

Mendonça, Miguel, Stephen Lacey, and Frede Hvelplund. 2009. Stability, Participation and Transparency in Renewable Energy Policy: Lessons from Denmark and the United States. *Policy and Society* 27 (4): 379–398.

Meyer, John W., John Boli, George M. Thomas, and Francisco O. Ramirez. 1997. World Society and the Nation-State. *American Journal of Sociology* 103 (1): 144–181.

Meyer, Niels I. 2007. Learning from Wind Energy Policy in the EU: Lessons from Denmark, Sweden and Spain. *European Environment* 17 (5): 347–362.

Meyer, Timothy. 2013. Epistemic Institutions and Epistemic Cooperation in International Environmental Governance. *Transnational Environmental Law* 2 (2): 15–44.

Michaelowa, Axel. 2005. The German Wind Energy Lobby: How to Promote Costly Technological Change Successfully. *European Environment* 15 (3): 192–199.

Milner, Helen V. 1997. *Interests, Institutions, and Information: Domestic Politics and International Relations*. Princeton, NJ: Princeton University Press.

Mitchell, C., D. Bauknecht, and P. M. Connor. 2006. Effectiveness Through Risk Reduction: A Comparison of the Renewable Obligation in England and Wales and the Feed-In System in Germany. *Energy Policy* 34 (3): 297–305.

Mitchell, Catherine, and Peter Connor. 2004. Renewable Energy Policy in the UK 1990–2003. *Energy Policy* 32 (17): 1935–1947.

Moe, Espen. 2015. *Renewable Energy Transformation or Fossil Fuel Backlash*. New York: Palgrave Macmillan.

Moe, Terry M. 2005. Power and Political Institutions. *Perspectives on Politics* 3 (2): 215–233.

Morris, Braig, and Martin Pehnt. 2012. *Energy Transition: The German Energiewende*. Berlin: Heinrich Böll Foundation.

Muller, Nicholas Z., Robert Mendelsohn, and William Nordhaus. 2011. Environmental Accounting for Pollution in the United States Economy. *American Economic Review* 101 (5): 1649–1675.

Müller-Kraenner, Sascha. 2010. *Energy Security: Re-Measuring the World*. Sterling, VA: Earthscan.

Müller-Rommel, Ferdinand. 1998. Explaining the Electoral Success of Green Parties: A Cross-National Analysis. *Environmental Politics* 7 (4): 145–154.

Narum, David. 1992. A Troublesome Legacy: The Reagan Administration's Conservation and Renewable Energy Policy. *Energy Policy* 20 (1): 40–53.

Navigant. 2013. World Market Update 2012: International Wind Energy Development *World Market Update 2012. International Wind Energy Development, Forecast 2013–2017*. http://www.navigantresearch.com/wp-content/uploads/2013/03/WWMU-13-Executive-Summary.pdf.

Nehring, Richard. 1975. *Oil and Gas Supplies for California: Past and Future*. Santa Monica, CA: Rand Corporation.

Nemet, Gregory F. 2012. Subsidies for New Technologies and Knowledge Spillovers from Learning by Doing. *Journal of Policy Analysis and Management* 31 (3): 601–622.

Nemet, Gregory F., and Daniel M. Kammen. 2007. U.S. Energy Research and Development: Declining Investment, Increasing Need, and the Feasibility of Expansion. *Energy Policy* 35 (1): 746–755.

Newbery, David M. 1998. Freer Electricity Markets in the UK: A Progress Report. *Energy Policy* 26 (10): 743–749.

Nguyen, Dzung D, and Brad Lehman. 2006. Modeling and Simulation of Solar PV Arrays under Changing Illumination Conditions. Paper presented at the 2006 IEEE Workshop on Computers in Power Electronics, Rensselaer Polytechnic Institute, Troy, NY, July 16-19. http://ieeexplore.ieee.org/stamp/stamp.jsp?arnumber=4097441.

Niemi, R., J. Mikkola, and P. D. Lund. 2012. Urban Energy Systems with Smart Multi-Carrier Energy Networks and Renewable Energy Generation. *Renewable Energy* 48:524–536.

Nigam, Poonam Singh, and Anoop Singh. 2011. Production of Liquid Biofuels from Renewable Resources. *Progress in Energy and Combustion Science* 37:52–68.

Nolden, Colin. 2013. Governing Community Energy: Feed-In Tariffs and the Development of Community Wind Energy Schemes in the United Kingdom and Germany. *Energy Policy* 63: 543–552.

Norberg-Bohm, Vicki. 2000. Creating Incentives for Environmentally Enhancing Technological Change: Lessons From 30 Years of U.S. Energy Technology Policy. *Technological Forecasting and Social Change* 65 (2): 125–148.

NSS. 2010. *Energy Sources of Indian Households for Cooking and Lighting*. New Delhi, India: National Sample Survey Organization, Ministry of Statistics and Programme Implementation, Government of India.

Nye, David E. 1998. *Consuming Power: A Social History of American Energies*. Cambridge, MA: MIT Press.

Ockwell, David G., Ruediger Haum, Alexandra Mallett, and Jim Watson. 2010. Intellectual Property Rights and Low Carbon Technology Transfer: Conflicting Discourses of Diffusion and Development. *Global Environmental Change* 20 (4): 729–739.

OECD. 2014a. OECD.StatExtract. Database. http://stats.oecd.org/

OECD. 2014b. STAN Database for Structural Analysis. Accessed May 25, 2014. http://oe.cd/stan.

Olson, Mancur. 1968. *The Logic of Collective Action*. Cambridge, MA: Harvard University Press.

Ostry, Sylvia, and Richard R. Nelson. 1995. *Techno-Nationalism and Techno-Globalism: Conflict and Cooperation*. Washington, DC: Brookings Institution Press.

Pegels, Anna. 2010. Renewable Energy in South Africa: Potentials, Barriers and Options for Support. *Energy Policy* 38 (9): 4945–4954.

Peltzman, Sam. 1976. Toward a More General Theory of Regulation. *Journal of Law & Economics* 19 (2): 211–240.

Peters, B. Guy, Jon Pierre, and Desmond S. King. 2005. The Politics of Path Dependency: Political Conflict in Historical Institutionalism. *Journal of Politics* 67 (4): 1275–1300.

Pew. 2013. *2012: Who's Winning the Clean Energy Race? The Clean Energy Economy*. http://www.pewtrusts.org/~/media/legacy/uploadedfiles/peg/publications/report/clen G20Report2012Digitalpdf.pdf.

Pew. 2014. *2013: Who's Winning the Clean Energy Race? The Clean Energy Economy*. http://www.pewtrusts.org/~/media/Assets/2014/04/01/clenwhoswinningthecleanenergyrace2013pdf.pdf.

Pierson, Paul. 2000. Increasing Returns, Path Dependence, and the Study of Politics. *American Political Science Review* 94 (2): 251–267.

Pillai, Indu R., and Rangan Banerjee. 2009. Renewable Energy in India: Status and Potential. *Energy* 34 (8): 970–980.

Pinchot, Gifford. 1910. *The Fight for Conservation*. New York: Doubleday.

Pousa, Gabriella P. A. G., André L. F. Santos, and Paulo A. Z. Suarez. 2007. History and Policy of Biodiesel in Brazil. *Energy Policy* 35 (11): 5393–5398.

Price, David. 1995. Energy and Human Evolution. *Population and Environment* 16 (4): 301–319.

Prodi, Romano, and Alberto Clô. 1975. Europe. *Daedalus* 104 (4): 91–112.

Putnam, Robert D. 1988. Diplomacy and Domestic Politics: The Logic of Two-Level Games. *International Organization* 44 (3): 427–460.

Quiggin, John. 2012. The End of the Nuclear Renaissance. *National Interest*. http://nationalinterest.org/commentary/the-end-the-nuclear-renaissance-6325/.

Rabe, Barry. 2004. *Statehouse and Greenhouse: The Evolving Politics of American Climate Change Policy*. Washington, DC: Brookings Institution Press.

Rabe, Barry. 2006. Race to the Top: The Expanding Role of U.S. State Renewable Portfolio Standards. *Sustainable Development Law and Policy* 7:10–16.

Rader, Nancy. 2000. The Hazard of Implementing Renewables Portfolio Standards. *Energy & Environment* 11 (4): 391–405.

Rai, Varun, David G. Victor, and Mark C. Thurber. 2010. Carbon Capture and Storage at Scale: Lessons From the Growth of Analogous Energy Technologies. *Energy Policy* 38 (8): 4089–4098.

Rasmussen, Morten Grud, Gorm Bruun Andresen, and Martin Greiner. 2012. Storage and Balancing Synergies in a Fully or Highly Renewable Pan-European Power System. *Energy Policy* 51:642–651.

Reece, Ray. 1979. *The Sun Betrayed: A Report on the Corporate Seizure of US Solar Energy Development*. Boston: South End Press.

REN21. 2012. *Renewables Global Status Report: 2012 Update*. Paris: REN21 Secretariat. http://www.ren21.net/renewables-2012-global-status-report/.

REN21. 2013. *Renewables Global Futures Report: 2013*. Paris: REN21 Secretariat. http://www.ren21.net/Global-Futures-Report-2013-EN.

REN21. 2014. *Renewables 2014: Global Status Report*. Paris: REN21 Secretariat. http://www.ren21.net/renewables-2014-global-status-report-full/.

REN21. 2015. *Renewables 2015: Global Status Report*. Paris: REN21 Secretariat. http://www.ren21.net/renewables-2015-global-status-report-full-report-en/.

RenewableUK. 2013. *Working for a Green Britain & Northern Ireland 2013–23: Employment in the UK Wind & Marine Energy Industries*. http://www.renewableuk.com/resource/resmgr/publications/reports/working-for-a-greener-britai.pdf.

Renn, Ortwin. 1994. Public Acceptance of Energy Technologies. European Strategy for Energy Research and Technological Development: Proceedings of a Seminar at Fondazione Giorgio Cini, Venice, November 18–20, 1993.

Rich, Daniel, and J. David Roessner. 1990. Tax Credits and US Solar Commercialization Policy. *Energy Policy* 18 (2): 186–198.

Richards, Deanna J., and Ann B. Fullerton. 1994. *Industrial Ecology: U.S.–Japan Perspectives*. Washington, DC: National Academy Press.

Righter, Robert W. 1996. Pioneering in Wind Energy: The California Experience. *Renewable Energy* 9 (1–4): 781–784.

Roman-Leshkov, Yuriy, Christopher J. Barrett, Zhen Y. Liu, and James A. Dumesic. 2007. Production of Dimethylfuran for Liquid Fuels from Biomass-Derived Carbohydrates. *Nature* 447 (7147): 982–985.

Romero-Lankao, Patricia, and David Dodman. 2011. Cities in Transition: Transforming Urban Centers from Hotbeds of GHG Emissions and Vulnerability to Seedbeds of Sustainability and Resilience: Introduction and Editorial Overview. *Current Opinion in Environmental Sustainability* 3 (3): 113–120.

Rosas, Guillermo. 2006. Bagehot or Bailout? An Analysis of Government Responses to Banking Crises. *American Journal of Political Science* 50 (1): 175–191.

Ruedig, Wolfgang. 2000. Phasing Out Nuclear Energy in Germany. *German Politics* 9 (3): 43–80.

Ruttan, Vernon W. 1997. Induced Innovation, Evolutionary Theory and Path Dependence: Sources of Technical Change. *Economic Journal (London)* 107 (444): 1520–1529.

RWE. 2015. *Fact Book*. Essen, Germany: RWE Innogy GmbH. http://www.rwe.com/web/cms/contentblob/108824/data/167072/RWE-Fact-Book-Renewable-Energy-April-2015.pdf.

Ryland, Elisabeth. 2010. Danish Wind Power Policy: Domestic and International Forces. *Environmental Politics* 19 (1): 80–85.

Sabatier, Paul A., and Hank C. Jenkins-Smith, eds. 1993. *Policy Change and Learning: An Advocacy Coalition Approach*. Boulder, CO: Westview Press.

Sadorsky, Perry. 1999. Oil Price Shocks and Stock Market Activity. *Energy Economics* 21 (5): 449–469.

Salje, Peter. 1998. *Stromeinspeisungsgesetz*. [*Electricity Feed-In Law*.] Cologne, Germany: Heymann.

Sandén, Björn A., and Christian Azar. 2005. Near-Term Technology Policies for Long-Term Climate Targets: Economy Wide Versus Technology Specific Approaches. *Energy Policy* 33 (12): 1557–1576.

Särkikoski, Tuomo. 2011. *Rauhan atomi, sodan koodi: Suomalaisen atomivoimaratkaisun teknopolitiikka 1955–1970*. [*The Peace Atom, the Code of War: The Technology Policy of the Finnish Atomic Energy Solution, 1955–1970*.] Helsinki, Finland: Unigrafia.

Satchwell, Andrew, Andrew Mills, and Galen Barbose. 2014. *Financial Impacts of Net-Metered PV on Utilities and Ratepayers*. Report for the Office of Energy Efficiency and Renewable Energy, US Department of Energy.

Sawin, Janet Laughlin. 2001. The Role of Government in the Development and Diffusion of Renewable Energy Technologies: Wind Power in the United States, California, Denmark and Germany, 1970–2000. Unpublished dissertation.

Scheer, Hermann. 2007. *Energy Autonomy: The Economic, Social and Technological Case for Renewable Energy*. Sterling, VA: Earthscan.

Schmalensee, Richard. 2012. Evaluating Policies to Increase Electricity Generation from Renewable Energy. *Review of Environmental Economics and Policy* 6 (1): 45–64.

Schmidt, Robert C., and Robert Marschinski. 2009. A Model of Technological Breakthrough in the Renewable Energy Sector. *Ecological Economics* 69 (2): 435–444.

Schrag, Daniel P. 2012. Is Shale Gas Good for Climate Change? *Daedalus* 141 (2): 72–80.

Schreuer, Anna, and Daniela Weismeier-Sammer. 2010. *Energy Cooperatives and Local Ownership in the Field of Renewable Energy Technologies: A Literature Review*. Vienna, Austria: Research Reports.

Schumacher, Ernest F. 1973. *Small Is Beautiful: Economics as If People Mattered*. London: Blond and Briggs.

Schwabe, Paul, Sander Lensink, and Maureen Hand. 2011. Multi-National Case Study of the Financial Cost of Wind Energy. IEA Wind Task 26. https://www.nrel.gov/docs/fy11osti/48155.pdf.

Searchinger, Timothy, Ralph Heimlich, R. A. Houghton, Fengxia Dong, Amani Elobeid, Jacinto Fabiosa, Simla Tokgoz, Dermot Hayes, and Tun-Hsiang Yu. 2008. Use of U.S. Croplands for Biofuels Increases Greenhouse Gases Through Emissions from Land-Use Change. *Science* 319 (5867): 1238–1240.

Seawright, Jason, and John Gerring. 2008. Case Selection Techniques in Case Study Research: A Menu of Qualitative and Quantitative Options. *Political Research Quarterly* 61 (2): 294–308.

Sebenius, James K. 1983. Negotiation Arithmetic: Adding and Subtracting Issues and Parties. *International Organization* 37 (2): 281–316.

Sekhon, Jasjeet S. 2004. Quality Meets Quantity: Case Studies, Conditional Probability, and Counterfactuals. *Perspectives on Politics* 2 (2): 281–293.

Seto, Karen C., Roberto Sánchez-Rodríguez, and Michail Fragkias. 2010. The New Geography of Contemporary Urbanization and the Environment. *Annual Review of Environment and Resources* 35:167–194.

Sharma, D. Parameswara, P. S. Chandramohanan Nair, and R. Balasubramanian. 2005. Performance of Indian Power Sector During a Decade under Restructuring: A Critique. *Energy Policy* 33 (4): 563–576.

Shrimali, Gireesh, David Nelson, Shobhit Goel, Charith Konda, and Raj Kumar. 2013. Renewable Deployment in India: Financing Costs and Implications for Policy. *Energy Policy* 62 (0): 28–43.

Simmons, Beth A., Frank Dobbin, and Geoffrey Garrett, eds. 2008. *The Global Diffusion of Markets and Democracy*. New York: Cambridge University Press.

Sine, Wesley D., and Brandon H. Lee. 2009. Tilting at Windmills? The Environmental Movement and the Emergence of the US Wind Energy Sector. *Administrative Science Quarterly* 54 (1): 123–155.

Singh, Anoop. 2006. Power Sector Reform in India: Current Issues and Prospects. *Energy Policy* 34 (16): 2480–2490.

Singh, G. K. 2013. Solar Power Generation by PV (Photovoltaic) Technology: A Review. *Energy* 53:1–13.

Skidmore, Mark, Chad Cotti, and James Alm. 2013. The Political Economy of State Government Subsidy Adoption: The Case of Ethanol. *Economics and Politics* 25 (2): 162–180.

Smil, Vaclav. 2010. *Energy Transitions: History, Requirements, Prospects*. Santa Barbara, CA: Praeger.

Smith, Adrian, Florian Kern, Rob Raven, and Bram Verhees. 2014. Spaces for Sustainable Innovation: Solar Photovoltaic Electricity in the UK. *Technological Forecasting and Social Change* 81:115–130.

Smith, Eric R. A. N. 2002. *Energy, the Environment, and Public Opinion*. Lanham, MD: Rowman and Littlefield.

Smith, Michael G., and Johannes Urpelainen. 2014. The Effect of Feed-in Tariffs on Renewable Electricity Generation: An Instrumental Variables Approach. *Environmental and Resource Economics* 57 (3): 367–392.

Snyder, Jack, Robert Y. Shapiro, and Yaeli Bloch-Elkon. 2009. Free Hand Abroad, Divide and Rule at Home. *World Politics* 61 (1): 155–187.

Solangi, K. H., M. R. Islam, R. Saidur, N. A. Rahim, and H. Fayaz. 2011. A Review on Global Solar Energy Policy. *Renewable & Sustainable Energy Reviews* 15 (4): 2149–2163.

Solar Foundation. 2014. National Solar Jobs Census 2013. Accessed May 13, 2014. http://www.thesolarfoundation.org/wp-content/uploads/2016/10/TSF-Solar-Jobs-Census-2013-Final_Reduced-1.pdf.

Solow, Robert M. 1974. The Economics of Resources or the Resources of Economics. *American Economic Review* 64 (2): 1–14.

Sørensen, Bent. 1991. A History of Renewable Energy Technology. *Energy Policy* 19 (1): 8–12.

Sovacool, Benjamin K. 2008. *The Dirty Energy Dilemma: What's Blocking Clean Power in the United States*. Westport, CT: Praeger.

Sovacool, Benjamin K. 2009. The Intermittency of Wind, Solar, and Renewable Electricity Generators: Technical Barrier or Rhetorical Excuse? *Utilities Policy* 17 (3–4): 288–296.

Spalding-Fecher, Randall, Anthony Williams, and Clive van Horen. 2000. Energy and Environment in South Africa: Charting a Course to Sustainability. *Energy for Sustainable Development* 4 (4): 8–17.

Sperling, Karl, Frede Hvelplund, and Brian Vad Mathiesen. 2010. Evaluation of Wind Power Planning in Denmark—Towards an Integrated Perspective. *Energy* 35 (12): 5443–5454.

Spiller, Pablo T., and Mariano Tommasi. 2003. The Institutional Foundations of Public Policy: A Transactions Approach with Application to Argentina. *Journal of Law Economics and Organization* 19 (2): 281–306.

Stainforth, D., A. Cole, P. Dolley, H. Edwards, J. Wilczek, and M. Wood. 1996. An Overview of the UK Department of Trade and Industry's (DTI's) Programme in Solar Energy. *Solar Energy* 58 (1–3): 111–119.

Stavins, Robert N. 1998. What Can We Learn from the Grand Policy Experiment? Lessons from SO2 Allowance Trading. *Journal of Economic Perspectives* 12 (3): 69–88.

Stefes, Christoph H. 2010. Bypassing Germany's Reformstau: The Remarkable Rise of Renewable Energy. *German Politics* 19 (2): 148–163.

Steinbach, Armin. 2013. Barriers and Solutions for Expansion of Electricity Grids: The German Experience. *Energy Policy* 63:224–229.

Steinberg, Theodore. 2004. *Nature Incorporated: Industrialization and the Waters of New England*. New York: Cambridge University Press.

Stenberg, Anders, and Hannele Holttinen. 2011. Tuulivoiman tuotantotilastot. [Wind Generation Production Statistics.] VTT Working Paper 178.

Stenzel, Till, and Alexander Frenzel. 2008. Regulating Technological Change: The Strategic Reactions of Utility Companies Towards Subsidy Policies in the German, Spanish and UK Electricity Markets. *Energy Policy* 36 (7): 2645–2657.

Stern, Jon, and Stuart Holder. 1999. Regulatory Governance: Criteria for Assessing the Performance of Regulatory Systems: An Application to Infrastructure Industries in the Developing Countries of Asia. *Utilities Policy* 8 (1): 33–50.

Stern, Jonathan. 2004. UK Gas Security: Time to Get Serious. *Energy Policy* 32 (17): 1967–1979.

Stern, Roger J. 2016. Oil Scarcity Ideology in US Foreign Policy, 1908–1997. *Security Studies* 25 (2): 214–257.

Stigler, George J. 1971. The Theory of Economic Regulation. *Bell Journal of Economics and Management Science* 2 (1): 3–21.

Sunell, Milka. 2001. Miten Suomen yksityinen metsäteollisuus hankki länsimaisen ydinvoimalan: Tutkimus taloudellisesta ja poliittisesta vallankäytöstä 1970-luvulla. [How Did the Private Forest Industry in Finland Acquire a Western Nuclear Power Plant?: A Study of Economic and Political Power in the 1970s.] Master's Thesis, University of Helsinki, Department of Social History.

Surrey, John, and Charlotte Huggett. 1976. Opposition to Nuclear Power: A Review of International Experience. *Energy Policy* 4 (4): 286–307.

Szarka, Joseph. 2007. Why Is There No Wind Rush in France? *European Environment* 17 (5): 321–333.

Tarr, Joel A., and Carl Zimring. 1997. The Struggle for Smoke Control in St. Louis. In *Common Fields: An Environmental History of St. Louis*, edited by Andrew Hurley. St. Louis: Missouri Historical Society Press.

Tarr, Joel A., and Karen Clay. 2014. Pittsburgh as an Energy Capital. In *Energy Capitals: Local Impact, Global Influence*, edited by Joseph A. Pratt, Martin V. Melosi, and Kathleen Brosnan. Pittsburgh, PA: University of Pittsburgh Press.

Taylor, Andrew. 1996. The Politics of Energy Policy in the United Kingdom. *Politics* 16 (3): 133–141.

Taylor, James W., Patrick E. McSharry, and Roberto Buizza. 2009. Wind Power Density Forecasting Using Ensemble Predictions and Time Series Models. *IEEE Transactions on Energy Conservation* 24:775–782.

Taylor, R. H., S. D. Probert, and P. D. Carmo. 1998. French Energy Policy. *Applied Energy* 59 (1): 39–61.

Teräväinen, Tuula, Markku Lehtonen, and Mari Martiskainen. 2011. Climate Change, Energy Security, and Risk: Debating Nuclear New Build in Finland, France and the UK. *Energy Policy* 39 (6): 3434–3442.

Tewari, Saurabh, Charles J. Geyer, and Ned Mohan. 2011. A Statistical Model for Wind Power Forecast Error and Its Application to the Estimation of Penalties in Liberalized Markets. *IEEE Transactions on Power Systems* 26 (4): 2031–2039.

Thelen, Kathleen. 1999. Historical Institutionalism in Comparative Politics. *Annual Review of Political Science* 2 (1): 369–404.

Thelen, Kathleen A. 2004. *How Institutions Evolve: The Political Economy of Skills in Germany, Britain, the United States, and Japan*. Cambridge: Cambridge University Press.

Thomas, Ronald L., and William H. Robbins. 1980. Large Wind-Turbine Projects in the United States Wind Energy Program. *Journal of Wind Engineering and Industrial Aerodynamics* 5 (3–4): 323–335.

Toke, Dave. 2002. Wind Power in UK and Denmark: Can Rational Choice Help Explain Different Outcomes? *Environmental Politics* 11:83–100.

Toke, David. 2011. The UK Offshore Wind Power Programme: A Sea-Change in UK Energy Policy? *Energy Policy* 39 (2): 526–534.

Toke, David, Sylvia Breukers, and Maarten Wolsink. 2008. Wind Power Deployment Outcomes: How Can We Account for the Differences? *Renewable & Sustainable Energy Reviews* 12 (4): 1129–1147.

Topçu, Sezin. 2007. Les Physiciens dans le Mouvement Antinucléaire: Entre Science, Expertise et Politique. [Physicians in the Antinuclear Movement: Between Science, Expertise and Politics.] *Cahiers d'Histoire* 102:89–108.

Torvanger, Asbjørn, and James Meadowcroft. 2011. The Political Economy of Technology Support: Making Decisions About Carbon Capture and Storage and Low Carbon Energy Technologies. *Global Environmental Change* 21 (2): 303–312.

Truman, David. 1951. *The Governmental Process: Political Interests and Public Opinion*. New York: Knopf.

Tsebelis, George. 2002. *Veto Players: How Political Institutions Work*. Princeton, NJ: Princeton University Press.

Tugwell, Franklin. 1988. *The Energy Crisis and the American Political Economy*. Stanford, CA: Stanford University Press.

Twidell, John, and Robert Brice. 1992. Strategies for Implementing Renewable Energy: Lessons from Europe. *Energy Policy* 20 (5): 464–479.

Tyner, Wallace E. 2008. The US Ethanol and Biofuels Boom: Its Origins, Current Status, and Future Prospects. *Bioscience* 58 (7): 646–653.

Ueckerdt, Falko, Lion Hirth, Gunnar Luderer, and Ottmar Edenhofer. 2013. System LCOE: What Are the Costs of Variable Renewables? *Energy* 63: 61–75.

Ummadisingu, Amita, and M. S. Soni. 2011. Concentrating Solar Power—Technology, Potential and Policy in India. *Renewable & Sustainable Energy Reviews* 15 (9): 5169–5175.

UNEP. 2012. *Feed-In Tariffs as a Policy Instrument for Promoting Renewable Energies in and Green Economies in Developing Countries*. Nairobi, Kenya: United Nations Environment Programme.

Unruh, Gregory C. 2000. Understanding Carbon Lock-In. *Energy Policy* 28 (12): 817–830.

Unruh, Gregory C. 2002. Escaping Carbon Lock-In. *Energy Policy* 30 (4): 317–325.

Unruh, Gregory C., and Javier Carrillo-Hermosilla. 2006. Globalizing Carbon Lock-In. *Energy Policy* 34 (10): 1185–1197.

Urpelainen, Johannes. 2009. Explaining the Schwarzenegger Phenomenon: Local Frontrunners in Climate Policy. *Global Environmental Politics* 9 (3): 82–105.

Urpelainen, Johannes. 2012. The Strategic Design of Technology Funds for Climate Cooperation: Generating Joint Gains. *Environmental Science & Policy* 15 (1): 92–105.

Urpelainen, Johannes. 2013. Can Strategic Technology Development Improve Climate Cooperation? A Game-Theoretic Analysis. *Mitigation and Adaptation Strategies for Global Change* 18 (6): 785–800.

Urpelainen, Johannes. 2014. Grid and Off-Grid Electrification: An Integrated Model with Applications to India. *Energy for Sustainable Development* 19:66–71.

Urpelainen, Johannes, and Antto Vihma. 2015. Soft Cooperation in the Shadow of Distributional Conflict? A Model-Based Assessment of the Two-Level Game between International Climate Change Negotiations and Domestic Politics. MIT Center for Energy and Environmental Policy Research (CEEPR). http://ceepr.mit.edu/publications/working-papers/620/.

Urpelainen, Johannes, and Thijs Van de Graaf. 2013. The International Renewable Energy Agency: A Success Story in Institutional Innovation? *International Environmental Agreement: Politics, Law and Economics* 15 (2): 159–177.

US Department of Energy. 2012. *2012 Renewable Energy Data Book*. Golden, CO: Energy Efficiency & Renewable Energy.

US Department of Energy. 2013. Revolution Now: The Future Arrives for Four Clean Energy Technologies. https://energy.gov/sites/prod/files/2013/09/f2/Revolution%20Now%20--%20The%20Future%20Arrives%20for%20Four%20Clean%20Energy%20Technologies.pdf.

Vad Mathiesen, Brian, Henrik Lund, and Kenneth Karlsson. 2011. 100% Renewable Energy Systems, Climate Mitigation and Economic Growth. *Applied Energy* 88 (2): 488–501.

van der Pligt, Joon. 1985. Public Attitudes to Nuclear Energy: Salience and Anxiety. *Journal of Environmental Psychology* 5 (1): 87–97.

van Est, Rinie. 1999. *Winds of Change: A Comparative Study of the Politics of Wind Energy Innovation in California and Denmark*. Utrecht, the Netherlands: International Books.

van Liere, Kent D., and Riley E. Dunlap. 1981. Environmental Concern: Does It Make a Difference How It's Measured? *Environment and Behavior* 13 (6): 651–676.

Vanvyve, Emilie, Luca Delle Monache, Andrew J. Monaghan, and James O. Pinto. 2015. Wind Resource Estimates with an Analog Ensemble Approach. *Renewable Energy* 74:761–773.

Vasi, Ion Bogdan. 2011. *Winds of Change: The Environmental Movement and the Global Development of the Wind Energy Industry*. New York: Oxford University Press.

Verbong, Geert, and Frank Geels. 2007. The Ongoing Energy Transition: Lessons from a Socio-Technical, Multi-Level Analysis of the Dutch Electricity System (1960–2004). *Energy Policy* 35 (2): 1025–1037.

Verbraucherzentrale Bundesverband. 2013. *Verbraucherinteressen in der Energiewende: Ergebnisse einer repräsentativen Befragung*. [*Consumer Interest in the Energy Transition: Results of a Representative Survey*.] https://www.vzbv.de/sites/default/files/downloads/Energiewende_Studie_lang_vzbv_2013.pdf.

Victor, David G. 2011. *Global Warming Gridlock*. Cambridge: Cambridge University Press.

Victor, David G., Amy M. Jaffe, and Mark H. Hayes. 2006. *Natural Gas and Geopolitics: From 1970 to 2040*. New York: Cambridge University Press.

Victor, David G., David R. Hults, and Mark Thurber, eds. 2012. *Oil and Governance: State-Owned Enterprises and the World Energy*. New York: Cambridge University Press.

Victor, David G., and Thomas C. Heller, eds. 2007. *The Political Economy of Power Sector Reform: The Experiences of Five Major Developing Countries*. New York: Cambridge University Press.

Vogel, Steven K. 1999. *Freer Markets, More Rules: Regulatory Reform in Advanced Industrial Countries*. Ithaca, NY: Cornell University Press.

Walker, Samuel J. 2004. *Three Mile Island: A Nuclear Crisis in Historical Perspective*. Berkeley, Los Angeles: University of California Press.

Wan, Yih-Huei, Erik Ela, and Kirsten Orwig. 2010. Development of an Equivalent Wind Plant Power-Curve. Conference Paper NREL/CP-550–48146.

Ward, Hugh. 2006. International Linkages and International Sustainability: The Effects of the Regime Network. *Journal of Peace Research* 43 (2): 149–166.

Warner, Ethan S., and Garvin A. Heath. 2012. Life Cycle Greenhouse Gas Emissions of Nuclear Electricity Generation. *Journal of Industrial Ecology* 16 (S1): S73–S92.

Watt, D. C. 1976. Britain and North Sea Oil: Policies Past and Present. *Political Quarterly* 47 (4): 377–397.

WDI. 2013. World Bank. Accessed December 1, 2013. http://data.worldbank.org/data-catalog/world-development-indicators/.

WDI. 2017. World Bank. Accessed January 24 2017. http://data.worldbank.org/data-catalog/world-development-indicators/.

Wellock, Thomas R. 2007. *Preserving the Nation: The Conservation and Environmental Movements 1870–2000*. New York: Wiley.

Wellock, Thomas Raymond. 1998. *Critical Masses: Opposition to Nuclear Power in California, 1958–1978*. Madison: University of Wisconsin Press.

Weyland, Kurt. 2002. *The Politics of Market Reform in Fragile Democracies: Argentina, Brazil, Peru and Venezuela*. Princeton, NJ: Princeton University Press.

Winkler, Harald. 2005. Renewable Energy Policy in South Africa: Policy Options for Renewable Electricity. *Energy Policy* 33 (1): 27–38.

Wiser, Ryan, Steven Pickle, and Charles Goldman. 1998. Renewable Energy Policy and Electricity Restructuring: A California Case Study. *Energy Policy* 26 (6): 465–475.

Wittman, Donald A. 1995. *The Myth of Democratic Failure: Why Political Institutions Are Efficient*. Chicago: University of Chicago Press.

Wolfram, Catherine D. 1999. Measuring Duopoly Power in the British Electricity Spot Market. *American Economic Review* 89 (4): 805–826.

Wu, Yanrui. 2003. Deregulation and Growth in China's Energy Sector: A Review of Recent Development. *Energy Policy* 31 (13): 1417–1425.

Wüstenhagen, Rolf, and Emanuela Menichetti. 2012. Strategic Choices for Renewable Energy Investment: Conceptual Framework and Opportunities for Further Research. *Energy Policy* 40:1–10.

Wüstenhagen, Rolf, Maarten Wolsink, and Mary Jean Bürer. 2007. Social Acceptance of Renewable Energy Innovation: An Introduction to the Concept. *Energy Policy* 35 (5): 2683–2691.

Wüstenhagen, Rolf, and Michael Bilharz. 2006. Green Energy Market Development in Germany: Effective Public Policy and Emerging Customer Demand. *Energy Policy* 34 (13): 1681–1696.

Yergin, Daniel. 1988. Energy Security in the 1990s. *Foreign Affairs* 67 (1): 110–132.

Yergin, Daniel. 2011a. *The Prize: The Epic Quest for Oil, Money and Power*. New York: Free Press.

Yergin, Daniel. 2011b. *The Quest: Energy, Security, and the Remaking of the Modern World*. New York: Penguin Press.

Yildiz, Özgür. 2014. Financing Renewable Energy Infrastructures Via Financial Citizen Participation. *Renewable Energy* 68:677–685.

Zachariadis, Theodoros. 2006. On the Baseline Evolution of Automobile Fuel Economy in Europe. *Energy Policy* 34 (14): 1773–1785.

Zahedi, A. 2011. Maximizing Solar PV Energy Penetration Using Energy Storage Technology. *Renewable & Sustainable Energy Reviews* 15 (1): 866–870.

Zerriffi, Hisham. 2011. *Rural Electrification: Strategies for Distributed Generation.* New York: Springer.

Zhang Peidong, Yang Yanli, Shi Jin, Zheng Yonghong, Wang Lisheng, and Li Xinrong. 2009. Opportunities and Challenges for Renewable Energy Policy in China. *Renewable & Sustainable Energy Reviews* 13 (2): 439–449.

Index